Node-RED and Raspberry Pi Pico W

From basics to flows for sensors, automation, motors, MQTT, and cloud services

Peter Dalmaris, PhD

elektor
design > share > earn

This is an Elektor Publication. Elektor is the media brand of
Elektor International Media B.V.
PO Box 11, NL-6114-ZG Susteren, The Netherlands
Phone: +31 46 4389444

Declaration

British Library Cataloguing in Publication Data
A catalogue record for this book is available from the British Library

ISBN 978-3-89576-588-9 Print
ISBN 978-3-89576-589-6 eBook

Prepress Production: D-Vision, Julian van den Berg
Printed in the Netherlands

Contents

Did you find an error?

Please let me know.

Go to **txplo.re/nodered**, and fill in the form.

I'll get it fixed right away

About the Author

Dr Peter Dalmaris is an educator, an electrical engineer, electronics hobbyist, and Maker. Creator of online video courses on DIY electronics and author of several technical books. Peter has recently released his book 'Maker Education Revolution', a book about how Making is changing the way we learn and teach in the 21st century.

As a Chief Tech Explorer since 2013 at Tech Explorations, the company he founded in Sydney, Australia, Peter's mission is to explore technology and help educate the world.

Tech Explorations offers educational courses and Bootcamps for electronics hobbyists, STEM students, and STEM teachers.

A lifelong learner, Peter's core skill lies in explaining difficult concepts through video and text. With over 15 years of tertiary teaching experience, Peter has developed a simple yet comprehensive style in teaching that students from all around the world appreciate.

His passion for technology and the world of DIY open-source hardware, has been a dominant driver that has guided his personal development and his work through Tech Explorations.

About Tech Explorations

Tech Explorations creates educational products for students and hobbyists of electronics who rather utilize their time making awesome gadgets instead of searching endlessly through blog posts and Youtube videos.

We deliver high-quality instructional videos and books through our online learning platform, txplore.com.

Supporting our students through their learning journey is our priority, and we do this through our dedicated online community and course forums.

Founded in 2013 by Peter Dalmaris, Tech Explorations was created after Peter realised how difficult it was to find high-quality definitive guides for the Arduino, written or produced by creators who responded to their reader questions.

Peter was frustrated having to search for YouTube videos and blog articles that almost never seemed to be made for the purpose of conveying knowledge. He decided to create Teach Explorations so that he could produce the educational content that he wished he could find back then.

Tech Explorations courses are designed to be comprehensive, definitive and practical. Whether it is through video, e-book, blog or email, our delivery is personal and conversational. It is like having a friend showing you something neat... the "AHA" moments just flow!

Peter left his career in Academia after his passion for electronics and making was rekindled with the arrival of his first Arduino. Although he was an electronics hobbyist from a young age, something the led him to study electrical and electronics engineering in University, the Arduino signalled a revolution in the way that electronics is taught and learned. Peter decided to be a part of this revolution and has never looked back.

We know that even today, with all the information of the world at your fingertips, thanks to Google, and all the components of the world one click away, thanks to eBay, the life of the electronics hobbyist is not easy. Busy lifestyles leave little time for your hobby, and you want this time to count.

We want to help you to enjoy your hobby. We want you to enjoy learning amazing practical things that you can use to make your own awesome gadgets. Electronics is a rewarding hobby. Science, engineering, mathematics, art, and curiosity all converge in a tiny circuit with a handful of components. We want to help you take this journey without delays and frustrations.

Our courses have been used by over 70,000 people across the world. From prototyping electronics with the Arduino to learning full-stack development with the Raspberry Pi or designing professional-looking printed circuit boards for their awesome gadgets, our students enjoyed taking our courses and improved their making skills dramatically. Here's what some of them had to say:

"*I'm about halfway through this course and I am learning so much. Peter is an outstanding instructor. I recommend this course if you really want to learn about the versatility of the amazing Raspberry Pi*" — Scott

"*The objectives of this course are uniquely defined and very useful. The instructor explains the material very clearly.*" — Huan

"*Logical for the beginner. Many things that I did not know so far about Arduino but easy to understand. Also the voice is easy to understand which is unlike many courses about microcontrollers that I have STARTED in the past. Thanks*" — Anthony

Please check out our courses at techexplorations.com and let us be part of your tech adventures.

Requirements

You will need a few things to make the most of this book. You probably already have most of all of them:

- A computer running Windows, Mac OS or Linux. If you have a spare Raspberry Pi Model B (any generation) or an old PC, you can use it as your Node-RED server.
- Access to the Internet.
- Hardware (all available in the Kepler kit from Sunfounder: https://txplo.re/kepler):
  - A Raspberry Pi Pico W
  - A breadboard and wires
  - Eight LEDs
  - Resistors (various Ω ratings)
  - Buttons and slide switches
  - Sensors: DHT11, HCSR04, motion PIR sensor, water level sensor, thermistor, analog light sensor.
  - A joystick
  - A 5 V relay
  - An RFID receiver and tag.
  - An infrared receiver and remote control.
  - A 2×16 LCD with the I²C backpack.
  - A 74HC595N THT integrated circuit.
  - A WS2812 RGB LED strip with eight RGB LEDs
  - A small servo motor.
  - A small DC motor

The Kepler kit from Sunfounder contains all the parts you will need for the projects in this book

The Book Errata Reporting and Resources Web Page

This book has a web page. This page contains an errata form to report bugs and access related resources as they become available.

Follow this URL to reach the book page: https://txplo.re/nodered.

Foreword

This book is a learning guide and a reference. You can use it to learn Node-RED, Raspberry Pi Pico W, and MicroPython. You can also use it as a source of information for these topics. I have organised this book into four parts to fulfil these dual roles.

Part 1 is dedicated to Node-RED for the absolute beginner. In Part 1, you will learn about Node-RED and event-driven systems, how to install an instance using the Docker option, the basics of nodes and flows, the dashboard and MQTT. If you are new to Node-RED, read the chapters in Part 1 carefully and complete the various projects.

Part 2 brings the Raspberry Pi Pico W into the mix. In the chapters of Part 2, you will learn how to use the Pico W as a Node-RED peripheral. You will learn to use MQTT to enable communications between the Pico and the Node-RED instance. You will also learn how to connect different hardware components to the Pico to implement simple circuits and use Node-RED (and its Dashboard) to control these components or view the data they produce. In Part 2, you will encounter motion, distance and water level sensors, motors, displays, relays, and joysticks, to mention a few. In all projects in Part 2, you will create Node-RED flows and write Raspberry Pi Pico W MicroPython scripts that typically implement an event-driven system.

Part 3 provides a primer to MicroPython. MicroPython is a language specifically designed for embedded systems using Python 3 syntax. Python 3 is one of the most successful programming languages ever. Python's syntax is straightforward, making it easy for beginners to learn. This simplicity encourages good programming practices and allows for a focus on problem-solving rather than syntax issues. MicroPython brings those attributes to Microcontroller programming. If you are not familiar with Python or MicroPython, the chapters in Part 3 will help you learn everything you need to be able to confidently write MicroPython programs for the Raspberry Pi Pico (and Pico W), as well as any other of the many Microcontroller boards that support MicroPython.

Finally, Part 4 provides additional Node-RED resources. These resources consist of content on important Nodes (all explained with the help of mini-projects), control structures, and ways to integrate your Node-RED flows with external services and APIs. In Part 4, you will learn how to create power flows that can be used in more advanced automation settings.

If you are new to Node-RED, Raspberry Pi Pico and MicroPython, I recommend reading this book in a linear fashion. Don't skip anything!

If you are familiar with Node-RED, you can quickly read Part 1 and continue with the projects in Part 2.

If you are unfamiliar with MicroPython, start with Part 3 and continue with other parts of the book. The Part 3 primer on MicroPython is a mini-book in this book and can be used independently of the rest of the content.

In this book, I chose Docker as the infrastructure technology on which I installed my Node-RED instance. You may prefer a different method (and there are several). You are free to choose any installation method you prefer. As long as, in the end, you have an accessible instance of Node-RED and MQTT broker running, you will be able to complete all of the projects in this book.

Enjoy!

Part 1: Node-RED Getting Started

Chapter 1.1 • What is Node-RED?

Node-RED is an open-source flow-based development tool that makes it easy to wire together devices, APIs, and online services. Imagine being able to drag and drop blocks on a screen to create a flowchart that does something—like turning on your lights at sunset or sending you an email when a sensor detects movement. That's what Node-RED lets you do, all without requiring you to write extensive code.

What is flow programming?

Flow programming in the context of Node-RED is a way to build applications by connecting different "nodes" in a flowchart-like manner. Each node performs a specific task, like reading from a sensor, performing a calculation, or sending an email. You create a "flow" by dragging and dropping these nodes onto a canvas and connecting them with "wires" to define the order of operations. The result visually represents your application logic, which you can deploy with a single click.

In traditional text-based programming, you write lines of code to define what your application should do. This often involves setting up loops, conditionals, and functions, and it requires a good understanding of the programming language you're using. In contrast, flow programming in Node-RED abstracts away much of this complexity. Instead of writing code, you're essentially drawing your program. This makes it easier to see the big picture, understand the data flow, and spot any potential issues.

Flow programming shines in scenarios where quick prototyping and iteration are essential. Because you can see the entire logic laid out visually, making changes or adding new functionalities is easier. You don't have to sift through lines of code to find the section you need to modify; you can rearrange or add nodes on the canvas.

It's also beneficial for people who may not have a strong background in programming. The drag-and-drop interface and pre-built nodes make it accessible for beginners, allowing them to focus on solving the problem rather than getting bogged down by syntax and language-specific rules.

Flow programming is compelling for IoT applications and automation tasks. When dealing with multiple devices, sensors, and APIs, the visual nature of flow programming makes it easier to manage the complexity. You can quickly see how data moves from your sensors to your logic nodes and output actions, making the development process more intuitive.

Where can you use Node-RED?

Before continuing, looking at some areas of everyday life where Node-RED is being used with great results is helpful. This will help you understand the impact that Node-RED can have on your projects. To keep this section short, I'll touch only on three areas where Node-RED makes a real difference: industry, education, and home automation.

In industrial settings, Node-RED is a game-changer. For instance, in manufacturing, it can be used to automate entire production lines. By connecting to PLCs (Programmable Logic

Controllers) and sensors, Node-RED can control the sequence of machinery operations. It's also valuable for energy management. Companies use Node-RED to monitor and control how much energy is being used in real time, allowing them to optimise consumption and reduce costs. Beyond that, it's used for predictive maintenance by collecting data from various machinery and using it to predict when a machine is likely to fail, thus scheduling timely maintenance.

Educational institutions find Node-RED to be an excellent tool for teaching and research. It's often used to introduce students to the concepts of IoT, networking, and automation in a hands-on manner. Because of its ease of use, students can quickly move from theory to practice. In research settings, Node-RED is a quick prototyping tool, enabling researchers to test their ideas without spending much time on initial setup and coding.

Node-RED offers endless possibilities for the DIY enthusiast or anyone interested in smart homes. You can set up intelligent lighting systems that adjust based on the time of day or occupancy. You can even create a home security system that sends you alerts based on sensor data. The possibilities are limited only by your imagination and the devices you have at hand.

The tech behind Node-RED

Node.js is an open-source, cross-platform JavaScript runtime environment that executes JavaScript code outside a web browser. Initially released in 2009, it has become a foundational technology for server-side development. Node.js allows you to build scalable network applications using JavaScript, a language traditionally used for client-side scripting in web browsers. It is incredibly lightweight and can run on various platforms, from a Raspberry Pi to a full-scale server.

It uses JSON to represent the flows, making it human-readable and machine-friendly. The logic is encapsulated in "nodes," essentially pre-written JavaScript functions you can string together to create your application.

Node.js is commonly used to build web servers and RESTful APIs. Its non-blocking, event-driven architecture makes it well-suited for handling multiple simultaneous connections, making it a popular choice for real-time applications. Applications like chat rooms and collaborative editing tools often use Node.js to manage real-time, bidirectional communication between the server and clients.

Node.js excels in scenarios where data streaming is crucial. For example, it can process files while uploaded, reducing the overall processing time. Node.js is frequently used in microservices architectures due to its ability to handle multiple small tasks concurrently. Companies like Netflix and Uber employ Node.js to manage their complex, distributed systems.

If you plan to use Node.js solely for Node-RED, you don't need to be an expert in Node.js. A basic understanding of JavaScript and how to install Node.js packages would be sufficient. Node-RED abstracts most of the complexities, allowing you to focus on the logic of your flows rather than the underlying code.

If you can only remember four facts about Node.js, these should be:

1. **Non-blocking, event-driven architecture**: Node.js uses a single-threaded, non-blocking event loop, allowing it to handle many connections simultaneously.

2. **NPM (Node Package Manager)**: Node.js has a rich ecosystem of libraries and frameworks available through NPM that can simplify the addition of new functionalities to your projects.

3. **Cross-Platform:** Node.js can run on various operating systems, including Windows, macOS, and Linux, making it highly versatile.

4. **Community support:** Being open-source and popular means that Node.js has a large, active community contributing to its development and a wealth of tutorials and resources.

Node.js is a powerful runtime environment with many applications beyond Node-RED. Its non-blocking architecture and rich ecosystem make it a go-to choice for many projects, from web servers to real-time communication apps. Even if you're only interested in using it for Node-RED, a minimal understanding will suffice, making it accessible for users with varying coding expertise.

And that's all you need to know about this (unless you plan to build Node.js applications).

Node-RED and microcontrollers

One of the most exciting aspects of Node-RED is its ability to interface with microcontrollers like Arduino and ESP8266. You can read sensor data from these devices and send commands back to control actuators, all through Node-RED's intuitive interface. This makes it an excellent tool for anyone looking to get into hardware hacking or create custom IoT devices.

Node-RED is a versatile and powerful tool with applications ranging from industrial automation to education and home DIY projects. Its drag-and-drop interface makes it accessible for people of all skill levels, while its underlying technology ensures it's robust enough for professional use. Whether you're a hobbyist looking to smarten up your home or an engineer aiming to optimise a production line, Node-RED has something to offer.

Chapter 1.2 • Node-RED in IoT and Event-driven Systems

The Internet of Things (IoT) has revolutionised how small devices are used to monitor and interact with their environment, and how they connect to other systems. As IoT networks grow, the need for an efficient, user-friendly, and powerful tool to manage these connected devices becomes apparent. Node-RED is an open-source, flow-based programming tool that simplifies the creation, deployment, and management of IoT applications and event-driven systems.
Node-RED's flexibility and ease of use make it a perfect fit for various IoT use cases. Here are some of the use cases where Node-RED is particularly well-suited.

Device management and control
Node-RED simplifies the process of connecting and managing IoT devices. It supports many communication protocols, such as MQTT, HTTP, WebSocket, and more, making it easy to collect data from sensors, control actuators, and interact with devices in real-time. For example, a Node-RED flow can be created to monitor a sensor's temperature and humidity data, process the data, and control a smart thermostat based on predefined conditions.

Data processing and analytics
IoT applications often require processing and analysing data from multiple sources. Node-RED provides various built-in nodes for data processing, such as Function nodes (for custom JavaScript code), Switch nodes (for conditional routing), and Change nodes (for modifying message properties). For instance, Node-RED can aggregate sensor data from multiple sources, filter out irrelevant information, and perform calculations to derive insights, such as detecting anomalies or predicting equipment failure.

Integration with cloud services and APIs
Node-RED enables seamless integration with popular cloud platforms, such as AWS, Azure, Google Cloud, and IBM Watson, as well as third-party APIs, like Twitter, Telegram, and Slack. This allows developers to build IoT applications that leverage cloud-based services for data storage, analytics, and machine learning.
An example use case involves using Node-RED to send sensor data to a cloud-based database, perform real-time analytics, and trigger notifications or alerts through third-party messaging services.

Node-RED in event-driven systems
Event-driven systems are designed to respond to specific events or changes in state. Node-RED's flow-based programming model and support for various communication protocols make it an excellent choice for building event-driven applications.

Home automation
Node-RED can create a custom home automation system that responds to events like motion detection, temperature changes, or voice commands. Integrating smart home devices and APIs, Node-RED can control lights, HVAC systems, security cameras, and more based on predefined rules and conditions.

Industrial automation and monitoring

In industrial settings, Node-RED can monitor and control equipment based on events or conditions, such as motors, valves, and conveyor belts. This can help optimise production processes, reduce downtime, and ensure equipment safety.

For example, a Node-RED flow can be created to monitor the status of a production line, detect when a machine malfunctions, and automatically shut down the affected equipment or send alerts to maintenance staff for immediate action.

Smart cities and infrastructure

Node-RED can be crucial in developing innovative city applications that respond to real-time events, such as traffic congestion, air quality, or energy consumption. By integrating with various sensors and services, Node-RED can be used to create intelligent systems that optimise urban infrastructure and enhance the quality of life for residents.

For instance, a Node-RED flow can be designed to analyse traffic data from multiple sources, detect congestion or accidents, and dynamically adjust traffic light timings or notify emergency services.

Node-RED's visual programming interface, extensive library of nodes, and support for a wide range of communication protocols make it a powerful tool for developing IoT and event-driven applications. Its flexibility and ease of use enable developers to create custom solutions for various use cases, from home automation and industrial monitoring to smart cities and infrastructure management.

As the IoT landscape continues to evolve, Node-RED is poised to play a significant role in developing and deploying connected systems that enhance our daily lives and drive innovation across industries

Chapter 1.3 • Communication in Node-RED: Protocols and Methods

Node-RED has an impressive ability to communicate with other systems thanks to its compatibility with various communication protocols. Its flexibility in this domain is key to its success. Let's delve deeper into the protocols and communication methods used by Node-RED.

HTTP and HTTPS

HTTP (HyperText Transfer Protocol) and its secure variant, HTTPS, are fundamental to the web as we know it. They serve as the basis for data communication on the World Wide Web. In Node-RED, you can use the "http" node to make HTTP requests to APIs on the web. For example, you could use an HTTP GET request to fetch data from a weather API or an HTTP POST request to send data to a database.

When working with Node-RED, using HTTP and HTTPS is one of the primary ways to communicate with clients. HTTP is easy to set up, and you don't need special certificates. It's great for quick tests and prototypes. HTTPS, on the other hand, adds a layer of security by encrypting the data. This is crucial if you're dealing with sensitive information.
However, with HTTP (notice the lack of the "S" from the acronym), your data is not encrypted, so transmitting confidential information is unsafe. Anyone can intercept it. HTTPS solves this problem but is a bit more complicated to set up. You'll need to get a security certificate, and sometimes, there might be compatibility issues with older devices or systems.
Later in this book, you will learn how to encrypt communications between your Node-RED server and the web clients using HTTPS.

MQTT

MQTT (Message Queuing Telemetry Transport) is a lightweight messaging protocol often used in IoT applications where bandwidth and power consumption are concerns. With MQTT, devices publish messages to topics, and other devices subscribe to these topics to receive the messages. Node-RED has built-in MQTT nodes, making integrating with MQTT-based systems straightforwardly. For instance, you could have a Raspberry Pi publish temperature sensor data to an MQTT topic and then use Node-RED to subscribe to this topic and react accordingly.

MQTT is a popular choice for sending and receiving data when using Node-RED. As I mentioned earlier, MQTT is lightweight. It doesn't use much bandwidth, which is great if you work with devices with limited resources. It's also good at handling intermittent connections, so MQTT can still work if your network is unreliable. In addition, MQTT is designed for real-time communication, so Node-RED can get updates as they happen.

On the flip side, MQTT has some limitations. It's not the best choice for large data sets because it's designed for small, frequent messages. Also, while MQTT itself is pretty secure, adding extra layers of security can be a bit complex. You might need to integrate it with other security protocols, which could be a hassle if you're new to this.

MQTT is great for real-time, lightweight data communication, but it might not be the best fit for every scenario, especially those requiring high security or large data transfers.

In this book, you will learn how to use MQTT to establish communications between Node-RED and the Raspberry Pi Pico W.

WebSockets

WebSockets is a protocol that provides full-duplex communication between a client and server over a long-lived connection. This is especially useful for applications that require real-time data updates. Node-RED provides a "websocket" node for client and server-side communication using WebSockets. This allows you to create interactive, real-time flows.

WebSockets can be a great way to handle data communications using Node-RED. One of the most interesting features of WebSockets is that they allow for two-way communication between the server and the client. This means you can send and receive data at the same time, making your applications more interactive and responsive. WebSockets are also fast because data can flow freely once the connection is established without repeatedly opening and closing connections.

On the other hand, WebSockets come with their own set of challenges. One issue is that they can be more complex to set up than other protocols like HTTP. You must ensure the client and server are configured correctly to handle WebSocket connections. Not all network configurations also support WebSockets, so you might run into issues if you're behind certain types of firewalls or proxies.

TCP and UDP

TCP (Transmission Control Protocol) and UDP (User Datagram Protocol) are two fundamental protocols of the Internet protocol suite. TCP is connection-oriented and ensures the reliable delivery of a stream of bytes, while UDP is simpler, connectionless, and suitable for scenarios where speed is more important than reliability. Node-RED provides TCP and UDP nodes for handling raw TCP and UDP communications, giving you even more flow flexibility.

If you're working with Node-RED, TCP is a reliable way to send your data. It makes sure all your data packets arrive in the correct order. This is great when you can't afford to lose data, like sending a file or updating a database. TCP also handles congestion well, so if your network is busy, it will adjust to ensure your data gets through.

The downside of TCP is that it can be slower than other methods. It takes time to establish a connection, and the checks it performs to ensure data integrity can add delays. So, TCP might not be the best choice for real-time applications where speed is crucial.

UDP, on the other hand, is all about speed. It sends data without worrying too much about whether it arrives or not. This is good for things like streaming video or audio, where you'd rather lose a few packets than have a delay.

The problem with UDP is that it's not reliable. If ensuring that every piece of data arrives is essential, then UDP is risky. It doesn't guarantee delivery or order. As a result, the recipient might get incomplete or jumbled data.

Serial

Node-RED also supports the serial communication protocol, commonly used for interfacing with hardware such as microcontrollers. With the serial node, you can read data from a serial port or write data to it. This is handy when working with devices like Arduino, which often use serial communication.

Using serial communications with Node-RED can be a straightforward way to communicate with hardware devices. It's a tried-and-true method that's been around for a long time, so it's well-supported and reliable. You'll often find it used in industrial settings or for connecting to older equipment.

Serial communications has its limitations. One of the main drawbacks is that it's not ideal for long-distance communications. The signal can degrade over distance, so it's mainly used for connections that are up to a few meters away. Also, it's generally slower than other modern data transfer methods, so it might not be the best choice for quickly transferring large amounts of data. Another thing to consider is that many modern computers don't come with a built-in serial communications port, so you might need an adapter if your Node-RED instance is running on more modern computer hardware.

In this book, as an example, I have set up my Node-RED instance to run inside a virtual machine, which, in turn, runs inside a modern personal computer without a serial port. Therefore, I cannot use this communication method without an adaptor.

On the other hand, if I had used a Raspberry Pi to host the Node-RED server, I would have been able to create flows that use the Pi's serial port to establish serial communication with microcontrollers such as the Raspberry Pi Pico and the Arduino UNO, both of which have support for serial communications.

Modbus

Modbus is a protocol often used in industrial applications for communication between electronic devices. If you're working in an industrial setting with Modbus equipment, you'll be pleased to know that Node-RED can handle this, too, thanks to community-contributed nodes.

Modbus can be a reliable choice for data communications using Node-RED, especially for industrial applications. One of the big advantages of Modbus is that it is widely used in industrial settings, so it's well-tested and reliable. It's also straightforward to set up. You can use it over various connections, like serial or TCP, giving you some flexibility. Modbus is also suitable for systems that require high reliability, as it has built-in error checking.

On the downside, Modbus has some limitations. It's not the fastest protocol out there, so it might not be the best fit if you need to quickly transfer large amounts of data. Also, while it's great for simple tasks, it's not as well-suited for more complex data structures. It might be limiting if you need to handle complex data types or commands. Another thing to consider is that it doesn't have built-in security features because it's an older protocol. You'll need to add those yourself if security is a concern.

Modbus is reliable and well-suited for industrial applications but may not be the best choice for high-speed or complex data communications.

Node-RED's extensive protocol support makes it an excellent tool for integrating diverse systems. Whether pulling data from a web API, communicating with an IoT device over MQTT, or reading sensor data from a microcontroller over a serial connection, Node-RED has you covered.

Chapter 1.4 • Node-RED Installation Options

To install and set up a Node-RED server, several options are available. Which one you choose depends on variables such as the resources that you have available and your goals. These options include installing Node-RED to your "regular" work computer as you would with any other program, using a Raspberry Pi or other Linux computers, setting up a Virtual Machine (VM), or leveraging Docker. Each choice has unique features and advantages, but as we move forward, we will focus on the benefits of Docker as a preferred choice.

Installing Node-RED on your local computer

Node-RED is very flexible when it comes to operating systems. You can install it on Windows, Mac OS, and various distributions of Linux.

Installing Node-RED locally has several advantages. For one, you have complete control over your environment. You can customise settings, add nodes, and test flows without worrying about external factors like network latency. It's also easier to integrate with other software and hardware on your computer, such as a webcam and microphone. Plus, you don't have to worry about monthly subscription fees or data limits that you might encounter with cloud-based solutions. Another advantage is that you do not need access to another computer with local installation. You already have a computer, and it's the one that you are (probably) working on right now!

However, the local installation has several significant disadvantages. At the top of the list is that your work computer can be a busy and constantly changing environment. If you are like me, your work computer is where you do all your programming, where you install, remove, reconfigure and experiment with software. It has multiple network connections and lots of connected devices. Once in a while, it will run out of disk space, or other issues will pop up. All this can affect the smooth operation of server software like Node-RED, and you may spend too much time troubleshooting self-inflicted problems rather than learning and using Node-RED. Another issue to consider is that since your flows are stored locally, you won't be able to access them from another computer unless you've set up remote access. And let's not forget, if your computer crashes, you risk losing all your work unless you've been diligent about backups.

For these reasons, I do not recommend installing Node-RED on your work computer unless you install it on a dedicated Virtual machine (see below) so that there is robust isolation between the host and the guest OS.

Raspberry Pi and other Linux computers

By deploying Node-RED on a dedicated Linux machine, such as a Raspberry Pi or an old re-purposed PC, you're positioning yourself for a highly efficient and customisable experience. Let's explore the nuances of this approach, from compatible Linux operating systems to the inherent pros and cons.

Node-RED offers compatibility with a range of Linux distributions. While the Raspberry Pi OS remains the go-to choice for Raspberry Pi users, other distributions like Ubuntu and

Fedora are also viable options. Your selection will likely hinge on the specific requirements of your project or your familiarity with a particular Linux distribution.

Linux's open-source nature is among its most compelling advantages, offering an unparalleled customisation level. Additionally, "headless" Linux distributions are generally lightweight, making them well-suited for hardware with limited computational resources, such as a Raspberry Pi. This results in a cost-effective yet powerful solution for running Node-RED. Furthermore, Linux's reputation for stability and security adds more reliability to your setup.

Perhaps you are unfamiliar with the term "headless", so let's take a moment to explain it. The term "headless Linux" refers to a Linux server or system run without a graphical user interface (GUI). In other words, it doesn't have the graphical desktop environment that is common on desktop computers. Instead, all interactions with the system are done through the command line interface (CLI). Running Linux in headless mode is resource-efficient, as it eliminates the need for the system to allocate resources to the GUI. This is particularly beneficial for servers or systems designed to run specific tasks and don't require user interaction via a graphical interface. It's also a typical setup for Linux instances running on cloud platforms or in data centres.

Despite its merits, Linux has challenges, particularly for those less acquainted with its environment. The operating system often necessitates command-line interactions, which can be daunting for newcomers. Moreover, Linux's robustness and stability come at the cost of limited software compatibility, especially for applications designed with Windows or macOS in mind. However, I recommend using your Linux machine as a Node-RED host only and not as a general-purpose computer; hence, the compatibility issue with applications you find on Windows and macOS computers is a non-issue.

Raspberry Pi, a low-cost, credit-card-sized computer, is one of the most popular choices for running a Node-RED server. Its affordability and accessibility make it a compelling choice for hobbyists and enthusiasts just starting with the Internet of Things (IoT) and Node-RED. Other Linux computers, with their robust and open-source nature, also serve as great hosts for Node-RED.

Installing Node-RED on Raspberry Pi or any other Linux system is straightforward. With a few commands in the terminal, your server will be up and running. However, while these systems are great for smaller, personal projects, they might not scale well for larger, more resource-intensive applications.

Virtual Machine

Deploying Node-RED on a Linux Virtual Machine (VM) offers a unique blend of flexibility and isolation. This approach allows you to run Node-RED in a controlled environment, separate from your primary operating system. The virtual machine can run on your local ("work") computer or another computer in your local network or accessible via the Internet.

In terms of virtualisation options, there are several to choose from. Which one you pick does not matter much since all options can run and manage Linux guests. For example, options consider VMware, VirtualBox, Parallels, QEMU, Citrix, and Zen (and there are more). These are among the most popular and user-friendly options. All provide a wide range of well-documented features, making it easier to find help if you run into issues. As I mentioned, all of these options support Linux guests; your choice will depend on your existing infrastructure and specific project requirements.

One of the most compelling benefits of using a VM is its isolation. Your Node-RED installation will be contained within the VM, separate from your primary operating system. This makes it easier to manage dependencies and avoid conflicts with other software. Additionally, VMs are highly portable. You can quickly move them between computers or upload them to cloud services. This portability simplifies backup procedures and makes it easier to share your Node-RED setup with others.

However, running Node-RED requires some planning and forward-thinking. Virtual machines consume additional system resources, including CPU, memory, and disk space, which could impact the performance of your primary operating system. You must be careful when you plan the deployment of VMs on a host so that the VM does not try to consume more resources (such as RAM, disk space and CPU cores) than are available on the host machine. Moreover, setting up a VM involves a series of configuration steps that might be daunting for those unfamiliar with virtualisation technology. Lastly, while VMs provide isolation, they can also add a layer of complexity to your setup, especially regarding networking and file sharing between the host and guest operating systems.

Installing Node-RED on a Linux Virtual Machine offers a flexible and isolated environment, ideal for testing and development. However, it does come with the trade-off of additional resource consumption and complexity. If you're comfortable navigating these challenges, a VM can be an excellent platform for your Node-RED projects.

My personal choice, and the one that I deployed for the production of this book, is to combine virtualisation with Docker (which I will discuss next). You will learn about the process I followed in the next chapter. The process involves setting up a Linux virtual machine on a small Windows personal computer (to act as a Docker container host) and installing Node-RED using the Docker method.

Docker: the preferred choice
While all these options are valid and have unique advantages, Docker has become a preferred choice for many when setting up a Node-RED server. Docker is a platform that allows you to package an application and its dependencies into a "container." This container can then be easily moved and run across different computing environments. Think of a Docker container as a lightweight, standalone, executable software package that includes everything needed to run a piece of software.

One of the primary benefits of using Docker for Node-RED is the ease of deployment. Docker containers encapsulate all the dependencies and configurations required, making the installation process remarkably straightforward. This eliminates the "it works on my machine" issue, where software behaves differently on different systems. Additionally, Docker containers are isolated from each other and the host system, enhancing security and reducing conflicts between software components. This isolation also makes it easier to manage different Node-RED versions or run multiple instances simultaneously.

While Docker offers numerous advantages, it's not without its drawbacks. There's a learning curve associated with understanding Docker concepts like containers, images, and Docker files. Getting comfortable might take some time if you're new to these. In this book, I used the Docker method to install my Node-RED instance, and I will show you how to install yours step-by-step.

Also, Docker containers do consume system resources, albeit less than traditional virtual machines. This could be a concern if you run on hardware with limited resources. Lastly, while Docker provides isolation, it can sometimes complicate networking configurations, especially for those not well-versed in networking principles.

Since Docker is the preferred method for Node-RED installation in this book, I take another moment to list the key takeaway advantages that convinced me to use this method:

1. **Portability:** Docker allows you to package an application with all its dependencies into a standardised unit called a Docker container. This container can run on any system that has Docker installed, irrespective of the underlying operating system. This high level of portability simplifies deployment and reduces potential issues caused by differences in system environments.

2. **Isolation:** Similar to VMs, Docker provides isolation. However, Docker containers are less resource-intensive than VMs, as they share the host system's kernel and do not require an entire operating system. This efficiency means you can run more containers on a given hardware combination than with VMs.

3. **Scalability:** Docker's design makes it easy to scale applications. You can quickly start, stop, replicate, or destroy containers with a single command or via Docker's APIs.

4. **Consistency:** With Docker, you can maintain consistency across multiple development, testing, and production environments. This consistency can significantly reduce developers' "it works on my machine" problem.

5. **Node-RED support**: Node-RED's Docker support and documentation are excellent. The Node-RED documentation includes a straightforward and detailed guide on how to install the software with Docker. This is the guide that I based my process on. See more here: https://nodered.org/docs/getting-started/docker.

While Raspberry Pi, other Linux computers, and VMs are all viable options for installing a Node-RED server, Docker's portability, resource efficiency, scalability, and consistency make it an excellent choice for development and production environments. With Docker, you can build, test, and deploy Node-RED applications quickly and confidently on any hardware that supports it, be it an embedded computer, a laptop, a local server, or a cloud server. In the next chapter, you will learn how to install a Node-RED server using the Docker option.

Chapter 1.5 • Set up Node-RED using Docker

Docker is an open-source platform that automates application deployment, scaling, and management. It uses containerisation technology to encapsulate an application and its dependencies into a self-sufficient unit known as a Docker container. This container can run uniformly and consistently on any infrastructure supporting Docker, irrespective of the underlying operating system.

Working with a virtual machine is like setting up a whole new computer inside your computer. This virtual computer has its own operating system, just like your computer. Think of a virtual machine as a house. The virtual machine acts as a separate house with its own rooms (applications) and utilities (like water and electricity, which are like the system libraries and binaries). It doesn't know it's not a real house and behaves like one for all intents and purposes.

Docker is different. If virtual machines are like houses, Docker is like an apartment complex. In this apartment complex, all apartments share the same utilities (like water and electricity). So, with Docker, all your applications share the same system libraries and binaries from your actual computer (the host). Each application lives in its own "apartment" or container, separate from the others, but they all share the same underlying system. This makes Docker lighter and faster than a virtual machine because it doesn't need to run a whole new operating system for each application.

However, both Docker and virtual machines have something in common. They both provide a way to isolate your applications from each other, making them more secure and less likely to interfere with each other. It's like having a private space, whether an entire house or an apartment.

This also means your applications can't directly communicate with each other. If they want to "talk", they have to do it in specific ways – through defined interfaces, just like you would have to call or text someone in another house or apartment.

In a nutshell, Docker and virtual machines provide isolated environments for running applications, but they do it differently. Docker is more lightweight and efficient because it shares resources with the host system, while a virtual machine is heavier but can provide a completely separate operating system.

Some example uses of Docker

Docker has found applications in many sectors, each leveraging its capabilities to solve unique challenges. Explore how Docker makes waves in industry, education, DIY electronics, and home automation.

Docker in Industry

Docker is often used in the industrial sector to create isolated environments for running various applications on factory floors. For instance, a manufacturing company might use Docker containers to run quality control software that inspects products as they come off the assembly line. The containerised approach ensures the software can be easily updated or rolled back without affecting other systems. This is crucial in an industrial setting where downtime can be costly.

Docker in education

Educational institutions also embrace Docker, especially in computer science and engineering courses. Instructors can package coding assignments or software tools within Docker containers, ensuring every student has the same working environment. This eliminates the need for lengthy setup procedures and resolves compatibility issues, allowing students to focus on learning rather than troubleshooting software problems.

Docker in DIY electronics

For people interested in DIY electronics, Docker offers a way to manage software for single-board computers like Raspberry Pi. Say you're building a weather station that collects and uploads data to the cloud. You could use a Docker container to package the data collection software, making deploying updates or adding new features easy. This modular approach simplifies the management of your DIY projects, allowing you to iterate and improve without hassle.

Docker in home automation

Home automation is another area where Docker is commonly used. If you're setting up a smart home, you might be dealing with various software solutions for controlling lights, thermostats, and security cameras. Docker allows you to run each of these software components in isolated containers, making it easier to manage and update them. This approach also enhances security by isolating each application, reducing the risk of one compromised system affecting others.

Docker's versatility makes it a valuable asset across multiple domains. Docker's containerised approach offers a robust solution for managing complex software ecosystems, from streamlining industrial processes to enhancing educational experiences and simplifying DIY electronics and home automation.

Key Docker terms and concepts

In the remainder of this chapter and book, I'll be referring to several of Docker's terms and concepts. I have listed and described these terms and concepts here to make it easy to find them. Remember that you can use Docker to install an instance of the Node-RED server without being a Docker expert. However, knowledge of the Docker fundamentals will make the installation and maintenance process a much smoother experience.

- **Docker Image:** A Docker image is a lightweight, standalone, executable package that includes everything needed to run the software, including the code, a runtime, libraries, environment variables, and config files.
- **Docker Container:** A Docker container is a runtime instance of a Docker image. It encapsulates the application and its dependencies, ensuring it runs uniformly across different environments.

- **Dockerfile:** A Dockerfile is a text file containing instructions for building a Docker image. It specifies the base image, the application code, and the dependencies.
- **Docker Hub:** Docker Hub is a cloud-based registry service that allows you to link to code repositories, build your images, store them, and deploy them to Docker hosts.
- **Docker Compose:** Docker Compose is a tool for defining and running multi-container applications. With Compose, you use a YAML file to configure your application's services and then start all the services with a single command.

Docker and Node-RED: a winning combination

Node-RED is a programming tool for wiring together hardware devices, APIs, and online services. Docker provides a robust and scalable environment for deploying Node-RED instances:

- **Portability:** Docker can encapsulate a Node-RED application and its dependencies into a container, ensuring it runs uniformly across different environments. This is particularly useful when deploying Node-RED flows across multiple devices or servers.
- **Isolation:** Each Node-RED instance runs in its container, isolating it from other applications and reducing the risk of conflicts.
- **Scalability:** Docker makes it easy to scale a Node-RED application by running multiple instances on the same host or distributing them across multiple hosts.
- **Management and Maintenance:** Docker simplifies updating Node-RED instances, managing dependencies, and handling configuration, making it easier to maintain your Node-RED applications over time.

Docker's ability to encapsulate an application and its dependencies into a portable, scalable, and isolated container makes it an ideal platform for running Node-RED, whether you're a hobbyist working on a single Raspberry Pi or a professional deploying a large-scale IoT application.

5.1. Docker containers: hardware options and considerations

A Docker container must be hosted on a computer. Let's consider a few options relating to Docker and installation and operation. This is a big topic because Docker is very flexible and can be used in many settings.

Host options

You can run Docker directly on a computer, whether it's your work desktop, a laptop or a dedicated computer if it has a supported operating system like Windows, Mac OS, or a Linux distribution. This is the simplest possible scenario, where Docker runs as yet another application on the machine, without an additional isolation layer as you would have if a virtual machine were used between the host operating system and Docker (see the following section for more details). This can be a good option if you're starting and want to try Docker without investing in new hardware.

I prefer not to install server software on my work computer because of the reasons I mentioned in an earlier chapter (see "Node-RED installation options"). To ensure that I get maximum reliability from server software like Node-RED, I always install it on dedicated hardware, even if that is a very old re-purposed PC or a low-cost (but entirely sufficient for low-load server applications) single-board computers like the Raspberry Pi.

Consider running Docker on a physical server to set up a dedicated environment. This computer is specifically designed to provide services to other computers. You could set up a server in your home or office or rent space in a data centre. Due to the cost involved in renting data centre space, this is an option you can consider if your computing and bandwidth requirements are very high, which is often the case for commercial applications.

You might consider running Docker on an embedded computer like a Raspberry Pi if you want something more compact and affordable. I run a few Docker containers in a Raspberry Pi 4, improved with the Pironman enclosure from Sunfounder, which provides additional SSD storage, a fan, and a power supply (you can read about this product on my blog post at https://txplo.re/pironman_case).

You can also run Node-RED on the Unihiker from DFRobot, a single-board computer with an integrated display. These small, low-cost devices are surprisingly powerful enough for server applications like Node-RED. An additional advantage of using a hardware host such as the Raspberry Pi or Unihiker is that these computers offer a full complement of input and output interfaces, including serial UART, unavailable on newer PCs. These interfaces are handy for applications that involve microcontroller peripherals. As a result, having your Node-RED instance running on a single-board computer such as one of those I mentioned earlier gives you additional connectivity options as opposed to running Node-RED in a virtual machine or a modern PC without (for example) a serial UART or I^2C interface.

Virtual machines

If you're interested in leveraging the benefits of Docker but want to keep it separate from your primary operating system, installing it in a virtual machine (VM) is an excellent approach. This setup offers a layer of isolation and flexibility that many find appealing. Let's delve into the specifics of this installation method.

A virtual machine allows you to run an operating system within another operating system, effectively isolating the two from each other. This is particularly useful for testing or development purposes. You can experiment with Docker without any risk to your primary system, and if something goes wrong, it's confined to the VM. Installing Docker in a VM is relatively straightforward; you will learn about it later in this chapter. You'll first need to set up a virtual machine using software like VMware, VirtualBox, or Hyper-V. Once the VM is up and running with a compatible operating system—usually a Linux distribution—you can install Docker just as you would on a regular machine. The installation steps typically involve a series of command-line instructions, which are well-documented on Docker's official website.

Installing Docker in a VM offers several benefits. The most obvious one is isolation. Your Docker containers will run in a controlled environment, isolated from your primary operating system, making managing dependencies easier and avoiding software conflicts. Additionally, VMs are portable. You can move your VM, complete with its Docker setup, to another machine or even upload it to a cloud service, offering a level of flexibility that's hard to beat.

You can create a virtual machine to host a Docker container on your work computer or any other computer in your local network. There are also low-cost virtual machine options on the cloud. Many cloud providers, like Amazon Web Services, Digital Ocean, Google Cloud, or Microsoft Azure, offer services that let you run Docker containers. This can be a great option if you need a lot of flexibility. You can create and destroy containers as required and only pay for what you use. Plus, you can take advantage of other cloud services, like databases or machine learning tools.

You can run Docker on almost any computer, from a tiny Raspberry Pi to a large physical server, or even in a virtual environment like a local virtual machine or the cloud. The best choice depends on your specific needs, your budget, and how much time and effort you want to put into managing your Docker environment.

In this book, I have opted to install and operate Docker in a local virtual machine. However, you may choose any other options I discussed in this chapter. Regardless of where you decide to install Docker, you can install a Node-RED instance using the instructions I describe later in this chapter.

5.2. Create the Ubuntu 22.04 VM

In this section, you will learn how to set up a VirtualBox virtual server running Ubuntu 22.04, on which you will install Docker and a Node-RED container. I have broken the process into a series of steps with comprehensive and detailed instructions to ensure you have a seamless setup experience.

I chose VirtualBox as my desktop virtualisation software for this book because it is free, open-source, cross-platform and reliable. It runs on Windows, Mac OS, Linux and Solaris hosts. VirtualBox is perfect for our goal of setting up a virtualised Docker environment for our Node-RED server, isolated from our regular work computer.

However, if you prefer to use other virtualisation software, you are certainly free to do it. If you know how to set up an Ubuntu virtual server using virtualisation software like VMWARE, Parallels, QEMU, Citrix or others, go ahead and do this now and skip this chapter. You can also use a cloud virtual machine provider and set up a Cloud virtual server. The instructions I have written in the next chapter and everything else that refers to the Node-RED server in this book will work regardless of the compute resource on which the Node-RED Docker container is running.

Let's continue with the setup of the Ubuntu server virtual machine.

Install VirtualBox

VirtualBox is an open-source virtualisation software package for x86 hardware that allows you to create virtual machines on your computer.

1. Access the official VirtualBox website (see https://txplo.re/virtualbox) to download the latest software version. As of the time of this guide, the newest version is VirtualBox 6.1.

2. Run the VirtualBox installer. Follow the on-screen instructions, clicking "Next" to proceed and "Install" to begin the installation process.

3. Click "Finish" to close the installer once the installation completes. VirtualBox is now installed on your machine.

You can install VirtualBox on your computer as you install other software, so I am not providing step-by-step instructions here. However, if you need help with installation, you should look at the official VirtualBox installation instructions at https://txplo.re/install_virtualbox.

Download Ubuntu Server 22.04 ISO

The next step is to download the ISO file for Ubuntu Server 22.04 from the Ubuntu website.

1. Navigate to the official Ubuntu website's download page (https://ubuntu.com/download/server) to download the ISO file for Ubuntu Server 22.04.

2. Click on the Ubuntu Server 22.04 ISO download link. The download should begin automatically.

3. Save the ISO file in an easily accessible location on your computer. You will need to locate this file during the virtual machine setup process.

Set up Ubuntu Server 22.04 in VirtualBox

Having installed VirtualBox and downloaded the Ubuntu Server 22.04 ISO, we can now set up our virtual machine.

1. Open VirtualBox and click "New" to create a new virtual machine.

2. Name your virtual machine, select "Linux" as the type, and "Ubuntu (64-bit)" as the version. Click "Next".

3. Allocate memory to your virtual machine. For a server setup, 2048 MB should be enough, but you can adjust it according to your system's capabilities. Click "Next".

4. Select "Create a virtual hard disk now" and click "Create". Choose "VDI (VirtualBox Disk Image)" and click "Next".

5. Select "Dynamically allocated" and click "Next". Set the size of the virtual hard disk. A minimum of 20GB is recommended for the server.

6. Set the Network option to "Bridged". This will set up the virtual machine networking so that the virtual machine appears as a regular host in your network and allows you to connect to it from your computer easily.

7. Click "Create" to finish setting up your virtual machine.

8. Select your newly created virtual machine and click "Start".

9. Navigate to the location of the downloaded Ubuntu Server 22.04 ISO file when prompted and select it. The server should now boot from the ISO file.

10. Follow the on-screen instructions to install Ubuntu Server 22.04 on your virtual machine.

You can see the configuration settings of my Ubuntu virtual machine in the screenshot below. I have marked the most important settings with an arrow.

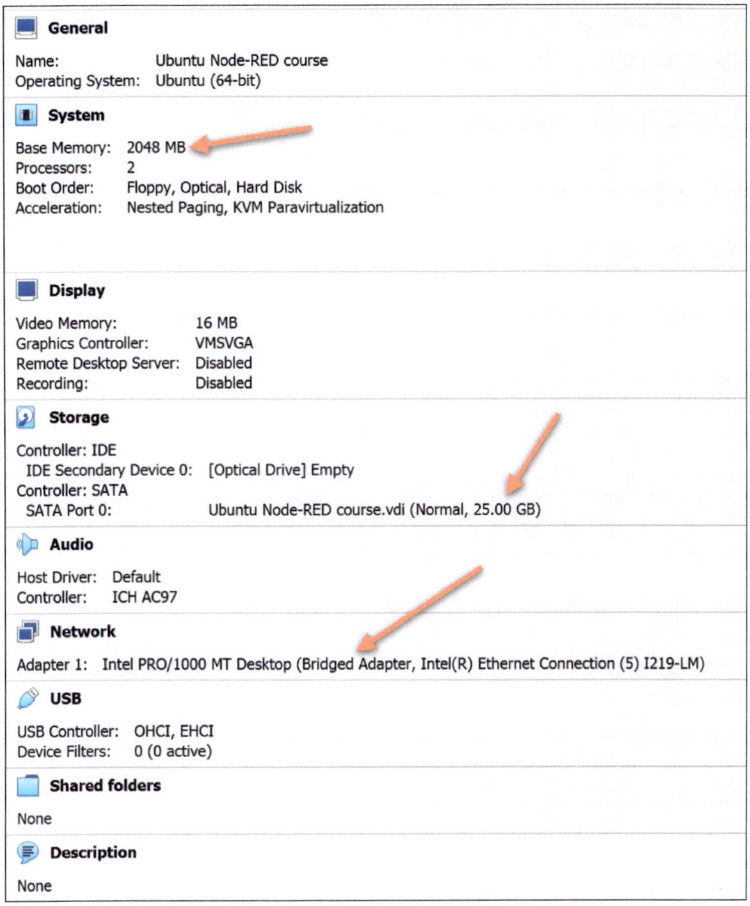

Figure 1.5.2.1: The settings of my Ubuntu virtual machine.

Once you complete this process, you will have a Ubuntu 22.04 virtual server ready to host Docker. Let's continue with Docker in the next section.

5.3. Install Docker on the Server

In this section, you will learn how to install Docker on a clean Ubuntu 22.04 server. Before you continue, ensure that you have a fresh and clean server running Ubuntu 22.04 server. By the end of this section, you will have Docker ready to accept the Node-RED container.

Preparing the Ubuntu Server

Before you install Docker, you need to ensure the Ubuntu server is up to date. This ensures that all the latest patches and updates are applied to your server, increasing its security and stability.

Here are the steps to follow:

1. Connect to your server through SSH or the console.

2. Once connected, first update the package lists for upgrades and new package installations by running the following command:

```
$ sudo apt-get update
```

3. Next, upgrade the system by executing the following command:

```
$ sudo apt-get upgrade
```

Your Ubuntu server is now up to date and ready for Docker's installation.

Installing Docker

Let's install Docker. Follow these steps:

1. Install Docker using the apt command:

```
$ sudo apt install docker.io
```

This command will install Docker from Ubuntu's repositories.

2. After installation, the Docker service should start automatically. Verify it's running with:

```
$ sudo systemctl status docker
```

3. If Docker is running, you should see output similar to:

```
docker.service - Docker Application Container Engine
        Loaded: loaded (/lib/systemd/system/docker.service; enabled; vendor
        preset: enabled)
        Active: active (running) since Tue 2023-06-16 19:20:30 UTC; 1min
        20s ago
        Docs: https://docs.docker.com
```

At this point, Docker is installed on your Ubuntu 22.04 server and running successfully. The server is ready to accept Docker containers.

Running Docker commands without sudo

By default, running Docker commands requires administrator privileges. If you wish to run Docker commands as a non-root user without prepending sudo, add your user to the Docker group.

1. Add your user to the docker group:

```
$ sudo usermod -aG docker $USER
```

2. To activate the changes to groups, log out and then log back in, or you can type:

```
$ su - $USER
```

Now, you can run Docker commands without prepending "sudo".

Testing Docker Installation

Let's verify that Docker is installed correctly and that you can download images and run them in containers.

Run a Docker container from the *hello-world* image:

```
$ docker run hello-world
```

The output will indicate that Docker is working correctly and pulled the hello-world image from the Hub.

At this point, you have successfully installed Docker on your Ubuntu 22.04 server. The server is now ready to receive and run Docker containers. You can find more details on managing your installation in the official Docker documentation. In the next section, you will learn how to install the Node-RED Docker container on your server.

5.4. Install Node-RED using Docker

In this section, you will learn how to set up a Node-RED Docker container on an Ubuntu 22.04 server. I assume you already have an Ubuntu server, and Docker is installed and running on it. Below are the steps for installing the Node-RED Docker image. Upon completing this section, you will have the Node-RED Docker image ready to run in a container on your Ubuntu server.

Pull the Node-RED image from Docker Hub

Before you can run the Node-RED container, you must pull the Docker image from Docker Hub. Docker Hub is a cloud-based repository where you can find, share, and manage Docker containers. Think of it as a library of pre-packaged applications that you can easily download and run using Docker. It's a platform that brings developers and IT professionals together, making distributing and deploying software easier.

On Docker Hub, you'll find containers spanning various use cases and technologies. Software companies, open-source projects, or individual developers often share these containers. Whether you're looking for a database like MySQL, a web server like Apache, or even complex software stacks, chances are you'll find a corresponding Docker container on Docker Hub.

One of the key advantages of Docker Hub is that it streamlines the software deployment process. Instead of going through complicated installation procedures, you can pull a container from Docker Hub and run it on your system. This ensures you're working with a version of the software tested and configured to run optimally in a Docker environment. Here's how to "pull" the image for Node-RED from the Docker Hub:

1. Connect to your server through SSH or the console (unless you are already connected from the previous section).

2. To pull the Node-RED image, use the following command:

```
$ docker pull nodered/node-red
```

Docker will reach out to Docker Hub, fetch the Node-RED Docker image, and store it locally.

Running the Node-RED container
Now, run the Node-RED Docker container on the Ubuntu 22.04 server.

1. Start a Node-RED container by executing the following command:

```
$ docker run -it -p 1880:1880 --name mynodered nodered/node-red
```

Here's what the command does:

- "docker run" is the command that starts ("runs") a new Docker container.
- "-it" makes the container interactive. This allows you to interact with the container. It allocates a terminal (-t) and keeps it open for interactive (-i) input. This is useful if you want to enter additional commands within the container later.
- "-p 1880:1880" maps port 1880 of the container to port 1880 of the host. This is essential for accessing the Node-RED interface, which runs on port 1880 by default. You can change the host port to something different, such as "2880". This is useful if you have multiple Node-RED Docker containers running on the same machine, as you can set each one to use a discreet host port number.
- "—name mynodered" names the container 'mynodered'. The "—name" flag assigns a custom name to the container, in this case, "mynodered." Naming your container makes managing it easier, especially if multiple containers run.
- "nodered/node-red" is the Docker image's name to create the container. This image contains the Node-RED software and all its dependencies. It's pulled from Docker Hub if not already on your system.

1. Node-RED should now be running and accessible at http://<your-server-IP>:1880.

2. To verify the Node-RED container is running, use the following command:

```
$ docker ps
```

This command will show you all the containers currently running. You should see your Node-RED container in the list. After completing this section, the Node-RED container runs on your Ubuntu server.

Managing the Node-RED Container

Knowing how to manage your Docker containers is crucial. Here are some basic commands:

1. To stop the Node-RED container, use the following command:

```
$ docker stop mynodered
```

2. To start the Node-RED container again, use the following command:

```
$ docker start mynodered
```

3. If you need to remove the Node-RED container, first stop the container and then use the following command:

```
$ docker rm mynodered
```

You have now installed and run a Node-RED Docker container on your Ubuntu 22.04 server. The server is now ready to run Node-RED flows. You can create your flows by accessing Node-RED at http://<your-server-IP>:1880.

5.5. Testing Your New Node-RED Server

After installing the Node-RED Docker container on the Ubuntu 22.04 server, you should verify the installation by conducting a few tests. This section will guide you through the steps to confirm that your Node-RED instance works correctly. By the end of this chapter, you will have a fully functional Node-RED instance ready to run your flows.

Accessing Node-RED web interface

One of the first steps in verifying your Node-RED instance is accessing the Node-RED web interface.

1. Open your preferred web browser.

2. Enter the following URL, and replace <your-server-IP> with your server's IP address:
   http://<your-server-IP>:1880

3. You should see the Node-RED flow editor. If you can access this, it indicates that the Node-RED instance is up and running successfully.

If you can't access your Node-RED server in your browser, you will need to take one or more steps back to discover and fix the problem. It would be best if you did not continue with the rest of this chapter until you have a fully working and accessible instance of Node-RED.

Creating a simple flow

Creating a simple flow in Node-RED is the best way to verify that Node-RED is functioning correctly. The labels in the screenshot below correspond to the step numbers in the instruction list.

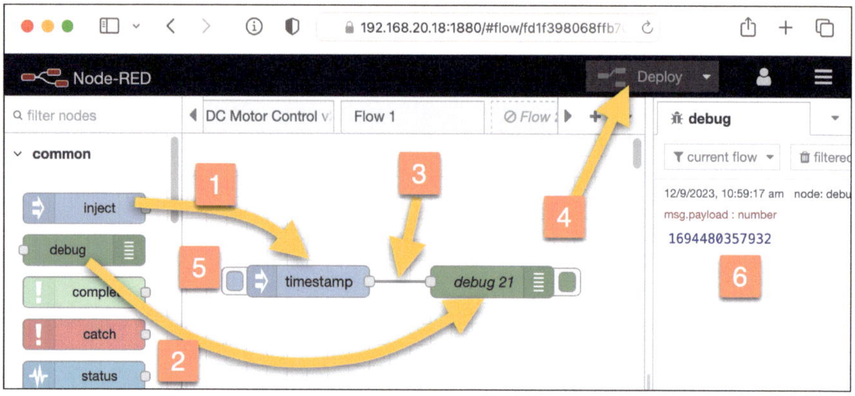

Figure 1.5.5.2: A very simple flow to test Node-RED.

Follow these steps to create this flow:

1. In your browser, open Node-RED by navigating to http://<your-server-IP>:1880.
2. Drag an 'inject' node (found under the 'input' category) and a 'debug' node (found under the 'output' category) onto the flow.
3. Click and drag from the edge of the 'inject' node to the 'debug' node to wire them together.
4. Click on 'Deploy' in the top right.
5. Click on the left side of the 'inject' node (a small square box) to send a timestamp to the debug node.
6. Click on the 'debug' tab on the right of the screen. You should see a message with the timestamp you just injected.

This indicates that the Node-RED is functioning correctly.

Inspecting Docker container logs

Inspecting the Docker container logs provides another way to verify the operation of your Node-RED instance.

1. Access the terminal of your Ubuntu server.

2. Use the following command to display the logs from the Node-RED Docker container:

```
$ docker logs mynodered
```

3. You should see logs from Node-RED indicating it is running correctly. Look for messages about the server starting, such as "Server now running at http://127.0.0.1:1880/". If you don't see this message, do not be alarmed. As long as you can access the web UI of your Node-RED server in your browser, you can be sure that the server is running.

You have successfully tested your Node-RED Docker container on the Ubuntu 22.04 server. You have verified its functioning by accessing the web interface, creating a simple flow, and inspecting the Docker container logs. Your Node-RED instance is ready for you to develop and run your applications.

5.6. Set up auto-start with Docker Compose

Docker Compose is a tool designed to make it easier to define and manage multi-container Docker applications. While Docker is great for running individual containers, things can get complicated when you have an application requiring multiple containers to work together. That's where Docker Compose comes in.

Imagine you're building a web application that needs a web server, a database, and a caching system. Each of these components could run in its own Docker container. Docker Compose allows you to define all these services in a single file called docker-compose.yml. This file outlines how the containers should interact, what ports should be open, and what resources they can use.

Once you've created this file, you can launch all the containers your application needs with a single command: "docker-compose up". Likewise, stopping all the containers is just as simple: "docker-compose down". This makes managing complex applications much more straightforward.

One of the main benefits of Docker Compose is its simplicity. You define everything in a single, human-readable YAML file. This makes it easy to share and version-control your application's architecture. It's also beneficial for development and testing environments, where you can mimic a multi-container production setup on a single machine.

In this section, you will learn how to configure your Node-RED Docker container to start using Docker Compose automatically. This can be particularly useful if your server restarts - ensuring your Node-RED instance is always available.

Install Docker Compose

Before using Docker Compose, you must ensure it is installed on the Ubuntu 22.04 server, as it is not part of the default Docker installation. Here is how to install Docker Compose:

1. If you haven't already done so, update the package database with the command:

```
$ sudo apt-get update
```

2. Install Docker Compose with the command:

```
$ sudo apt-get install docker-compose
```

3. Verify the installation by checking the version of Docker Compose:

```
$ docker-compose --version
```

The installation of Docker Compose should now be complete, and you should be able to create your Docker Compose configuration file.

Create the Docker Compose configuration

Now that Docker Compose is on your server, you can create a Docker Compose configuration file that tells Docker to run a container with the Node-RED image. Here is how to do this:

1. First, create a directory where you will store your Docker Compose configurtion:

```
$ mkdir ~/node-red && cd ~/node-red
```

2. Next, create a new Docker Compose file named "docker-compose.yml":

```
$ nano docker-compose.yml
```

3. In the text editor, insert the following Docker Compose configuration:

```
version: "3"
services:
  mynodered:
    image: nodered/node-red
    ports:
```

```
            - 1880:1880
          restart: unless-stopped
```

The "restart: unless-stopped" command tells Docker to automatically start the Node-RED container when the Docker daemon starts, which happens at system boot. Ensure the image name is correct and identical to the name you used when you installed Node-RED using the "docker run" command earlier in this chapter.

4. Save and close the file (Ctrl+X, Y, Enter).

Now, you have a Docker Compose configuration file set to start the Node-RED container automatically at boot.

Run Docker Compose

With the Docker Compose file created, you can use it to start the Node-RED container. Here is how to do this:

1. Stop and remove the running Node-RED Docker container to release the port that it currently occupies. We will reuse the same port "1883" in the container that Docker-Compose will manage in the next step:

```
$ sudo docker stop mynodered
$ sudo docker rm mynodered
```

2. From the directory containing your "docker-compose.yml" file, start the Node-RED service with Docker Compose:

```
$ sudo docker-compose up -d
```

The "-d" option tells Docker to run the container in detached mode, which means it runs in the background and doesn't occupy your terminal session. This means that once you start the Docker container in your terminal, you can close the terminal without affecting the container operation.

At this point, the Node-RED Docker container is running and configured to start automatically at boot.

Verify the setup

The final step is to verify that everything has been set up correctly.

1. Check that Docker Compose is running your Node-RED service:

```
$ sudo docker-compose ps
```

This command should show the Node-RED service running.

2. Open your web browser and navigate to http://<your-server-IP>:1880. You should see the Node-RED interface, indicating the service is running correctly.

The configuration of the Node-RED Docker container to automatically start using Docker Compose on the Ubuntu 22.04 server has been completed. Docker Compose provides an efficient way to manage your containers and ensure their proper execution order. Your Node-RED instance is now resilient to server restarts and will be automatically available when your server boots up.

Remember that while Docker Compose is powerful and convenient, it's primarily aimed at development and testing rather than production use. You should look into more advanced orchestration tools like Kubernetes for more complex production environments.

5.7. Set up Node-RED for Data Persistence

Data persistence in Docker involves storing data generated by your Docker container outside the container itself. By default, Docker containers are stateless, and all data generated inside the container is lost when the container is deleted. For Node-RED, this includes all your flows and configurations. Data persistence is essential because without it, all your Node-RED data, including flows and configuration, will be lost when you do something as routine as upgrading Node-RED to the latest version available. Data persistence has another significant benefit: it allows you to back up your Node-RED data only (ignoring the container itself) using simple server environment tools such as the basic "cp" and "rsync" commands. Without data persistence, you would need to make backups of the entire container, yielding bloated and cumbersome to manage backup files.

Fortunately, Docker provides a mechanism to ensure data persistence using volumes, which maps a directory inside a Docker container to a directory on the host system.

In this section, you will learn how to set up a Node-RED container to use a host directory for data persistence. This allows the data to be retained even if the container is deleted.

I will show you two ways to set up data persistence. The first way involves using Docker only. In the second way, you'll set up data persistence with Docker Compose. Because in this book, we use Docker Compose to manage the auto-start of our Node-RED container, you should implement method #2. You should still read and understand how data persistence is set up with Docker, as this will help you understand how to do the same with Docker Compose.

Data persistence with Docker

As you are about to find out, this is very easy. Here are the two steps involved:

1. Create a directory on the host system where Node-RED data will be stored. This can be any directory you choose. For this example, let's use "/home/user/nodered-data":

```
$ mkdir -p /home/user/nodered-data
```

2. Run the Node-RED Docker container, mapping the host directory to the data directory in the Node-RED container using the "-v" (volume) flag:

```
$ docker run -it -p 1880:1880 -v /home/user/nodered-data:/data
--name mynodered nodered/node-red
```

In this command, "-v /home/user/nodered-data:/data" tells Docker to map the "/home/user/nodered-data" directory on the host system to the "/data directory" inside the Node-RED container. Any changes made inside the "/data" directory inside the Node-RED container will be saved in "/home/user/nodered-data" on the host system.

The Node-RED data is stored on the host system and will remain intact even if the Docker container is removed. The next time you run a new Node-RED container with the same volume mapping, it will have access to the previously saved flows and configurations.

Note: Be sure to replace "/home/user/nodered-data" with the actual path on your system where you want to store Node-RED data.

Data persistence with Docker Compose

Let's set up data persistence with Docker Compose. This is the preferred method because, in this book, we use Docker Compose to manage the automatic start of our Node-RED container when the server starts. Here are the steps:

1. Create a directory on the host system where Node-RED data will be stored. This can be any directory you choose. For this example, let's use "/home/user/nodered-data":

```
$ mkdir -p /home/user/nodered-data
```

2. Go to the Node-RED Docker Compose directory, where the "docker-compose.yml" file is stored, and use "nano" to edit this file. We'll add new configuration lines that point to the data persistence directory you created in Step 1.

```
$ cd ~/node-red
$ nano docker-compose.yml
```

1. Edit the file to contain the contents you can see below:

```
version: "3.7"

services:
```

```
      node-red:
        image: node-red/node-red:latest
        container_name: mynodered
        environment:
          - TZ=Australia/Sydney
        ports:
          - "1880:1880"
        volumes:
          - /home/peter/nodered-data:/data
        restart: unless-stopped

    volumes:
      node-red-data:

    networks:
      node-red-net:
```

2. Stop, remove, and restart the container using Docker Compose (remember, this command must be issued in the Docker Compose directory, that is, the directory that contains the "yml" file):

```
$ docker-compose stop
$ docker-compose rm
$ docker-compose up -d
```

At this time, the Node-RED container will start, and files will begin to appear in the persistence directory "~/nodered-data". Navigate to this directory with "~/nodered-data" and use "ls -al" to view the files. One of them will be "flows.json", which is the file that contains all the flows running on the Node-RED server.

By using Docker volumes for data persistence, you can ensure the longevity of your Node-RED data, making your container setup more resilient and robust. For more information on Docker volumes, refer to the official documentation at https://docs.docker.com/storage/volumes/. You can also find information on setting up data persistence for Node-RED in the Node-RED documentation at https://nodered.org/docs/getting-started/docker.

5.8. Maintaining Your Instance of Node-RED

Regular maintenance is the key to running a thriving server environment, including Docker and its containers like Node-RED. Maintenance tasks ensure your system is secure, efficient, and up-to-date. These tasks include performing updates, taking backups, and restoring backups when necessary.

In this section, you will learn how to perform these basic maintenance tasks for Docker and your Node-RED container. Please note that there are many other ways to perform maintenance tasks, particularly taking backups. I am discussing the base-case scenarios in this section to help you get started. Specifically, the backup and restoration method I describe below will work on any Node-RED instance, whether within a Docker container, a virtual machine, or even directly on "metal" (i.e. on your computer without any isolation).

Backing Up Node-RED flows

Having a backup of your Node-RED flows is critical. In case of a system failure or data corruption, a backup can restore your work without starting from scratch. In Node-RED, backing up all your flows with just a few clicks is easy. Here is how to do this (the labels in the screenshot below correspond to the step numbers in the list):

Figure 1.5.8.3: A simple way to backup all your flows

1. Navigate to your Node-RED flow editor at http://<your-server-IP>:1880.

2. Click on the 'hamburger' menu in the top-right corner.

3. Select 'Export'. This will open the export window.

4. Choose whether to export the current or all flows. If you want to backup all your flows, choose "all flows". However, if you want to export the current flow only (for sharing, copying to another instance of Node-RED, etc.), choose "current flow".

5. Select the 'JSON' tab to make the JSON document for the flow (or all flows) available for copy. If you want to copy specific nods, select the "Export nodes" tab.

6. Click 'Copy to clipboard' to copy the selected JSON document to your clipboard for immediate paste to a text editor or the Import window of another instance of Node-RED. If you want to download a file of the JSON document for storage, choose the "Download" option.

Through these steps, you now have a backup of your Node-RED flows. It would be best to regularly back up your flows, especially before and after making significant changes.

Restoring Node-RED flows from backup

Restoring your Node-RED flows from a backup is just as straightforward. Here is how to restore the Node-RED flows from a JSON backup file:

1. Navigate to your Node-RED flow editor at http://<your-server-IP>:1880.

2. Click on the 'hamburger' menu in the top-right corner and select 'Import'.

3. Open your backup file, copy its content, and paste it into the import dialogue box.

4. Click 'Import' to restore your flows.

It would be best if you did a practice run of the backup and restoration process of your Node-RED flows so that should you ever need to recover your work in case of data loss or corruption, you will know how to do so.

Keeping Docker up-to-date

Docker, like any other software, is regularly updated by its developers. These updates often provide security patches, bug fixes, and new features. To ensure the highest level of security and performance, it's essential to update Docker to the latest version regularly.

Before you attempt the steps below, you should fully back up your flows, as you learned in the previous two sections of this chapter. When you are sure you have a good backup, continue. The upgrade process includes downloading the latest version of the Node-RED Docker image and replacing the old image with the new image. In the instructions below, I assume you have set up your Node-RED container with persistence, as described in the previous chapter. By the end of this section, you will have updated Docker to its latest version, ensuring a secure and efficient environment.

Here are the steps involved:

1. To update Docker, first update the package database with the following command:

```
$ sudo apt-get update
```

2. Next, upgrade Docker to the latest version using the following command:

```
$ sudo apt-get upgrade docker.io
```

Now, you have the latest version of the Docker runtime environment. Let's continue with the upgrade of the Node-RED container.

Updating Node-RED container

Like Docker, Node-RED is also regularly updated. These updates can bring new features, bug fixes, and security enhancements. It's essential to keep your Node-RED container updated to ensure it is secure, stable and has the latest features. Here is the process:

1. To update the Node-RED Docker container, first pull the latest image:

```
$ docker pull nodered/node-red
```

2. Then stop and remove the current Node-RED container. The result of this step is equivalent to deleting the data (all code and configuration) that makes up the Node-RED container. Because you set up data persistence earlier, your Node-RED flows and configuration are safe outside the container and will be used by the new and upgraded Node-RED container you will install shortly. To remove the current Node-RED contain, execute these two instructions:

```
$ docker stop mynodered
$ docker rm mynodered
```

3. Run a new Node-RED container with the latest image. This command connects the container to the persistence directory. This command should be familiar to you, as it is the same command you used to start the container in the previous section:

```
$ docker run -it -p 1880:1880 -v /home/user/nodered-data:/data --name mynodered
nodered/node-red
7 Sep 07:49:33 - [info]

Welcome to Node-RED
===================

7 Sep 07:49:33 - [info] Node-RED version: v3.1.0
7 Sep 07:49:33 - [info] Node.js  version: v16.20.2
7 Sep 07:49:33 - [info] Linux 5.15.0-79-generic x64 LE
7 Sep 07:49:34 - [info] Loading palette nodes
7 Sep 07:49:37 - [info] Dashboard version 3.4.0 started at /ui
7 Sep 07:49:37 - [info] Settings file  : /data/settings.js
7 Sep 07:49:37 - [info] Context store  : 'default' [module=memory]
7 Sep 07:49:37 - [info] User directory : /data
…
```

Wait for around 30 seconds before you try to load the web UI in your browser as Docker's services start. In the console output you see above, my updated instance of Node-RED is version 3.1.0, and Node.js is version 16.20.2. Both are the latest at the time I write these lines. When I started writing this book, Node-RED was at version 3.0.1.

Maintaining Docker and the Node-RED container on your Ubuntu 22.04 server involves regularly performing updates and managing backups. Keeping your software up-to-date ensures that you are running a secure and efficient environment. Periodically backing up and knowing how to restore your Node-RED flows protects your work from data loss.

To learn more about Docker maintenance, refer to the official documentation (see https://docs.docker.com/engine/install/ubuntu/). Remember, maintenance is not a one-time task but a routine part of managing a healthy server environment. Following these steps, your Docker and Node-RED container should be current and up-to-date.

5.9. Security

Node-RED's flexibility and functionality make it a powerful tool for developing applications. However, if you're using Node-RED, particularly in a production environment, it's essential to implement security measures to safeguard your installation and flows. This section will teach you how to secure your Node-RED setup, including using HTTPS, enabling authentication, and restricting access to sensitive nodes and configuration settings.

Using HTTPS for secure communication

The standard Node-RED installation communicates over HTTP, which is unencrypted. This means any data sent between the browser and Node-RED server is in clear text and can be intercepted by an attacker. Setting up HTTPS is essential to protect your data, which encrypts communication.

To set up HTTPS on your Node-RED server, you will need access to its command line and to issue a few commands. It is an excellent time to take a backup of your server now to make it easy to recover if you make a mistake.
Here's how to enable HTTPS:

1. **Generate a self-signed certificate**: Node-RED requires a private key and a certificate to enable HTTPS. The second command will ask you questions as part of the certificate-creation process. You can accept the defaults or add as many details as you want. It is essential to leave the last challenge password question to the default empty string to avoid having to enter this password every time the server starts. You can create the private key and certificate with OpenSSL like this:

   First, login to your Node-RED server via SSH:

```
$ ssh peter@192.168.20.18
```

Then, navigate to your Node-RED Docker container's persistence folder, which you (hopefully) set up earlier in this chapter.

```
$ cd nodered-data/
```

Create a new directory titled "ssl", where you will save the key and certificate, and navigate to it:

```
$ mkdir ssl
$ cd ssl
```

Continue with the OpenSSL commands:

```
$ openssl genrsa -out privatekey.pem 2048
$ openssl req -new -sha256 -key privatekey.pem -out csr.pem
$ openssl x509 -req -in csr.pem -signkey privatekey.pem -out certificate.pem
```

Here is my session:

```
peter@noderedvm:~/nodered-data/ssl$ openssl genrsa -out privatekey.pem 2048
peter@noderedvm:~/nodered-data/ssl$ openssl req -new -sha256 -key
privatekey.pem -out csr.pem
You are about to be asked to enter information that will be incorporated
into your certificate request.
What you are about to enter is what is called a Distinguished Name or a DN.
There are quite a few fields but you can leave some blank
For some fields there will be a default value,
If you enter '.', the field will be left blank.
-----
Country Name (2 letter code) [AU]:
State or Province Name (full name) [Some-State]:NSW
Locality Name (eg, city) []:Sydney
Organization Name (eg, company) [Internet Widgits Pty Ltd]:TechExplorations
Organizational Unit Name (eg, section) []:
Common Name (e.g. server FQDN or YOUR name) []:
Email Address []:

Please enter the following 'extra' attributes
to be sent with your certificate request
A challenge password []:
An optional company name []:
peter@noderedvm:~/nodered-data/ssl$ openssl x509 -req -in csr.pem -signkey
privatekey.pem -out certificate.pem
Certificate request self-signature ok
subject=C = AU, ST = NSW, L = Sydney, O = TechExplorations
```

The "ssl" directory now contains these three files:

```
peter@noderedvm:~/nodered-data/ssl$ ls -al
total 20
drwxrwxr-x 2 peter peter 4096 Jun 21 04:37 .
drwxrwxr-x 6 peter peter 4096 Jun 21 04:36 ..
-rw-rw-r-- 1 peter peter 1131 Jun 21 04:37 certificate.pem
-rw-rw-r-- 1 peter peter  960 Jun 21 04:37 csr.pem
-rw------- 1 peter peter 1704 Jun 21 04:36 privatekey.pem
```

Continue with the Node-RED configuration.

2. **Configure Node-RED to use the certificate**: In your Node-RED settings file, you'll find a section commented out for https. As I am using the Docker persistence directory for my Node-RED instance, the location of the settings file is "~/nodered-data/settings.js". Uncomment it and provide the paths to your private key and certificate:

```
https: {
    key: require("fs").readFileSync('/data/ssl/privatekey.pem'),
    cert: require("fs").readFileSync('/data/ssl/certificate.pem')
},
```

Before restarting the Docker container to make these changes effective, let's take a minute to understand why these changes work. Remember that earlier in this chapter, you created a persistence directory for the Node-RED Docker container with this command:

```
$ docker run -it -p 1880:1880 -v /home/user/nodered-data:/data --name
mynodered nodered/node-red
```

In this command, I marked the path "/data" in bold font. This is the path that the Docker container "sees" as its installation directory when it operates. Therefore, when we start the container and Node-RED, the Node-RED configuration file exists in the "/data" directory and not in the "~/nodered-data" we see when working with the container's host virtual machine.

If you would like to ensure that calls to HTTP are redirected to HTTPS automatically, uncomment this line in the configuration file:

```
requireHttps: true,
```

With this, the new configuration is ready. Restart the Node-RED Docker container to make the change effective:

```
$ docker restart mynodered
```

To verify that the container has restarted, use "docker ps":

```
$ docker ps
CONTAINER ID        IMAGE     COMMAND          CREATED          STATUS          PORTS
NAMES
f7271f95980e   nodered/node-red:latest   "./entrypoint.sh"   6 weeks ago   Up 35
seconds (healthy)   0.0.0.0:2880->1880/tcp, :::2880->1880/tcp   mynodered
```

Now, you should be able to access Node-RED over HTTPS at https://localhost:1880.

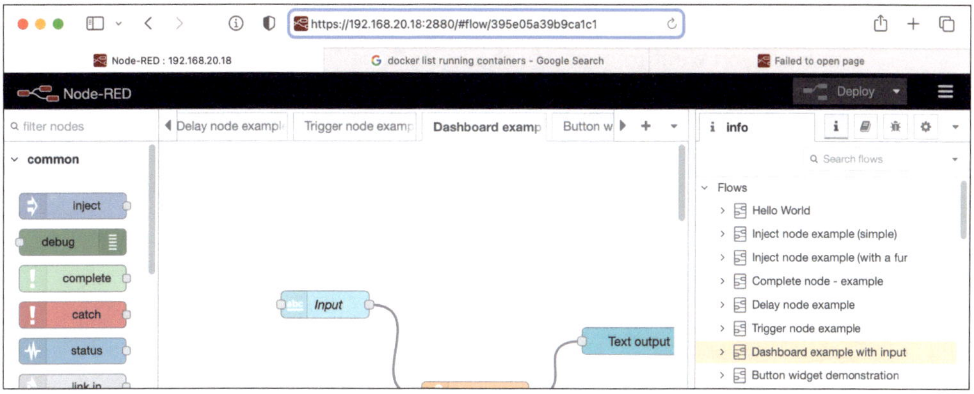

Figure 1.5.9.4: Securing Node-RED with HTTPS.

Enabling authentication in Node-RED

By default, Node-RED's editor doesn't require a login. This leaves your flows and configuration data open to anyone accessing the Node-RED server. To add a layer of security, you can enable username/password authentication.

To set up authentication, you must first create a user within Node-RED (not the operating system user). As in this book, I am working with a Docker container for Node-RED; we will make use of information found in the "Running under Docker" page of the Node-RED documentation (see https://nodered.org/docs/getting-started/docker).

Here is how to create a new user in Node-RED:

1. Login to the server where the Docker container is running using SSH (replace the IP address and username in the example below to match those in your server):

```
$ ssh peter@192.168.20.18
```

2. Navigate to the Node-RED persistence folder:

```
$ cd ~/nodered-data
```

3. Generate a password hash. For this purpose, Node-RED provides a utility named "node-red-admin". However, this utility is not exposed outside the Docker container, so we need an alternative way to create the password hash. Conveniently, Docker offers the "exec" command, which allows us to invoke the bash command prompt of the container and use it. Go ahead and use the Docker "exec" command like this:

```
$ docker exec -it mynodered /bin/bash
```

4. The command Node-RED container command prompt will appear. Use it to issue the node-red-admin command. This command will prompt you for the password you want to generate the hash. Type "exit" to return to the regular server prompt. I will create a hash for the password "noderedpass":

```
bash-5.1$ node-red-admin hash-pw
Password: <type in your password>
$2b$08$L4ajUDxj9G243YapMFoC5uItQ4K1Z0l7IA8qPIzWv44bTENB6Wxhm
bash-5.1$ exit
exit
```

Copy the hash string as you will need it in the next step.

5. Add the user to the settings file. In the settings.js file, find the adminAuth section. Uncomment it and add a user, replacing "admin" and my generated password hash with your username and the hashed password:

```
adminAuth: {
    type: "credentials",
    users: [{
        username: "admin",
        password: "$2b$08$L4ajUDxj9G243YapMFoC5uItQ4K1Z0l7IA8qPIzWv44bTENB6Wxhm",
        permissions: "*"
    }]
},
```

Restart Node-RED with "docker restart nodered", and you'll be prompted for the username and password when you access the editor. Here is the login prompt:

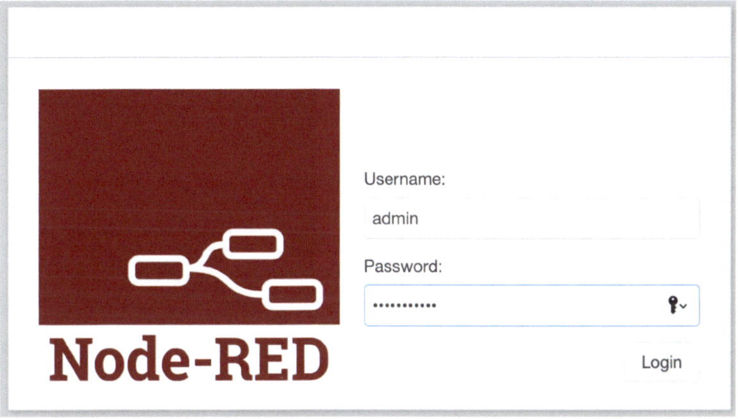

Figure 1.5.9.5: Node-RED secured with user credentials.

Once you log in, you will also notice that a user icon appears in the top-right corner of the Node-RED editor. Click on that icon to reveal the Logout link (see below):

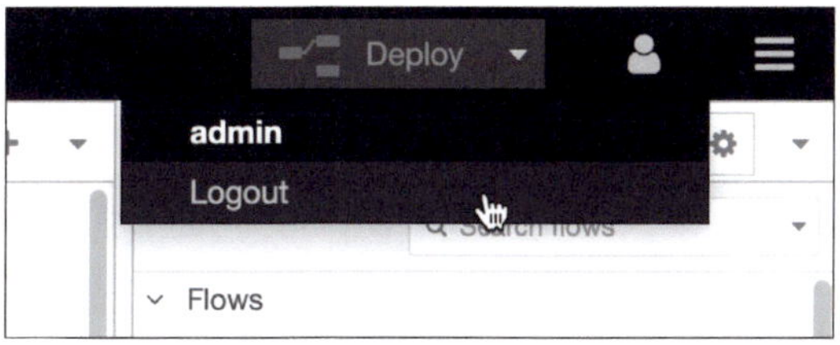

Figure 1.5.9.6: Logout by clicking on the user icon.

Implementing Node-Level Permissions

While user authentication protects your Node-RED editor at a high level, you may also want to control access more granularly. Node-RED allows you to assign permissions on a per-node basis, enabling you to restrict which users can view or edit specific nodes.

To do this, you must modify the adminAuth section of the "settings.js" file. Here's an example where different permissions are assigned to different users:

```
adminAuth: {
    type: "credentials",
    users: [
        {
            username: "admin",
            password: "YOUR_ADMIN_PASSWORD_HASH",
            permissions: "*"
        },
        {
            username: "user",
            password: "YOUR_USER_PASSWORD_HASH",
```

```
                permissions: "read"
            }
        ]
    },
```

In this example, the 'admin' user has full permissions, while the 'user' only has read permissions.

Limiting access to Node-RED Dashboard

The Node-RED Dashboard is a powerful tool for creating user interfaces for your flows. However, you should restrict who can access it. By enabling HTTP authentication for the dashboard, you can require a username and password to access it.

To enable HTTP authentication for the dashboard, add the following to the "settings.js" file:

```
httpNodeAuth: {user:"user", pass:"YOUR_PASSWORD_HASH"},
```

After saving the changes and restarting Node-RED, you'll be prompted for the username and password when accessing the dashboard.

Securely Managing Node-RED Configuration Settings

Node-RED stores configuration settings in the "settings.js" file, including essential security configurations. It's crucial to protect this file from unauthorised access.

Here are a few steps to secure your configuration settings:

1. **File permissions**: Ensure the "settings.js" file has appropriate permissions, limiting who can read and write to the file. For instance, you could use the following command only to allow the file owner to read and write to it:

```
chmod 600 settings.js
```

2. **Sensitive data**: Avoid storing sensitive data, like passwords or API keys, directly in the "settings.js" file. Consider using environment variables or Node-RED's built-in credential management system.

By implementing these measures, you can significantly enhance the security of your Node-RED installation and protect your flows from potential threats. Security is an ongoing process; you should always stay informed about the latest best practices and updates.

Chapter 1.6 • Node-RED Basics

In the previous chapters, you learned about Node-RED at a high level and installed a server instance in a virtual machine with Docker. It is now time to roll up your sleeves and start using it.

In the following parts and chapters of this book, you will learn a wide range of Node-RED features and capabilities that will enable you to create any Node-RED project you can imagine. To help you focus your learning, I have listed the five core skills that any Node-RED programmer must have. Remember, you will not build those skills with the help of any single experiment in this book. But you will build them by working on all the experiments, as each will help you gradually learn and improve.

As you work more with Node-RED, there are a few essential skills you should focus on developing:

1. Understanding JSON: Many nodes work with messages in JSON format, so understanding how to work with JSON is crucial.

2. Working with Function Nodes: These nodes allow you to write JavaScript code to manipulate messages and offer a powerful tool to customise your flows.

3. Understanding Asynchronous Operations: Since Node-RED is built on Node.js, understanding how asynchronous operations work will be beneficial.

4. Debugging: Learn how to use the debug node and debug sidebar to troubleshoot your flows.

5. Securing Your Node-RED: Understanding how to secure your Node-RED instance is essential, especially when exposing it to the internet.

Mastering these skills will significantly enhance your proficiency with Node-RED.

You will start to develop those skills right now. In this chapter, you will learn the following:

1. The Node-RED editor allows you to build your flows by dragging and dropping "nodes" onto your workspace and wiring them together.

2. Nodes, which consist of the building blocks of Node-RED flows that allow you to process and route data between various sources and destinations.

3. How to create and deploy flows. Flows are collections of nodes analogous to a program written in a text-based language like Python or Ruby.

4. How to use several fundamental nodes such as "debug", "function", "inject", and "trigger".

5. How to fine-tune your Node-RED instance by customising its settings and configuration.

6. The necessary resources outside of this book, where you can look for more information and help.

Let's continue with an introduction to the Node-RED editor.

6.1. Understanding the Node-RED Editor

Node-RED offers a visual editor that allows you to build your flows by dragging and dropping "nodes" onto your workspace and wiring them together. It provides a great deal of flexibility and functionality that can seem overwhelming at first. This section will help you become familiar with the Node-RED editor and comfortable using its most essential functions.

Accessing the Node-RED editor

The Node-RED editor is a web application you can access through a web browser. It serves as the primary interface for creating and managing your flows. Follow these steps to use the Node-RED editor:

1. Open your preferred web browser. The editor will work with all modern web browsers.

2. Type the URL for your Node-RED instance in the address bar. This is usually your server's IP address or domain, followed by the port number (1880 by default). If you specified a different port when you launched the Node-RED server with the "docker run" command or in the Docker Compose configuration file, remember to use the specified port instead of the default "1880". For example:

```
http://<your-server-IP>:1880
```

3. Hit the "Enter" key to load the Node-RED editor.

At the end of this section, you should be able to access your Node-RED editor from your web browser, marking your entry point into the Node-RED environment.

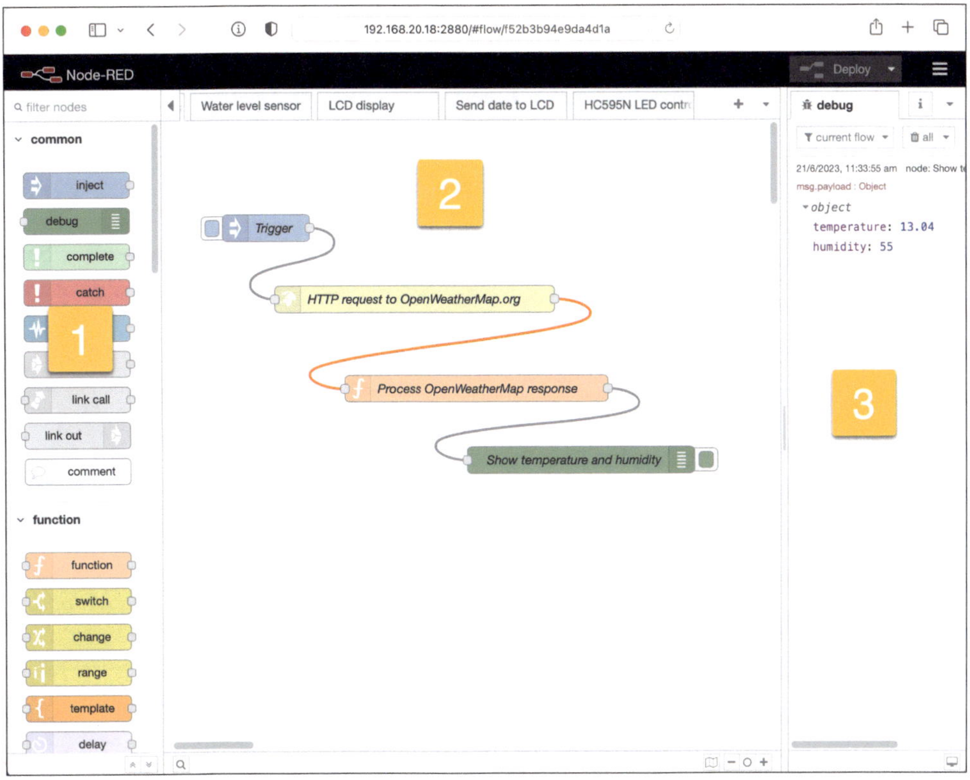

Figure 1.6.1.1: The Node-RED editor.

Node-RED Editor overview

The Node-RED editor is divided into three main sections: the Node Palette, the Flow Workspace, and the Sidebar.

1. **Node Palette**: Located on the left-hand side (marked "1" in Figure 1.6.1.1), the Node Palette contains different nodes that you can drag onto the workspace to build your flows. Nodes are categorised into input, output, function, and social. You can install additional nodes via the Palette Manager, which you will learn about later.

2. **Flow Workspace**: This is the central area where you build your flows (marked "2" in Figure 1.6.1.1). You can drag nodes from the Node Palette onto the workspace, wire them together, and arrange them as you see fit. In the example of Figure 1.6.1.1, you can see that the flow workspace contains multiple tabs. Each tab includes a flow that you can edit and deploy.

3. **Sidebar**: On the right-hand side is the sidebar (marked "3" in Figure 1.6.1.1), which contains several tabs for Info, Debug, and others. The Debug tab is helpful for viewing debug output messages from your flows. You will also use the Sidebar to access the Dashboard.

Understanding the layout of the Node-RED editor is the first step in becoming proficient in using it.

Creating a flow

Creating a flow involves dragging nodes from the Node Palette onto the Flow Workspace and wiring them together to perform a specific task.

1. Create a new flow workspace by clicking the "+" button. You will find this button in the top-right corner of the editor (see below).

Figure 1.6.1.2: Create a new flow.

2. Drag a node from the Node Palette onto the Flow Workspace. In the example below, I am dragging an Inject node, followed by a debug node.

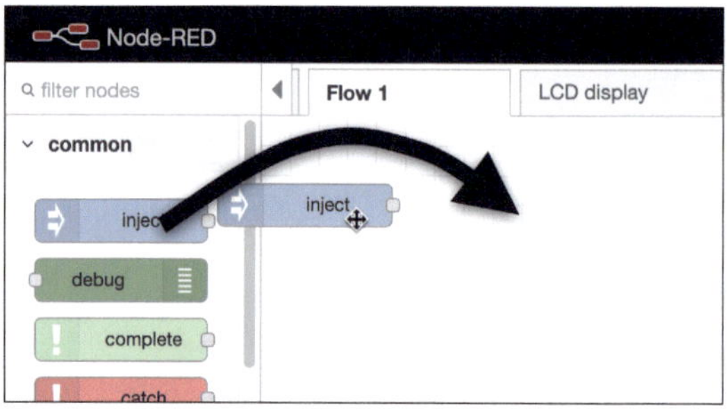

Figure 1.6.1.3: Drag a node to the workspace.

3. Click and drag the handle on the edge of a node to wire it to another node. Nodes are connected from the output (right side) to the input (left side). I am wiring the Inject node's output with the debug node's input in the example below. When the wiring is complete, the colour of the wire will be grey.

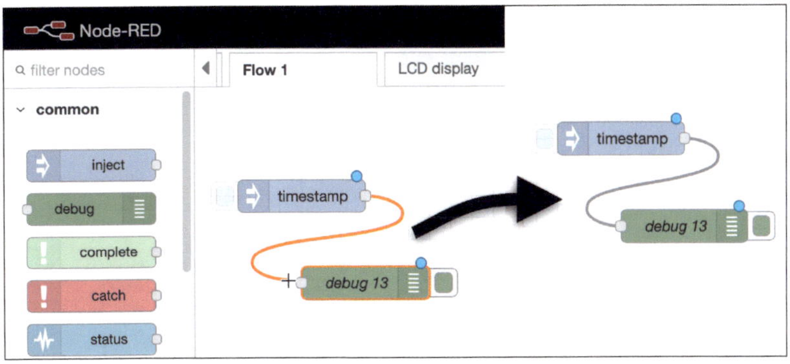

Figure 1.6.1.4: Wire the output of the Inject node with the input of the debug node.

4. To edit a node's properties, double-click on it. This opens a configuration window where you can set different options based on the node type. Every node has a different set of properties that you can edit. The Inject node's properties in the example below contain the two text fields "payload" and "topic".

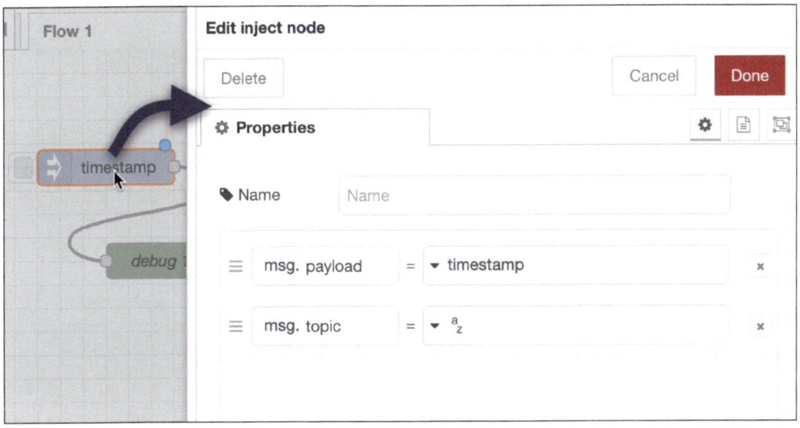

Figure 1.6.1.5: Double-click on a node to show its properties.

5. Once you finish your flow, click the 'Deploy' button in the top-right corner. This will save and activate your flow. Deploying a flow is similar to "running" a program. Every time you change the flow, you must redeploy it for the changes to become effective.

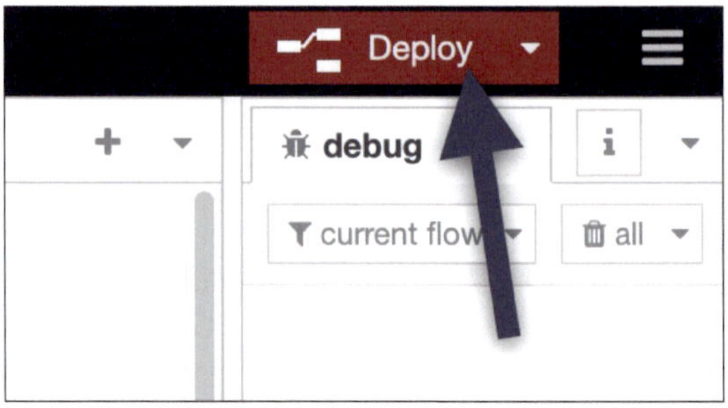

Figure 1.6.1.6: To deploy (run) a flow, click the Deploy button.

6. Use the flow. In this first example, you can click on the button of the Inject node. This will cause the Inject node to send a message that contains the current timestamp from its output. The Inject node message will then be received by the Debug node, which will print it in the Debug window.

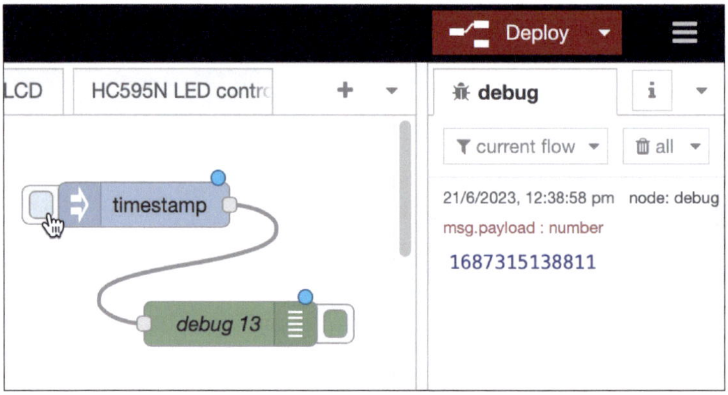

Figure 1.6.1.7: Running the deployed flow.

You can create and deploy a basic flow in Node-RED with these steps.

Importing and exporting flows

Node-RED allows you to export and import flows. This is useful when you want to back up your flows or share them with others. In this book, I have used the export feature to export all example flows and share them with you on the book's GitHub repository. From there, you can copy and import them to your Node-RED instance to play with.

- **Exporting Flows**: To export a flow with the flow's tab active, click on the hamburger menu (≡) in the top right corner and then click "Export". The Export window will appear. The Export window gives you various export options. For example, you can export all flows (an excellent way to do a full server backup) or the current flow only. You can also choose where to export to a file or the clipboard. I am exporting the JSON code for the current flow to the clipboard in the example below. Later, I can paste the JSON code into a text file on my computer.

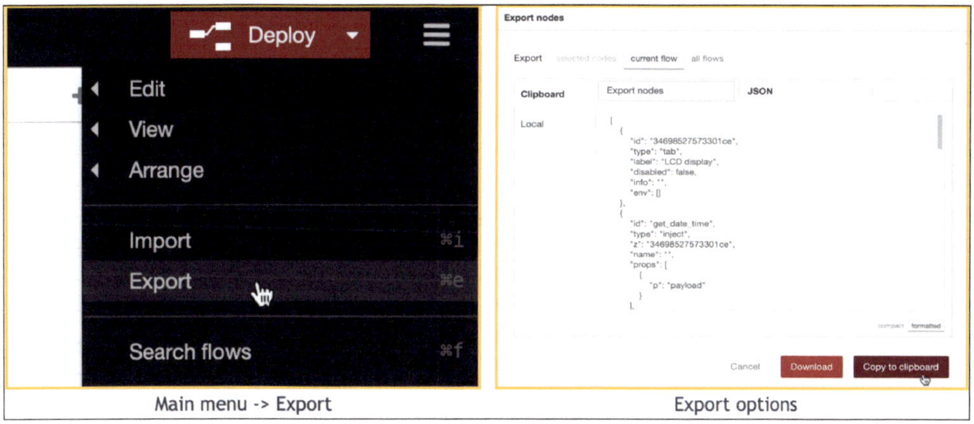

Figure 1.6.1.8: Export function and options

- **Importing Flows**: To import a flow, select the flow tab you'd like to import to. Then, click on Import from the hamburger menu (≡) in the top right corner. This will open the Import window. This window gives you several import options. I am pasting the JSON code I copied from the previous export function in the example below. You can also import JSON code stored in a file. When you click the "Import" button, the imported flow will appear in the editor.

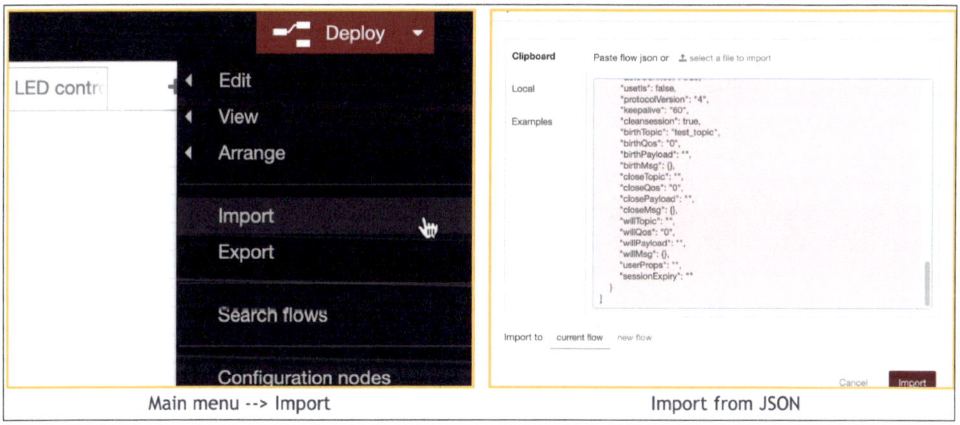

Figure 1.6.1.9: Import function and options

You can use the import and export flow functions to share your flows with other people, move them between different Node-RED instances, or keep backups.

Zoom and Pan

Navigating your workspace is essential to managing your Node-RED flows, especially as they increase in complexity. Here, we cover the zoom and pan features.

- **Zoom**: You can zoom in and out of the workspace using the slider in the right-hand corner.

Figure 1.6.1.10: Zoom buttons

- **Pan**: To pan around your workspace, click and drag anywhere on the workspace that isn't occupied by a node. If your mouse has three buttons (or a scroll wheel with a button), press the middle button to activate panning.

This ability to zoom and pan will enable you to maintain a good overview of your workspace, no matter how large or complex your flows become.

Grid and Align/Distribute

The grid and align features help you to manage and organise your nodes and flows in an orderly fashion.

- **Grid**: You can enable the grid from the User Settings window (≡ > Settings > Grid). When the grid is enabled, you can set the nodes to snap to the nearest grid point, helping you keep your workspace tidy.

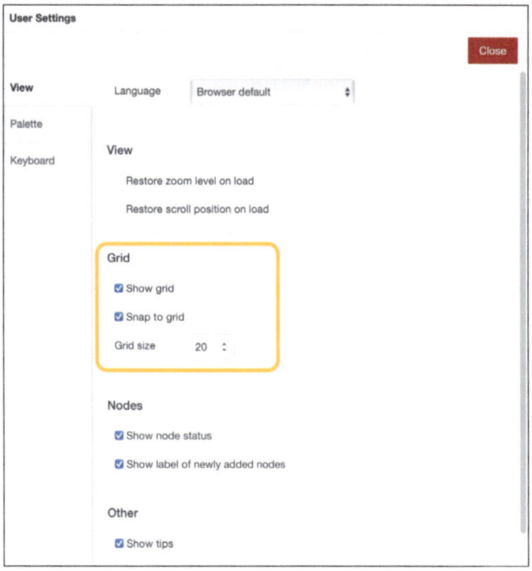

Figure 1.6.1.11: Grid settings

- **Align/Distribute**: To align nodes, select them with your mouse, click on the main menu and choose the desired arrangement function from the Arrange sub-menu. This helps to align your nodes for a cleaner look and easier understanding of your flows.

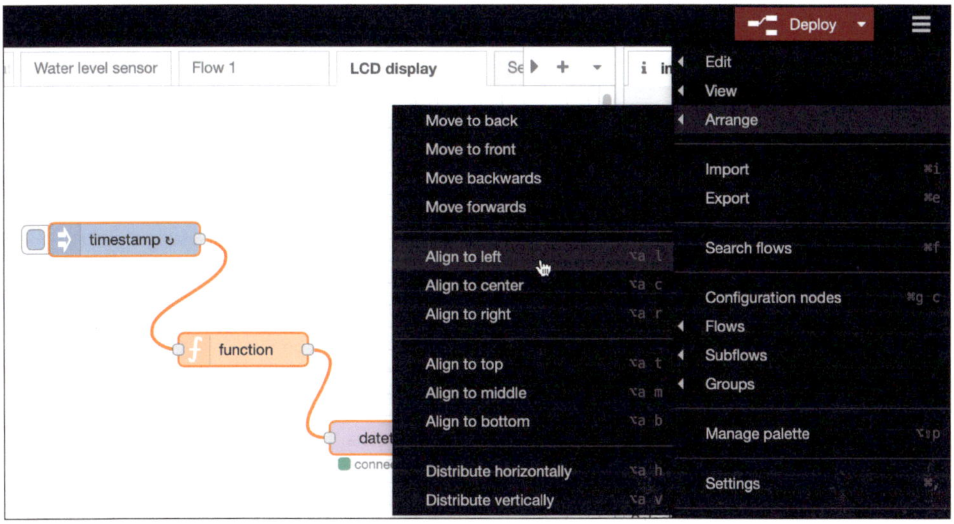

Figure 1.6.1.12: Align and distribute nodes.

Using the grid and align tools will help you keep your workspace organised, making it easier to design and understand your flows.

Search

The search feature helps you to find nodes quickly and flows in your workspace.

- **Step 1**: Access the search by clicking on the magnifying glass icon in the bottom-left corner of the editor or use the keyboard shortcut Ctrl + F (or Command + F on Mac).

Figure 1.6.1.13: Search

- **Step 2**: Type your search term into the search bar that appears. This will highlight any matching nodes or flows in your workspace. This search feature is handy as your flows grow in size and complexity. Use this feature to find flows and nodes across your entire Node-RED instance. You can also narrow your searches to subsets of nodes and flows. For example, you can use filters to search among modified or invalid nodes only.

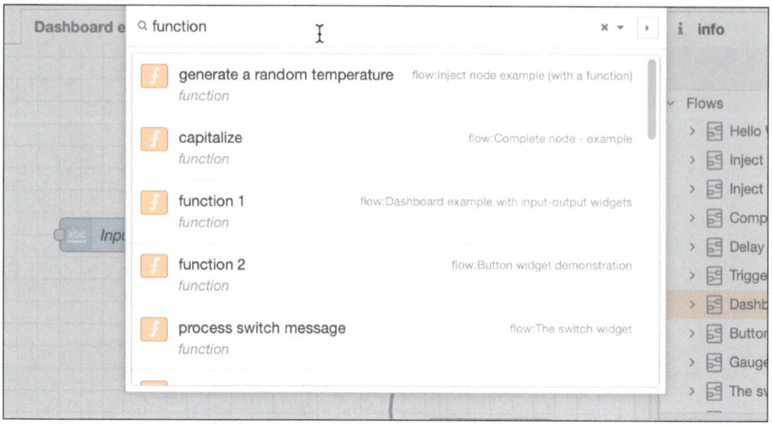

Figure 1.6.1.14: Example search

The "search" feature is a powerful tool for quickly finding and navigating to parts of your flows, saving you from manually locating them.

Keyboard Shortcuts

Node-RED offers a variety of keyboard shortcuts to increase your efficiency in the editor.

- **Ctrl + I**: Import nodes.
- **Ctrl + E**: Export nodes.
- **Ctrl + D**: Delete selected nodes.
- **Ctrl + Z**: Undo last action.
- **Ctrl + Shift + Z**: Redo last action.

A complete list of keyboard shortcuts can be found by going to the main menu "☐", then Keyboard Shortcuts, or using the shortcut "Shift + ?".

Using keyboard shortcuts can significantly improve your efficiency and speed when creating and managing your Node-RED flows.

These additional sections on zooming, panning, aligning, using the grid, searching, and using keyboard shortcuts further enhance your proficiency in the Node-RED editor.

6.2. Nodes

Nodes are the building blocks of Node-RED flows. Nodes can process and route data between various sources and destinations. In this section, you will learn the concept of "nodes" in Node-RED and learn about some of the most essential built-in nodes. Later in this book, you will learn how to use the nodes discussed here "hands-on" in the projects.

What is a Node?

In Node-RED, a node represents a specific functionality, such as receiving or sending data, calculating, or controlling a device. Each node has input and output ports, which can be connected to other nodes using wires. Nodes can have properties and configurations that

determine their behaviour, and they can exchange messages with other nodes through their ports.
Nodes in Node-RED are based on flow-based programming, where data flows through a series of connected nodes that process and transform the data as needed.

Built-in nodes
Node-RED comes with a set of built-in nodes that cover a wide range of functionalities, such as:

- **Input nodes** receive data from external sources, such as HTTP requests, MQTT messages, or file inputs. Examples include the "http in", "mqtt in", and "file in" nodes.
- **Output nodes** send data to external destinations, such as HTTP responses, MQTT messages, or file outputs. Examples include the "http response", "mqtt out", and "file out" nodes.
- **Function nodes** provide specific processing or transformation capabilities, such as parsing JSON data, performing calculations, or making decisions based on conditions. Examples include the "json", "switch", and "change" nodes.
- **Debug nodes** help you inspect and debug your flows by displaying messages and data in the Debug sidebar. The "debug" node is the most commonly used.

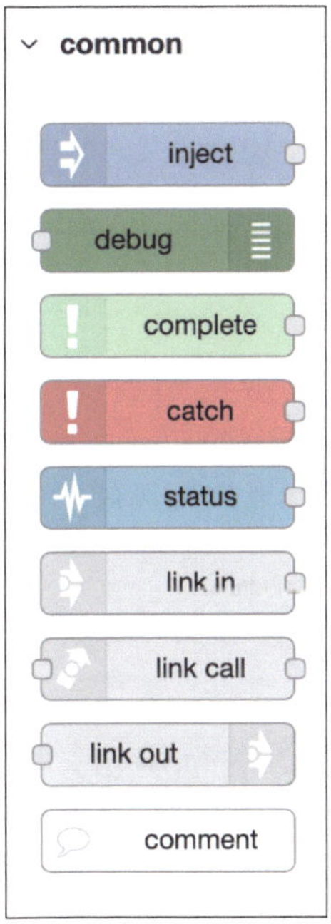

Figure 1.6.2.15: Example common nodes.

Third-party nodes
In addition to the built-in nodes, Node-RED supports third-party nodes that can be installed from the Node-RED palette manager or by using the npm package manager. These nodes extend Node-RED's functionality by supporting specific devices, protocols, or services.

For example, you can find third-party nodes to interact with databases, control smart home devices, or integrate with cloud services such as AWS, Google Cloud, and Microsoft Azure.

To install a third-party node, follow these steps:

1. Open the Node-RED editor and click the menu button in the upper-right corner.
2. Select "Manage Palette" from the menu.
3. Click the "Install" tab in the Palette Manager.
4. Search for the desired node package and click the "Install" button next to the package.

In the example below, I have installed two third-party nodes that allow my flows to interact with Amazon Echo hubs and devices. To find these nodes, search for "amazon" in the Install tab of the Palette Manager. The name of the palette that contains these nodes is "node-red-contrib-amazon-echo-theukshadow".

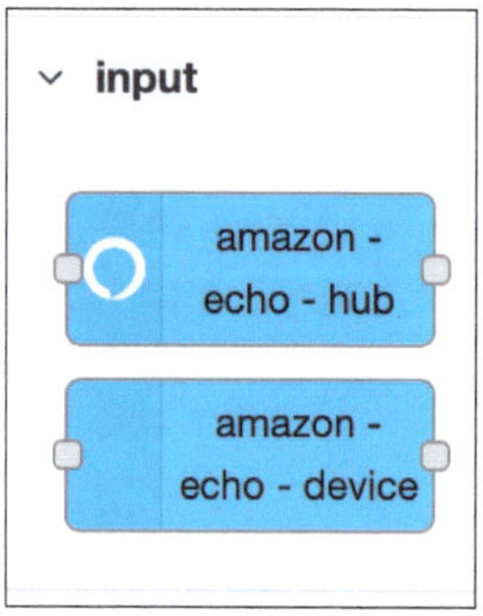

Figure 1.6.2.16: Example of third-party nodes.

Best practices for working with nodes

When working with nodes in Node-RED, consider the following best practices to create maintainable and efficient flows:

1. **Organize your flows**: Use multiple tabs (sheets) in the workspace to separate different project parts, making them easier to understand and maintain. Group related nodes together and use comments to provide context and explanations.

2. **Reuse nodes**: If you have a set of nodes that perform a common task, consider creating a subflow to encapsulate that functionality. Subflows can be used multiple times within your project, making it easier to maintain and update your flows.

3. **Choose appropriate nodes**: Select the most suitable node for the task. While it's possible to accomplish many tasks using the built-in function node with custom code, a specific node designed for the task can result in more readable and maintainable flows.

4. **Optimize performance**: Be mindful of the performance implications of your flows, especially when processing large amounts of data or working with resource-intensive nodes. Use nodes that process data efficiently and consider using rate limiting or buffering strategies to manage resource usage.

5. **Secure your flows**: Protect your Node-RED installation and flows by following security best practices, such as using HTTPS, enabling authentication, and restricting access to sensitive nodes and configuration settings.

By following these best practices and understanding the role of nodes in Node-RED, you can create powerful and efficient flows to tackle a wide range of challenges and automation tasks.

6.3. Creating and Deploying Flows

In this section, you will focus on flows and how to create and operate them. After this you will learn about the best practices for designing and maintaining flows in Node-RED.

What are flows?

In Node-RED, a flow represents a collection of connected nodes that work together to perform a specific task or solve a particular problem. Flows are created by adding nodes to the workspace, connecting them using wires, and configuring their properties and behaviour.

Flows can be organised into separate tabs (also known as "sheets") within the Node-RED editor, allowing you to divide your project into logical sections or group related functionality together.

The screenshot below shows a flow with multiple nodes and connections I designed for my terrarium controller video course (see https://techexplorations.com/so/nodered/). This flow receives sensor input from my terrarium and uses this information to control a water pump. The flow generates email notifications and logs its activities in a Google Sheet.

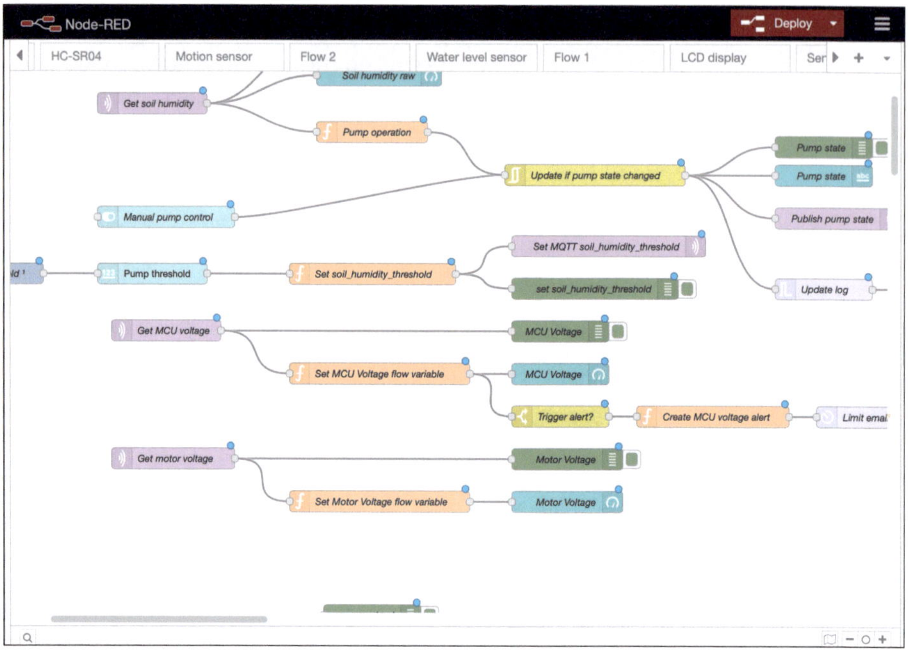

Figure 1.6.3.17: An example flow.

Flow programming

Flows in Node-RED are based on flow-based programming, where data flows through a series of connected nodes that process and transform the data as needed. Nodes exchange messages through their input and output ports, allowing you to create complex logic and behaviour by chaining together different nodes.

In traditional programming, you write a series of instructions, and the computer executes them line by line. This procedural approach has been a mainstay of software development for decades. But with Flow Programming, things take a different turn.

Flow Programming, also known as Dataflow Programming, is a software design paradigm focused on the flow of data rather than the control flow. In this model, your program is a network of "black box" processes, called nodes, which exchange data along connections called edges. Each node operates independently and executes its computation when data arrives on its inputs.

The processes don't know about each other. They don't need to know where the data is coming from or where it's going. This makes the system highly modular and flexible. You can add, remove, or change processes without disrupting the system as a whole.

Several programming languages and tools support the Flow Programming paradigm. One of the most notable is Node-RED, the topic of this book. It allows you to wire devices, APIs, and online services together by dragging and dropping nodes onto a workspace and connecting them with wires to represent data flow.

There's also NoFlo, an open-source Flow-Based Programming environment for JavaScript. It lets you build applications as graphs of processes that exchange data.

Working with flows

When a flow is deployed, the nodes are instantiated and start processing messages based on their configuration and connections. As messages flow through the nodes, they are processed, transformed, or routed to other nodes as needed.

Starting, stopping, and pausing flows

In Node-RED, you can control the execution of your flows using the following methods.

Starting flows

When you first deploy a flow, the nodes within the flow are automatically started and begin processing messages based on their configuration and connections. Clicking the "Deploy" button in the Node-RED editor to start a flow.

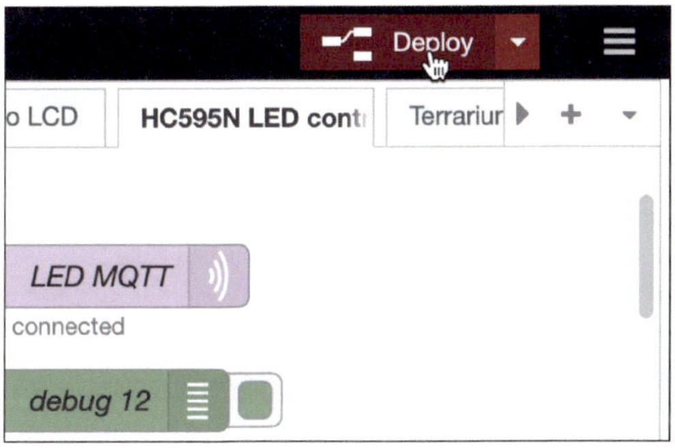

Figure 1.6.3.18: Click Deploy to start a flow.

Enabling and disabling flows

You can enable and disable a flow by clicking the Enable/Disable button in the Information sidebar next to the flow in question. When a flow is disabled, none of its nodes will be created when deployed. If you deploy a disabled flow, the flow will not work. See the screenshot below for the location of the enable/disable button in the information sidebar.

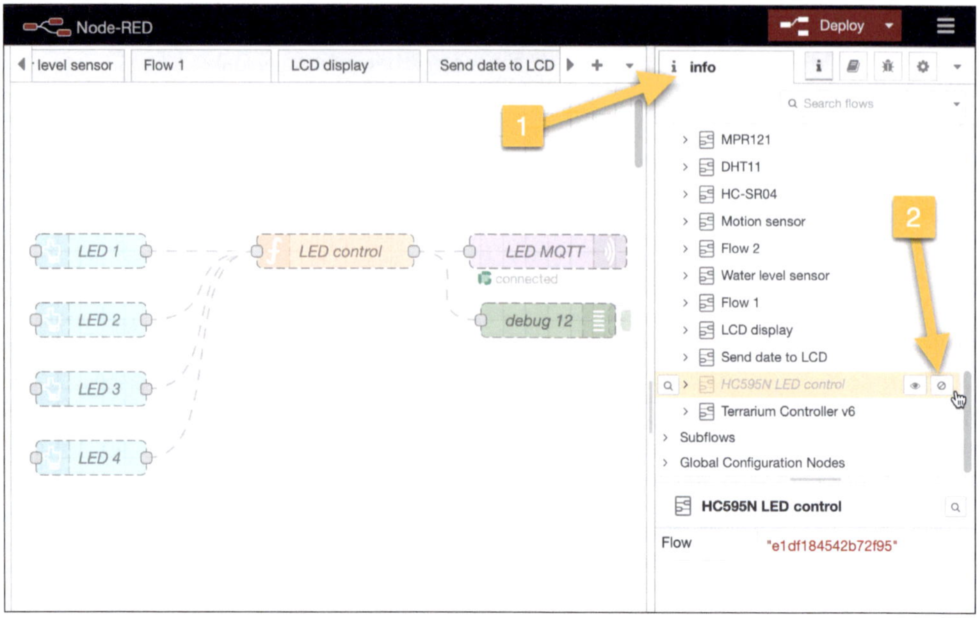

Figure 1.6.3.19: Enable and disable a flow.

Node-RED does not have a built-in feature to pause flows directly. However, you can achieve a similar effect by using nodes that control the flow of messages within your flows, such as the "delay" node or the "trigger" node. By configuring these nodes to buffer, rate limit, or conditionally release messages, you can effectively pause and resume the data flow through your flows.

Hiding and showing a flow

If your editor is becoming crowded with flow tabs, you can hide the ones you don't need. You can show hidden tabs at any time. To hide a flow, click the "eye" button next to the flow name in the tab box (see screenshot below). To show a hidden tab, click the "eye" button next to the flow name in the Information sidebar.

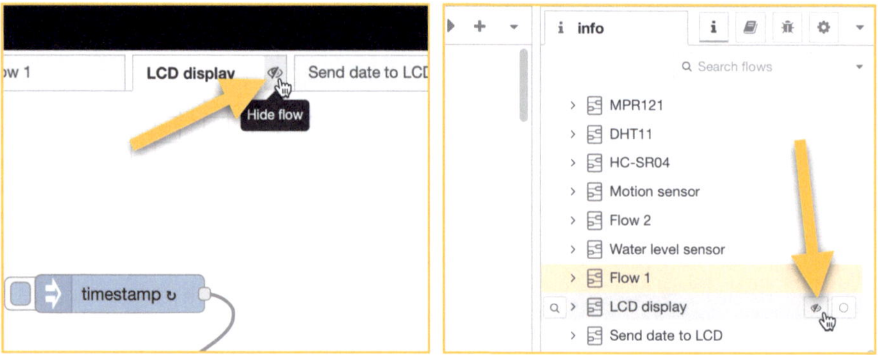

Hide flow Show flow

Figure 1.6.3.20: Enable and disable a flow.

6.4. Best Practices for Working with Flows

Like any other software development activity, following best practices when working with Node-RED is crucial for multiple reasons:

1. **Efficiency**: Best practices optimize the development process, making it more efficient. They ensure that you're taking full advantage of the features and capabilities of Node-RED, thereby saving time and resources.

2. **Reliability**: By adhering to best practices, you reduce the likelihood of encountering bugs or other unexpected behaviours in your applications.

3. **Maintainability**: Best practices enhance the maintainability of your flows, making it easier for others (or even you in the future) to understand, update, and manage them.

4. **Security**: Adherence to best practices often means considering security measures in your application design, thus reducing the potential for vulnerabilities that can lead to breaches or data loss.

Here are some best practices when working with Node-RED:

1. **Keep Flows Simple and Readable**: The best flows are those that are easy to understand and troubleshoot. Avoid overly complex flows and strive to make your flows self-explanatory. Use comments and labels to describe what each node and flow does. Remember, you're not just coding for now, but also for future you or other team members who may need to understand your flows.

2. **Use Subflows and Link Nodes**: When you have repetitive tasks, consider using subflows. Subflows allow you to create a set of nodes that accomplish a specific task, and then reuse that set in multiple places. Link nodes can be used to send messages from one part of a flow to another, making it easier to manage complex flows.

3. **Plan for Error Handling**: Every flow should include some form of error handling. Whether you're retrieving data from an API or writing to a database, things can go wrong, and your flow needs to be able to handle those situations gracefully.

4. **Secure Your Node-RED Instance**: Security is a critical aspect of any software development project. This includes setting up an admin password, using HTTPS where possible, and limiting who can deploy changes. Keep in mind that Node-RED's default settings are designed for development, not production.

5. **Optimize Performance**: Be mindful of how you handle messages and consider the resources your flows consume. For example, don't leave debug nodes turned on in production, as they can consume a lot of resources.

6. **Regular Backups**: Ensure you have a backup of your flows. You can export your flows to a JSON file and keep them backed up. This way, if something goes wrong, you can easily restore your work.

7. **Use Version Control**: It's a good practice to use version control systems like Git for your flows. This allows you to keep track of changes, collaborate with others more efficiently, and revert back to earlier versions if necessary.

You should apply these best practices, to leverage the full potential of Node-RED, create high-quality applications, and make your development process smoother and more efficient.

6.5. The "Trigger" Node

In this section, you will learn about the 'Trigger' node, which provides the most common way to start the execution of a flow.

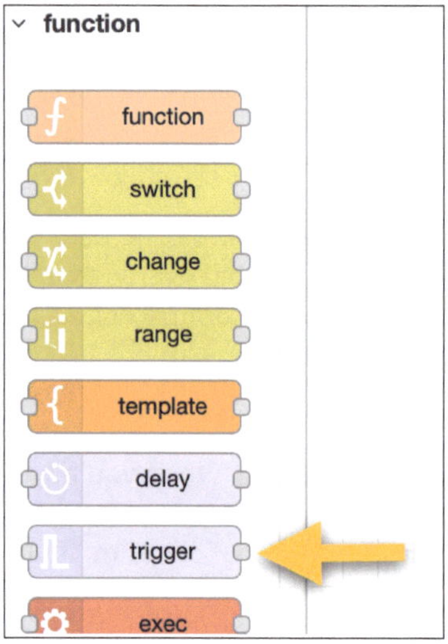

Figure 1.6.10.21: The Trigger node in the palette.

What is the 'Trigger' node?

The 'Trigger' node is designed to generate two messages with a controllable delay in between. When a message arrives at the input of the 'Trigger' node, it sends a predefined message immediately and then another one after a certain delay.

How the 'Trigger' Node works

Three parameters define the operation of the 'Trigger' node:

1. **"Send"**: This defines the payload of the first message to be sent immediately when a message is received at the node's input. You can choose to send the incoming payload or a custom payload.

2. **"Then"**: This sets the delay between the first and the second message. This can be set in seconds, minutes, or even hours.

3. **"Then send"**: This defines the payload of the second message after the delay. It can be a custom payload or nothing, where nothing stops ongoing delays and resets the node.

Using the 'Trigger' Node in Flows

The 'Trigger' node's ability to control the timing and content of messages makes it useful in various scenarios. Here are a few examples:

- **Timeouts**: The 'Trigger' node can implement a timeout mechanism. For instance, if you send a command to a device and expect a response within a certain time, you can use the 'Trigger' node to send an error message if no response is received within the expected timeframe.
- **Pulse Generation**: The 'Trigger' node can generate a pulse-like sequence of messages, helpful in controlling devices that require a short activation period followed by deactivation.
- **Polling Mechanisms**: You can use the 'Trigger' node to create a polling mechanism that sends a request, waits for a while, and then sends another request.
- **State Resets**: If you have a flow that maintains some state, the 'Trigger' node can be used to reset that state after a certain period of inactivity.

By effectively understanding and utilising the 'Trigger' node, you can add a new layer of control and sophistication to your Node-RED flows, enhancing the functionality of your applications.

An example

Here's a simple example of a flow using the 'Trigger' and 'Debug' nodes:

1. Drag and drop an 'Inject' node onto the workspace.

2. Double-click the 'Inject' node to open its configuration settings. Set the following properties:
   - **Payload**: Choose "string" from the dropdown menu, and enter a string like "Start" in the field below.
   - **Topic**: Leave it blank.

3. Next, drag and drop a 'Trigger' node onto the workspace. Double-click it to open its configuration settings and set the following properties:
   - **Send**: Choose String and enter "Processing started" in the text field.
   - **then wait for**: Set this to "5 seconds".
   - **then send**: Choose "string", and in the field, enter "Processing finished".

Here is the configuration of the example Trigger node:

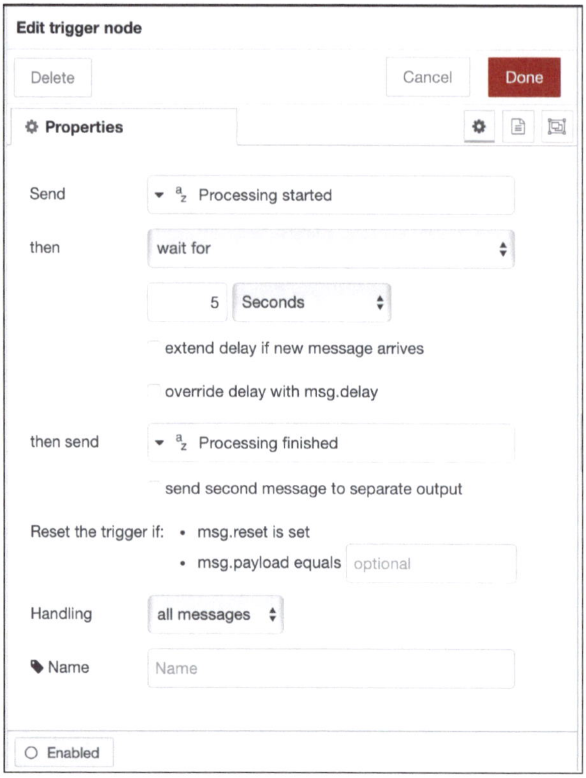

Figure 1.6.10.22: The configuration of the example Triger node.

4. Connect the output port of the 'Inject' node to the input port of the 'Trigger' node.

5. Finally, drag and drop a 'Debug' node onto the workspace and connect the output port of the 'Trigger' node to the input port of the 'Debug' node.

6. Click 'Deploy' to deploy the flow.

Here is the example flow in the editor:

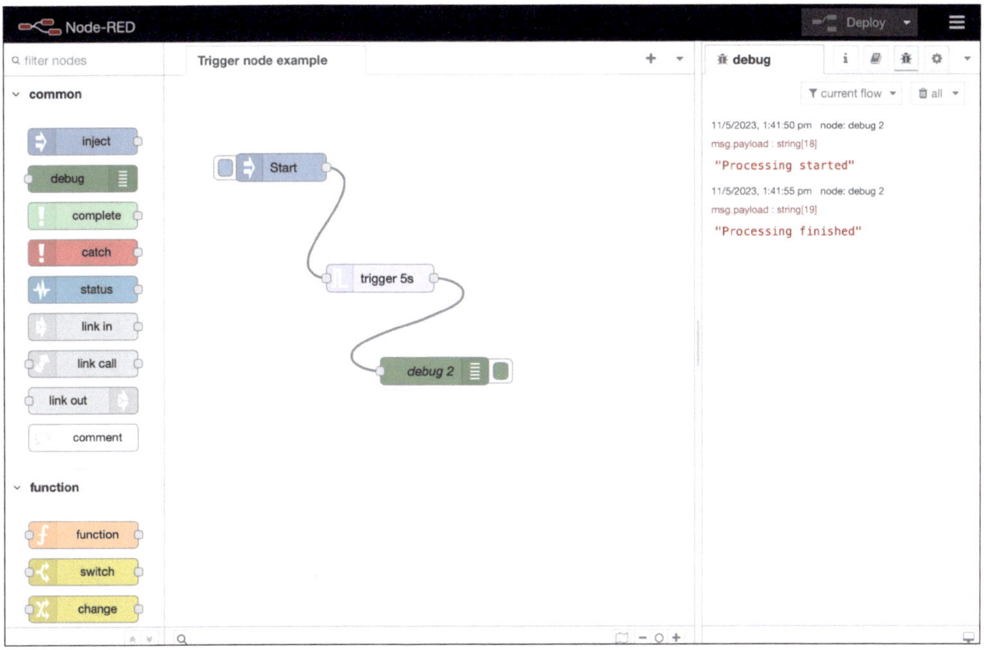

Figure 1.6.10.23: The example flow.

In this flow, when you click the button on the 'Inject' node, it will send a message to the 'Trigger' node. The 'Trigger' node will immediately send a message with the payload "Processing started"; after a delay of 5 seconds, it will send a second message with the payload "Processing finished". The 'Debug' node will log both of these messages.

6.6. The "Inject" Node

One of the most fundamental and versatile nodes in Node-RED is the 'Inject' node. It's often used as a starting point in many flows due to its straightforward yet powerful functionality. In this section, we will discuss the 'Inject' node, how it works, and how you can use it in your Node-RED flows.

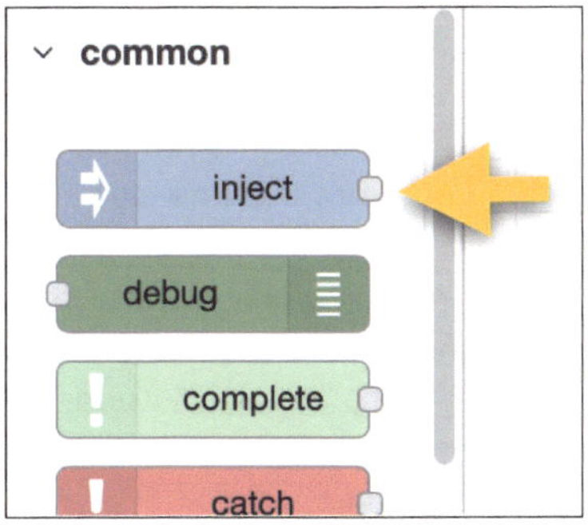

Figure 1.6.8.24: The Inject node in the palette

What is the 'Inject' Node?

The 'Inject' node in Node-RED is an input node that lets you manually or automatically inject messages into a flow. This can be very helpful for various purposes, such as triggering events, testing other nodes in your flow, or sending predefined data.

The 'Inject' node can be configured to send various payload types, including strings, numbers, booleans, timestamps, JSON objects, and more. You can also set the "msg.topic" property to help route messages in your flow.

How the 'Inject' node works

Once you've added an 'Inject' node to your flow, you can double-click it to open its configuration settings.

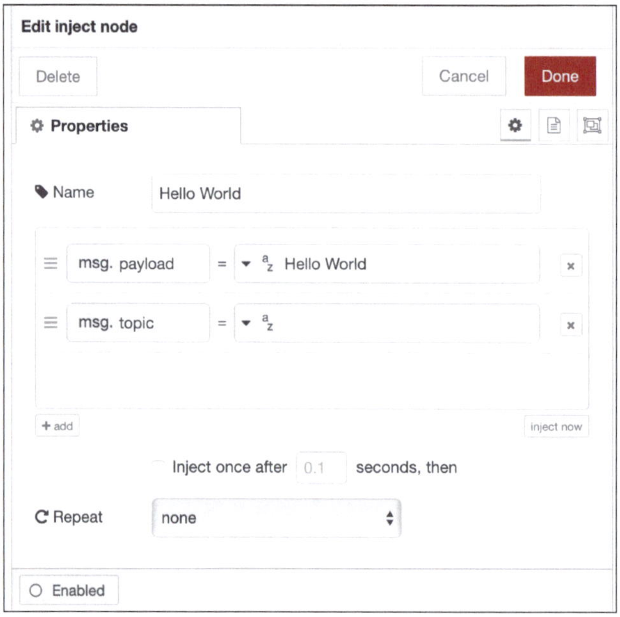

Figure 1.6.8.25: The Inject node configuration window.

Here, you can set the following properties:

- "**Payload**": This is the primary data that the node will inject into the flow. You can choose the payload type from the dropdown menu and enter the payload data in the field below.
- "**Topic**": This optional property can be used to set the "msg.topic" field of the message. This can be useful for routing or filtering messages in your flow.
- "**Repeat**": This allows you to configure the node to inject messages at regular intervals automatically. You can set the interval in seconds, minutes, hours, or days or choose a specific time.
- "**Inject once after**": If you enable this option and set a delay, the node will automatically inject a message after your specified number of seconds since the node button is clicked.

Once you've configured the 'Inject' node, you can connect its output port to the input port of another node in your flow. To manually inject a message, click the button on the left side of the node in the workspace.

Using the 'Inject' node in flows

The 'Inject' node can be used in various ways in your Node-RED flows. Here are a few examples:

- **Testing**: You can use the 'Inject' node to send test data to other nodes in your flow. This is particularly useful when developing or debugging your flow.
- **Scheduling**: Using the repeat functionality, you can use the 'Inject' node to schedule events or tasks. For instance, you could set the node to inject a message every morning to trigger a flow that fetches the weather forecast.
- **Manual control**: You can use the 'Inject' node to manually trigger events or actions in your flow. For example, you could set up a flow where clicking the 'Inject' node turns on a connected light.

The 'Inject' node is a powerful tool in your Node-RED toolkit, allowing you to manually or automatically trigger events, test your flows, and control the data flow. By understanding and utilising the 'Inject' node, you'll be well-equipped to build a wide range of effective flows in Node-RED.

A simple example

Let's consider a simple flow where we use the Inject node to simulate sensor data that is then logged by the Debug node. This flow simulates a temperature sensor that sends a temperature reading every 5 seconds.

1. Drag and drop an 'Inject' node onto the workspace.

2. Double-click the 'Inject' node to open its configuration settings. Set the following properties:
   - **Payload**: Choose "Number" from the dropdown menu, and enter a value like "25" in the field below. This will be our simulated temperature reading.
   - **Topic**: Leave it blank.
   - **Repeat**: Choose "interval" and set it to "5" seconds. This will make the node automatically inject the message every 5 seconds.
   - **Inject once at start:** Disable this option.

3. Next, drag and drop a 'Debug' node onto the workspace.

4. Connect the output port of the 'Inject' node to the input port of the 'Debug' node.

5. Click 'Deploy' to deploy the flow.

In this flow, the 'Inject' node will automatically send a message with a payload of "25" every 5 seconds, simulating a temperature sensor sending readings. The 'Debug' node will receive these messages and log them to the debug sidebar, allowing you to see the simulated sensor data (see screenshot below).

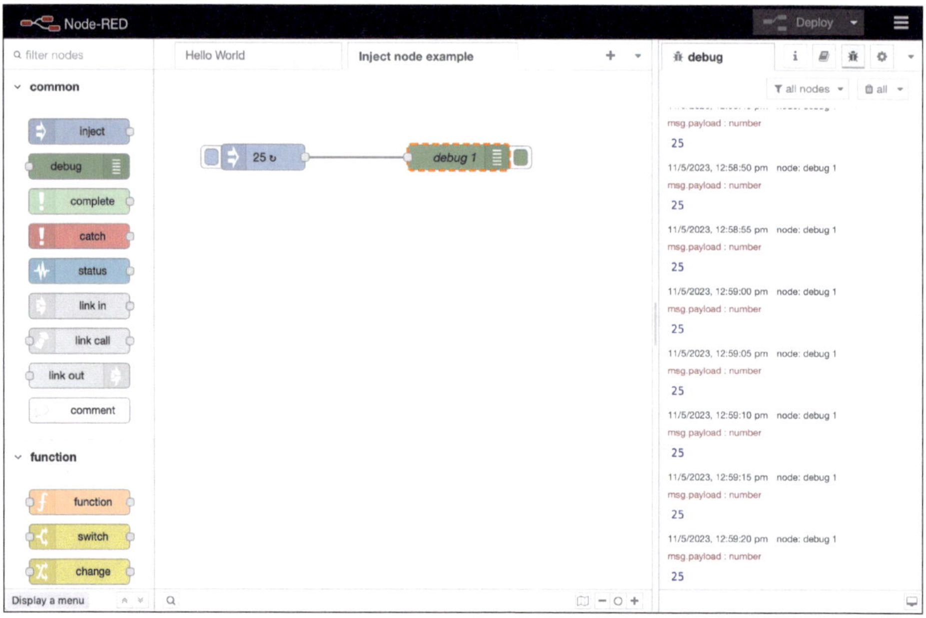

Figure 1.6.8.26: A simple Inject node demo.

Please note that this is a simplistic example. In a more realistic scenario, you might use a 'Function' node or a 'Random' node to generate varying temperature data or a 'Change' node to add more properties to the message.

An example of a function

Let's extend the previous example using a "function" node to generate a random temperature value between 10 °C and 30 °C.

1. Drag and drop an 'Inject' node onto the workspace.

2. Double-click the 'Inject' node to open its configuration settings. Set the following properties:
   - **Payload**: Choose "timestamp" from the dropdown menu. This will inject the current timestamp, but we will not use this value. The purpose of the 'Inject' node in this example is to trigger the flow at regular intervals.
   - **Topic**: Leave it blank.
   - **Repeat**: Choose "interval" and set it to "5" seconds. This will make the node automatically inject the message every 5 seconds.
   - **Inject once at start**: Disable this option.

3. Next, drag and drop a 'Function' node onto the workspace.

4. Double-click the 'Function' node to open its configuration settings. In the "Function" field, enter the following JavaScript code:

```
msg.payload = Math.floor(Math.random() * (30 - 10 + 1) + 10);
return msg;
```

This code generates a random integer between 10 and 30 and sets it as the message's payload. The 'Function' node will output this message when it receives a message from the 'Inject' node.

5. Connect the output port of the 'Inject' node to the input port of the 'Function' node.

6. Finally, drag and drop a 'Debug' node onto the workspace and connect the output port of the 'Function' node to the input port of the 'Debug' node.

7. Click 'Deploy' to deploy the flow.

In this flow, the 'Inject' node will automatically send a timestamp message every 5 seconds. The 'Function' node will receive this message, generate a random temperature between 10° C and 30° C, and output this value. The 'Debug' node will receive the temperature value and log it to the debug sidebar.

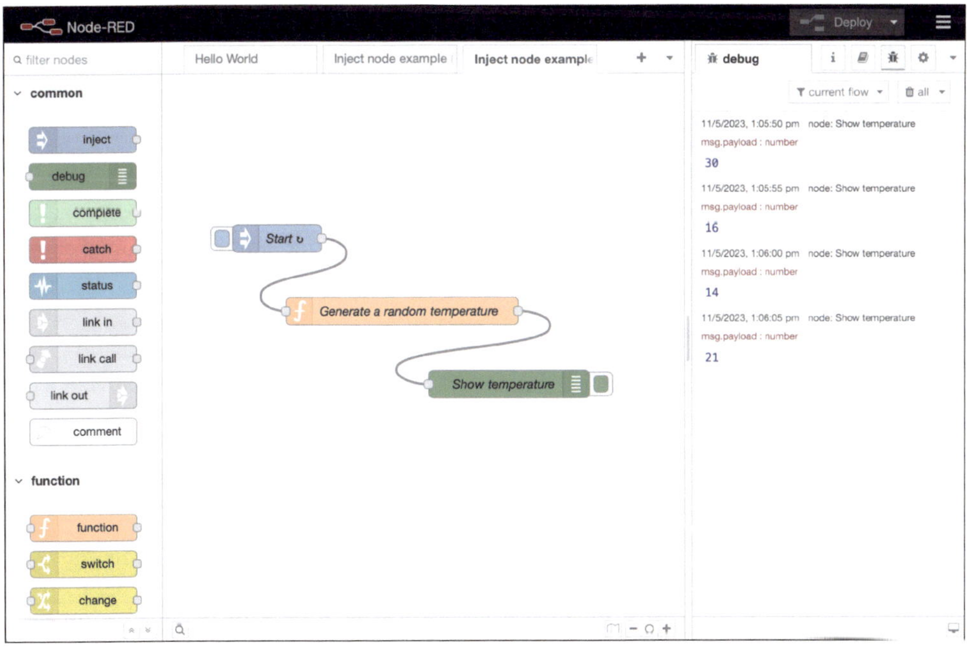

Figure 1.6.8.27: An Inject node combined with a Function node.

6.7. The "Debug" Node

In this section, you will explore the Debug node in Node-RED and learn how to use it effectively for monitoring and troubleshooting your flows. The Debug node is a powerful tool for understanding how data flows through your nodes and identifying any issues that may arise during development.

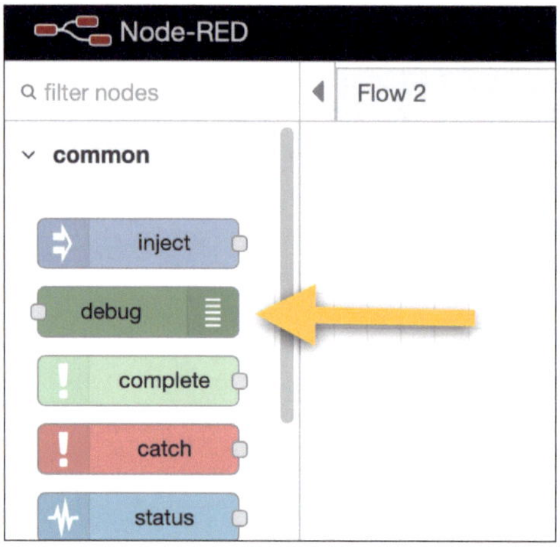

Figure 1.6.5.28: The Debug node in the palette

Understanding the Debug node

The Debug node in Node-RED displays messages, data, and other information during development and testing. By default, the Debug node outputs its messages to the Debug sidebar, which can be accessed by clicking on the bug icon in the top-right corner of the Node-RED editor.

Adding a Debug node to your flow

To add a Debug node to your flow, follow these steps:

1. In the Node-RED editor, locate the Debug node in the nodes palette on the left side of the screen. The Debug node is typically found under the "function" category and has a bug icon.

2. Drag and drop the Debug node onto the flow canvas.

3. Wire the Debug node to the output of any node you want to monitor.

Configuring the Debug node

By default, the Debug node will output any message payload it receives. However, you can configure the Debug node to output specific message properties or custom JavaScript expressions.

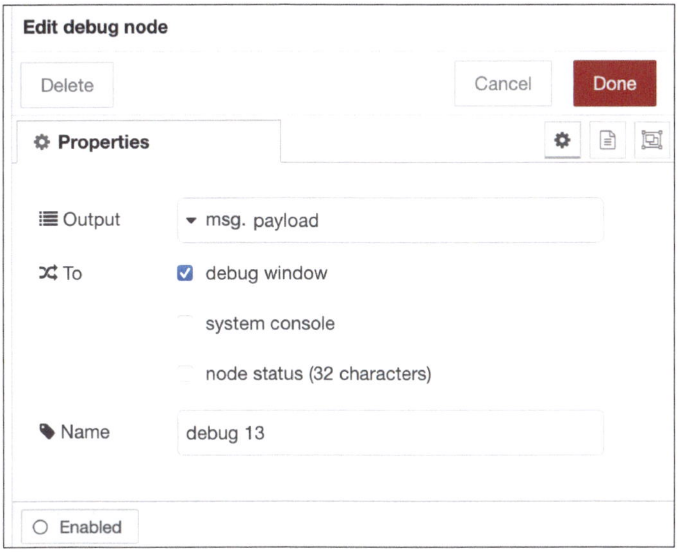

Figure 1.6.5.29: The Debug node configuration options.

To configure the Debug node, follow these steps (also see screenshot above):

1. Double-click the Debug node to open its configuration window.

2. In the "Output" field, you can choose from the following options:
   - "msg.payload": Displays the message payload (default setting).
   - "msg.": Allows you to specify a custom message property (e.g., "msg. topic").
   - "full msg object": Displays the entire message object, including all properties.
   - "expression": You can enter a custom JavaScript expression that will be evaluated and displayed (e.g., "msg.payload * 2").

3. Optionally, you can add a custom label to the Debug node to help identify its purpose in the flow.

4. Click "Done" to save your changes.

Viewing Debug output

The debug node produces text output printed in the Debug tab of the sidebar, as you can see in the screenshot below.

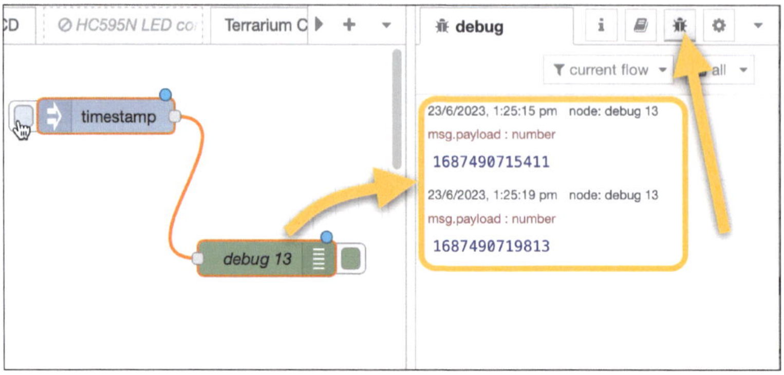

Figure 1.6.5.30: Debug node output.

To view the output from your Debug nodes, follow these steps:

1. Click on the bug icon in the top-right corner of the Node-RED editor to open the Debug sidebar.
2. Deploy your flow by clicking the "Deploy" button in the top-right corner of the editor.
3. Interact with your flow by triggering events or sending data to connected nodes.
4. Observe the output from the Debug nodes in the Debug sidebar. Messages will be timestamped and displayed with the label you specified for the Debug node.

Filtering Debug output

If you have multiple Debug nodes in your flow, you should filter the output in the Debug sidebar to focus on specific nodes. To filter the Debug output, follow these steps:

1. In the Debug sidebar, click on the search icon at the top.
2. Enter a search term that matches the label or output of the Debug node(s) you want to focus on.
3. The Debug sidebar will now only display messages from Debug nodes that match your search criteria.

Clearing Debug output

Click the trash can icon at the top to clear the Debug output in the Debug sidebar. This will remove all messages and allow you to start with a clean slate.

The Debug node is an essential tool for understanding how data flows through your nodes, identifying issues during development, and ensuring the correct operation of your flow.

6.8. The "Function" Node

In this section, you will focus on the Function node and learn how to use it to create custom functionality within your flows. The Function node allows you to write JavaScript code to manipulate message payloads, control flow execution, and interact with other nodes. We will also walk through a simple example to demonstrate the power and flexibility of the Function node.

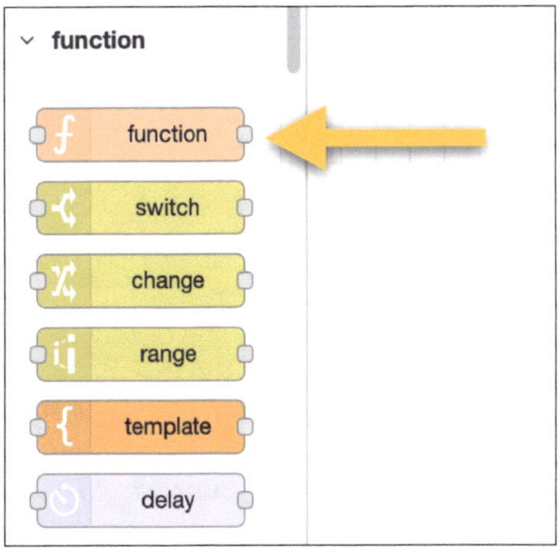

Figure 1.6.6.31: The Function node in the palette

Understanding the Function node

The Function node in Node-RED is a versatile tool that lets you write custom JavaScript code to perform various tasks within your flows. It can process data, make decisions, and even call external APIs. The Function node operates on message objects passed between nodes in a flow.

Adding a function node to your flow

To add a Function node to your flow, follow these steps:

1. In the Node-RED editor, locate the Function node in the nodes palette on the left side of the screen. The Function node is typically found under the "function" category and has a "*f*" icon.
2. Drag and drop the Function node onto the flow canvas.
3. Wire the Function node between the input and output nodes where you want to add custom functionality.

Configuring the function node

In the screenshot below, you can see the configuration window for the Function node.

Figure 1.6.6.32: The Function node configuration window.

To configure the Function node and write custom JavaScript code, follow these steps:

1. Double-click the Function node to open its configuration window.

2. Enter your custom JavaScript code in the "Function" field. Your code should follow Node-RED's coding conventions, such as using the "msg" object to pass data between nodes.

3. Optionally, you can add a custom name to the Function node to help identify its purpose in the flow.

4. Click "Done" to save your changes.

Simple example: converting temperature units

In this example, let's create a flow that receives a temperature value in Celsius and converts it to Fahrenheit using a Function node. This flow looks like this:

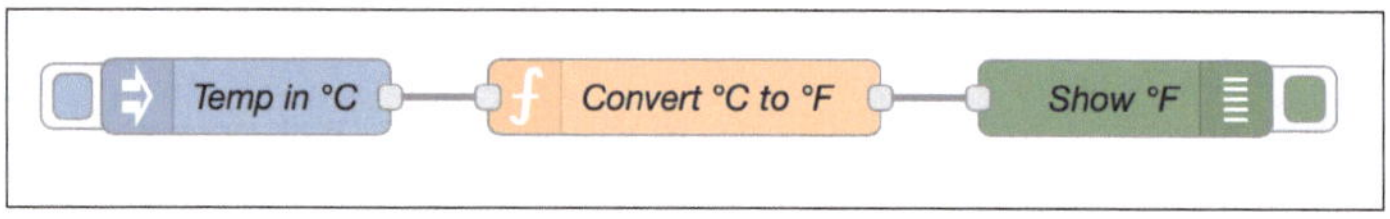

Figure 1.6.6.33: A simple flow that converts a temperature.

1. Create a new flow and add the following nodes:
   - An "inject" node to enter the temperature value in Celsius. Set the payload type to "number". In the value field, type "15". The properties for this node look like this:

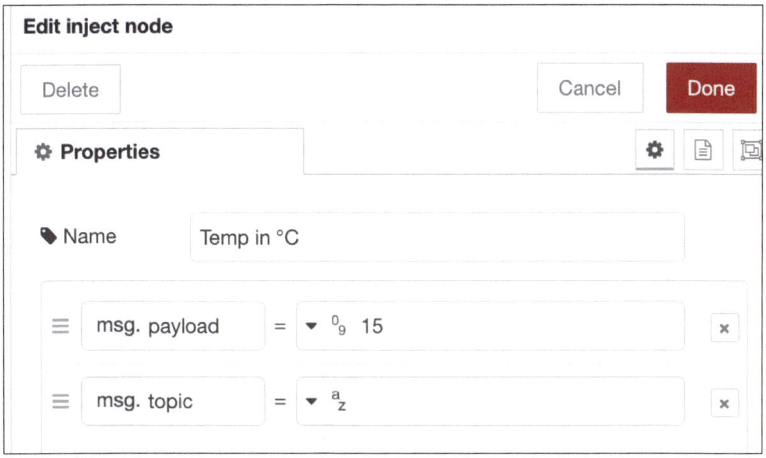

Figure 1.6.6.34: The Inject node properties.

- A Function node to convert the Celsius value to Fahrenheit. Use the following code in the "Function" field:

```
msg.payload = (msg.payload * 9/5) + 32;
return msg;
```

The Function node properties look like this:

Figure 1.6.6.35: The Function node properties.

- A "Debug" node to display the converted temperature value in Fahrenheit.

1. Wire the nodes together: "ui_text_input" → "Function" → "ui_text".

2. Deploy the flow and open the Node-RED debug tab in your browser.

3. Trigger the Inject node by clicking on its button, and notice the converted value in Fahrenheit be displayed in the debug tab. To convert a different temper-

ature, enter the new temperature in the Inject node's payload field, re-deploy the flow, and trigger the Inject node.

The Function node allows you to create nodes with custom functionality. As your flows become more complex, you will find that the Function node is indispensable.

6.9. The "Complete" Node

In this section, you will focus on the "Complete" node. This node allows you to monitor the completion of a node or an entire flow, which can be helpful in many scenarios. You can find the "Complete" node in the "common" palette group.

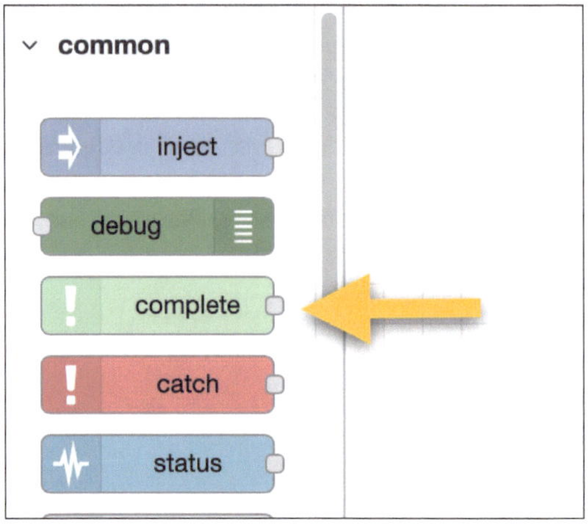

Figure 1.6.9.36: The Complete node in the palette

What is the "Complete" node?

The "Complete" node in Node-RED is an output node triggered when another node (or nodes) has finished processing a message. It does not matter whether the node outputs a message; once it completes its task, the "Complete" node will be triggered. The node can also monitor the completion of an entire flow. This feature is often used when you want to execute something after completing a task in your Node-RED flow.

The "Complete" node can be set to monitor the completion of a particular node, or it can be set to watch for the completion of any node in the same flow.

How the "complete" node works

The working of a "Complete" node is straightforward. Select the node you want to monitor from a drop-down list in the "Complete" node settings. When the monitored node finishes processing a message, the "Complete" node sends a message. This message contains the entire message object.

The "Complete" node output message contains details about the node that completed its task. The "msg.payload" will contain the payload of the original message, while the "msg. status" field will contain a copy of the status object from the watched node.

Using the "Complete" node in flows

The "complete" node can be used in various ways in your Node-RED flows. Here are a few examples:

- **Monitoring task completion**: Use the "Complete" node to monitor the completion of a task. For example, you can use it to watch a "File" node and be notified when a file read or write operation has been completed.
- **Error handling**: If a node encounters an error during its execution, it will still complete its operation. Using a "Complete" node, you can be alerted when a node has finished processing, even if it encounters an error. This can be useful for error handling and debugging your flows.
- **Post-processing**: You can use the "Complete" node to trigger some post-processing tasks after completing a particular task. For example, after writing data to a database, you might use a "Complete" node to send a confirmation message or log the completion of the task.

The "Complete" node is an essential part of Node-RED, allowing you to monitor the completion of tasks and execute further actions accordingly. You can create more robust, reliable, and efficient Node-RED applications by leveraging your flows'" Complete" node.

An example

Here is an example of a flow that uses the "Complete" node to monitor the completion of a "Function" node. The "Function" node processes a message from the "Inject" node, and the "Complete" node logs a message when the processing is done.

1. Drag and drop an "Inject" node onto the workspace.

2. Double-click the "Inject" node to open its configuration settings. Set the following properties:
   - **Payload**: Choose "string" from the dropdown menu, and enter a string like "Hello, Node-RED!" in the field below.
   - **Topic**: Leave it blank.

3. Next, drag and drop a "Function" node onto the workspace. Double-click it to open its configuration settings and set the following properties:
   - **Name**: Enter a name like "Process Message".
   - **Function**: In the function field, enter the following JavaScript code:

```
msg.payload = msg.payload.toUpperCase();
     return msg;
```

   This function will convert the string to uppercase.

4. Connect the output port of the "Inject" node to the input port of the "Function" node.

5. Drag and drop a "Complete" node onto the workspace.

6. Double-click the "Complete" node to open its configuration settings. In the "Watch" field, select the "Function" node that you just added (see below).

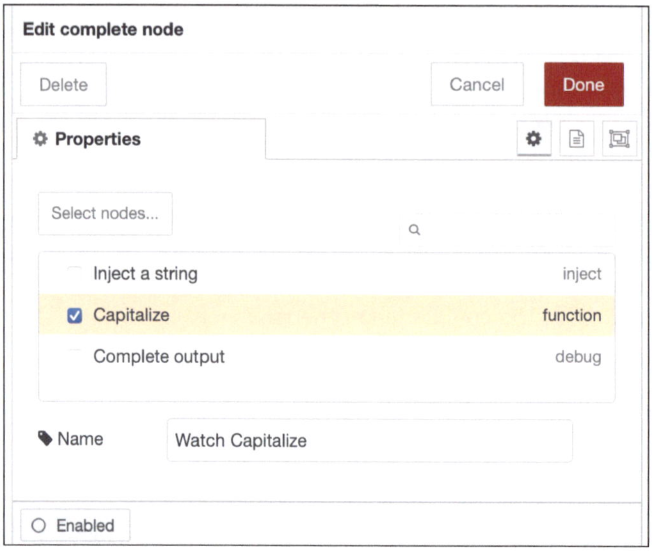

Figure 1.6.9.37: The Complete node set to monitor the Capitalise node.

7. Finally, drag and drop a "Debug" node onto the workspace and connect the output port of the "Complete" node to the input port of the "Debug" node.

8. Click "Deploy" to deploy the flow.

In this flow, when you click the button on the "Inject" node, it will send a message with the payload "Hello, Node-RED!" to the "Function" node. The "Function" node will convert the string to uppercase and output the result. Regardless of the processing result, the "Complete" node will be triggered when the "Function" node finishes processing the message. The "Complete" node will then send a message, which will be logged by the "Debug" node.

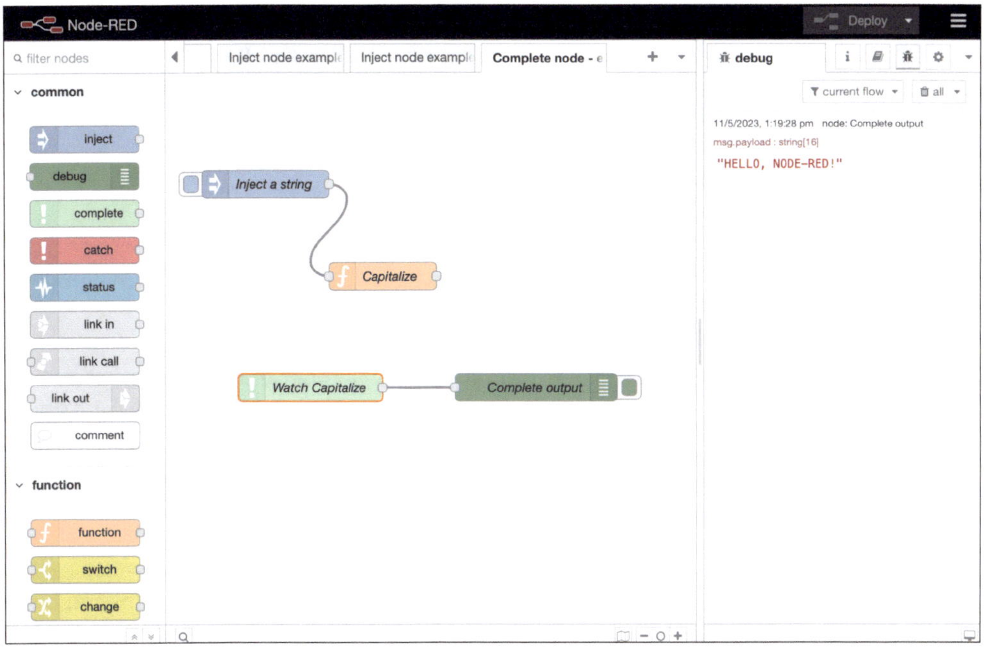

Figure 1.6.9.38: The Complete node as part of a flow.

This example illustrates how the "Complete" node can be used to monitor the completion of any node, not just input or output nodes.

6.10. The "Delay" Node

In this section, you will explore the 'Delay' node. This node introduces a delay in the message flow, proving useful in various scenarios, from simple pauses to rate limiting. Here, you will learn the 'Delay' node, how it works, and how it can be utilised in your flows.

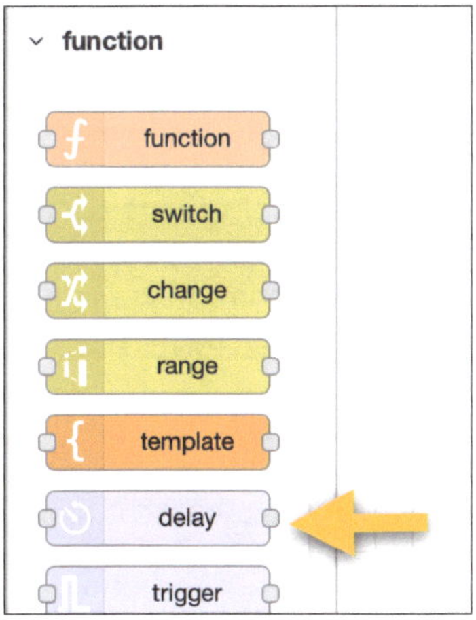

Figure 1.6.10.39: The Delay node in the palette

What is the 'Delay' node?

The 'Delay' node in Node-RED is used to introduce a delay in the flow of messages. This can control the rate at which messages are processed, add a pause between operations, or stagger the execution of different flow parts.

How the 'Delay' Node Works

The operation of the 'Delay' node can be configured in several ways:

1. **Delay each message**: This is the simplest mode, where each incoming message is delayed by a specified amount of time before being sent.

2. **Rate limit**: This mode allows you to limit the number of messages that can pass through the node per unit of time. Additional messages within the same time window can be discarded or queued for later processing. This can be useful when interacting with services imposing rate limits or dealing with traffic bursts.

3. **Delay sequence of messages**: In this mode, a sequence of messages can be delayed by a specified amount of time. The sequence can be identified by a specific 'topic' property or a message flow rate.

In each mode, the 'Delay' node can also be configured to handle delays longer than a certain threshold by dropping intermediate messages or sending the latest message at regular intervals.

Using the 'Delay' Node in Flows

The 'Delay' node is quite versatile and can be used in various ways in your Node-RED flows. Here are a few examples:

- **Throttling API Calls**: If you are interacting with third-party APIs with rate limits, you can use the 'Delay' node to ensure you do not exceed these limits.
- **Staggering Operations**: In some scenarios, you may need to stagger the execution of different operations. For example, if you control a set of devices, avoid all devices being activated simultaneously.
- **Simulating Network Latency**: If you are testing how your flows handle network latency, you can use the 'Delay' node to introduce artificial delays.
- **Message Queueing**: In a scenario where many messages are being produced, you can use the 'Delay' node to queue messages and process them at a manageable rate.

The 'Delay' node is a powerful tool in your Node-RED toolkit, allowing you to control the flow of messages and build more robust and efficient applications. By understanding and leveraging the 'Delay' node, you can add a new level of sophistication to your Node-RED flows.

An example

Here's an example of a flow that uses the 'Delay' node to rate limit messages from an 'Inject' node:

1. Drag and drop an 'Inject' node onto the workspace.

2. Double-click the 'Inject' node to open its configuration settings. Set the following properties:
   - **Payload**: Choose "timestamp" from the dropdown menu.
   - **Repeat**: Select "interval" and set it to "1 seconds". This will make the 'Inject' node send a message every second.

3. Next, drag and drop a 'Delay' node onto the workspace. Double-click it to open its configuration settings and set the following properties:
   - **Action**: Choose "Rate limit".
   - **Rate**: Set the "msg(s) field to "1", set the period field to "5", and select "seconds" from the dropdown menu. This will limit the rate to 1 message per 5 seconds.
   - **If the rate is exceeded**, Choose "drop intermediate messages". This will discard all but the first message in any given minute.

   Here is the delay node configuration window:

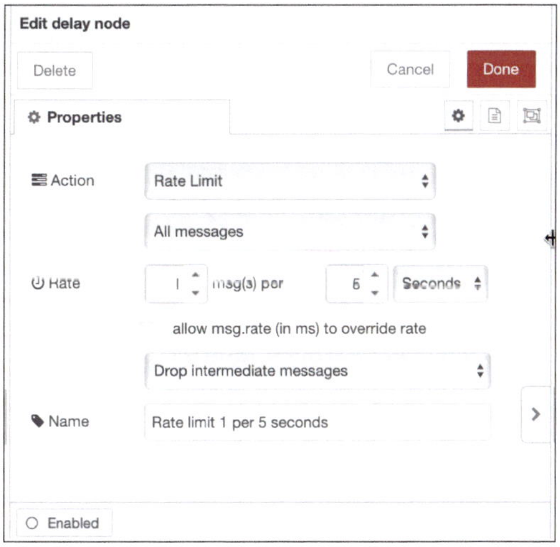

Figure 1.6.9.40: Example Delay node configuration.

4. Connect the output port of the 'Inject' node to the input port of the 'Delay' node.

5. Finally, drag and drop a 'Debug' node onto the workspace and connect the output port of the 'Delay' node to the input port of the 'Debug' node.

6. Click 'Deploy' to deploy the flow.

   Here is the example flow in the editor:

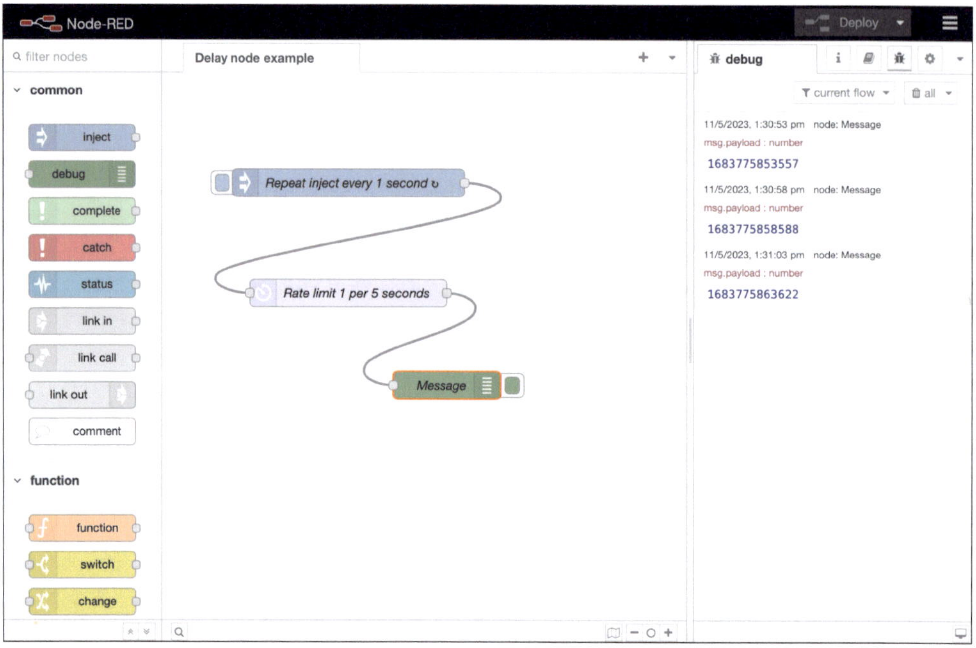

Figure 1.6.9.41: Delay node example flow.

In this flow, the 'Inject' node will try to send a message every second, but the 'Delay' node will throttle this only to allow one message every five seconds to pass through. The discarded messages will not appear in the debug sidebar. This demonstrates how the 'Delay' node can rate limit messages in a Node-RED flow.

6.11. Node-RED Settings and Configuration

In this section, you will learn about the Node-RED configuration file, which controls various aspects of Node-RED's behaviour, appearance, and security. You can configure Node-RED to meet your specific needs and preferences by customising this file.

The Node-RED configuration file, usually named "settings.js", is a JavaScript file that contains a set of key-value pairs defining the configuration options for your Node-RED instance. The configuration file is in your Node-RED user directory, typically ".node-red" in your home directory. If you followed the instructions on enhancing Node-RED security in the previous chapter, you are probably already familiar with this file.

You can find the documentation for the configuration file in the Node-RED user guide at https://nodered.org/docs/user-guide/runtime/configuration.

Below are the main features of the Node-RED configuration file:

Flow file

You can specify the flow file name where your Node-RED flows are stored. By default, this is set to "flows_<hostname>.json", where "<hostname>" is your computer's hostname.

```
flowFile: 'flows.json',
```

User interface theme

Node-RED allows you to customise the appearance of its editor by specifying a user interface theme. The theme is "light" by default, but you can change it to "dark" or another custom theme.

```
editorTheme: {
        theme: 'dark',
},
```

You can find a list of available themes in the Node-RED themes Github repository: https://github.com/node-red-contrib-themes/theme-collection. Before you use one of the themes in this repository, you will need to install it.

Context storage

Node-RED provides context storage for storing flow, global, and node-specific variables. By default, memory storage is used, but you can configure other storage options, such as file-based storage or custom storage modules.

```
contextStorage: {
        default: {
        module: 'memory',
        },
        file: {
        module: 'localfilesystem',
        config: {
                dir: 'context',
        },
        },
},
```

Palette editor

You can enable or disable the palette editor, which allows you to manage the node modules installed in your Node-RED instance directly from the editor.

```
editorTheme: {
        palette: {
        editable: true,
        },
},
```

HTTP Node Configuration

You can configure the HTTP nodes (i.e., the "http in" and "http response" nodes) to behave differently, such as using a custom path prefix or enabling CORS (Cross-Origin Resource Sharing).

```
httpNode: {
        root: '/api',
        cors: {
        origin: '*',
        methods: 'GET,PUT,POST,DELETE',
        },
},
```

Admin authentication

We have covered this security option in the previous chapter, but here's a recap. You can enable admin authentication to protect your Node-RED editor with a username and password for added security. The built-in "node-red-admin" tool can generate a password hash.

```
adminAuth: {
        type: 'credentials',
        users: [
        {
                username: 'admin',
                password: '<generated-password-hash>',
                permissions: '*',
        },
        ],
},
```

Node-RED library

You can configure the path to the Node-RED library, which stores your custom nodes, flows, and other resources.

```
userDir: '/path/to/your/custom/directory',
```

Custom nodes

You can specify custom nodes you've created or downloaded from external sources. To add a custom node, include its file path in the "nodes" array.

```
nodes: [
        'path/to/your/custom/node.js',
],
```

For this book, you will not be changing any configuration settings I have mentioned above, except for "adminAuth" which is useful for strengthening the server security.

6.12. Node-RED Documentation and Resources

The official Node-RED documentation is an excellent resource for learning about and working with Node-RED. You can find high-quality documentation at the following sources:

1. **Official Node-RED Documentation** The official Node-RED documentation is available on the Node-RED website at https://nodered.org/docs/. This documentation covers installation, configuration, creating flows, working with nodes, and more.

2. **Node-RED GitHub Repository**: The Node-RED GitHub repository (https://github.com/node-red/node-red) contains the source code and additional information and resources related to Node-RED. The repository's README file and Wiki pages can provide insights into various aspects of the project.

3. **Node-RED Discourse Forum**: The Node-RED Discourse forum (https://discourse.nodered.org/) is a community-driven platform where you can ask questions, share experiences, and find solutions to problems related to Node-RED. This forum is invaluable for learning from other Node-RED users and developers.

4. **Node-RED Cookbook**: The Node-RED Cookbook (https://cookbook.nodered.org/) provides practical examples and recipes for common tasks and scenarios in Node-RED. This resource is excellent for learning best practices and solving specific problems.

5. **Node-RED Blog**: The Node-RED Blog (https://nodered.org/blog/) features news, updates, and articles related to Node-RED development and use cases. This is an excellent source of information to stay up-to-date with the latest changes and improvements in Node-RED.

These resources should give you a comprehensive understanding of Node-RED, its features, and best practices for using it in your projects.

Chapter 1.7 • Node-RED Dashboard

The Node-RED Dashboard is a powerful tool that enables the creation of interactive graphical user interfaces for your Node-RED applications. It provides a set of nodes that allow you to create live graphs, charts, gauges, form inputs, and other UI components. This chapter introduces the Node-RED Dashboard, its main features, and its use.

The screenshot below shows a dashboard containing various widgets from my course, "Node-RED ESP32: Make a Terrarium Controller". In this example, you can see these types of widgets:

- Switch
- Numeric
- Gauge
- Chart
- Text

Figure 1.7.0.1: An example Node-RED dashboard

What is the Node-RED Dashboard?

The Node-RED Dashboard is an add-on module that provides a set of nodes in Node-RED to create live dashboards. It's a visual tool for wiring the Internet of Things (IoT), but it can also be used for other applications to visualise data and interact with your Node-RED server.

The Node-RED Dashboard offers a range of UI components:

- **Buttons**: Buttons are simple UI elements that can send a message when clicked.

- **Sliders & Numeric Inputs**: These allow users to input numeric values in various ways.
- **Text Input & Dropdown**: Text input fields and dropdown selectors provide more ways for users to interact with your flows.
- **Charts & Gauges**: These nodes allow you to display data in various graphical formats.
- **UI Control**: This node can be used to control the layout and content of the dashboard dynamically.

Setting up the Node-RED Dashboard

To start using the Node-RED Dashboard, you first have to install it. You can do this by navigating to 'Manage palette' from the menu, then searching for `node-red-dashboard` in the 'Install' tab.

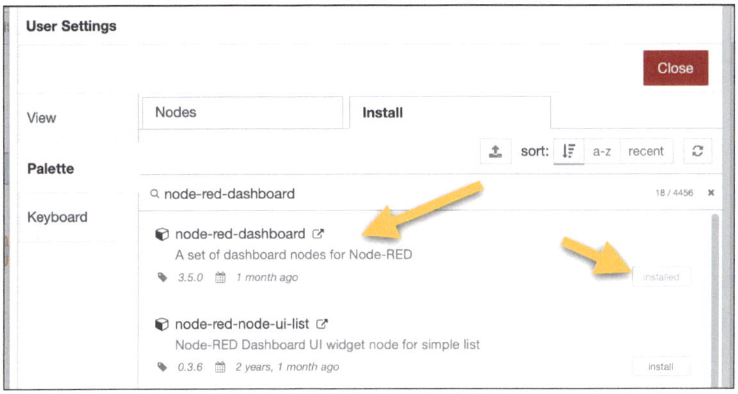

Figure 1.7.0.2: Install the Dashboard nodes.

After installation, a new set of nodes labelled 'dashboard' will be available in the palette. You can create your dashboard by dragging and dropping these nodes onto your flow. You can see some of the installed Dashboard nodes in the screenshot below.

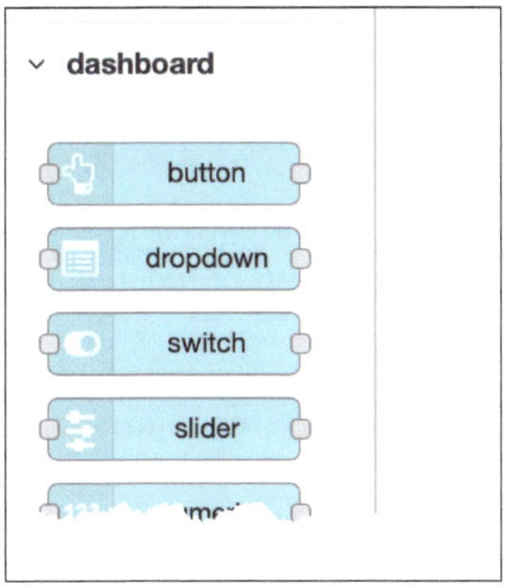

Figure 1.7.0.3: Some of the installed Dashboard nodes.

Using the Node-RED Dashboard

Creating a dashboard involves creating a flow (or adding a dashboard to an existing flow) and adding input and output nodes that interact with other dashboard nodes. Once you've created a flow with dashboard nodes and deployed it, you can view your dashboard by appending `/ui` to your Node-RED URL in your browser (e.g., `http://localhost:1880/ui`). You can also click on the Dashboard button that you can find in the Dashboard tab of the Information in the sidebar (see below).

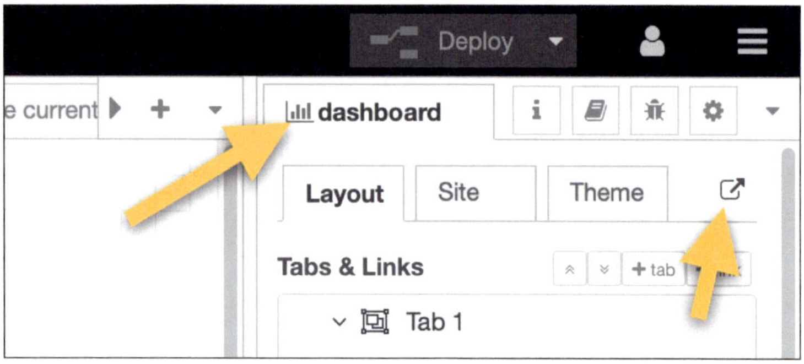

Figure 1.7.0.4: You can open the. Dashboard from the sidebar.

Each dashboard node provides different configuration options, such as the group and tab it belongs to, size and labels. Experimenting with these options will help you create a unique and interactive dashboard. The dashboard is designed to create a single-page application, which means all interactions occur on a single web page.

Dashboard Groups and Tabs

You can organise the widgets in the Node-RED dashboard using Groups and Tabs.

A Node-RED Dashboard group comprises UI widgets or components, such as buttons, charts, and text inputs. It helps organise and group similar or related user interface elements on the Node-RED Dashboard.

A Node-RED Dashboard tab is a primary navigation element in the Node-RED Dashboard. It represents a distinct page or section within the user interface and can contain multiple Dashboard groups and associated UI widgets.

Each group is associated with a specific tab in the Node-RED Dashboard. This allows for better navigation and categorisation of different functionalities.

In the experiments through this book with flows that include dashboard widgets, I have organised these widgets in a tab and one or more groups.

Let's look at an example. Take any dashboard node from the Dashboard group. In this example, I will use a text node widget (you will learn more about this in the next section). This node generates a text widget in the dashboard. Double-click on the text dashboard node to bring up its properties window. You can see this window below.

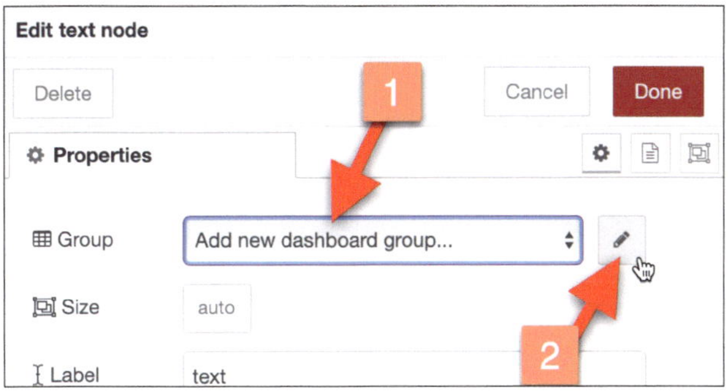

Figure 1.7.0.5: The properties window of the text widget node (top part only).

Notice that the properties contain a Group drop-down ("1"). In this dropdown, you can select one of the existing groups or the option to create a new group. Let's assume there are no existing groups and opt to create a new dashboard group. After selecting "Add new dashboard group..." from the drop-down, click on the pencil button to edit the new Group's properties ("2").

A new dialogue titled "Add new dashboard group config node" will appear. It looks like this:

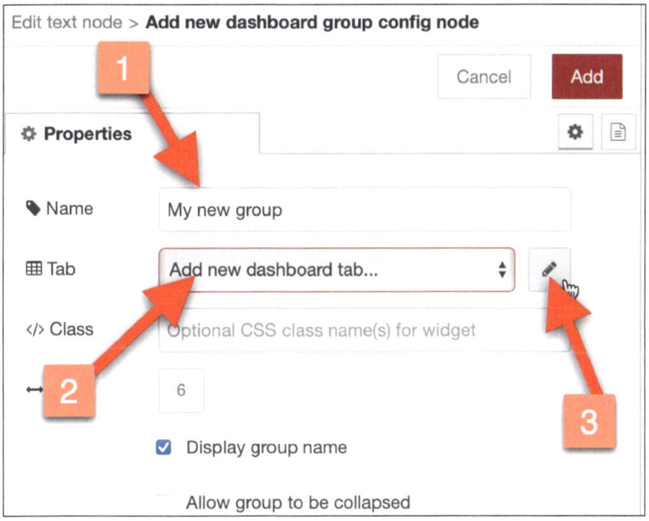

Figure 1.7.0.6: The properties window of the group.

In this dialogue, give the new group a name ("1") and select a tab where the group (and the widgets assigned to it) will be displayed. For this example, let's assume that there is no existing tab. Select the option "Add new dashboard tab..." from the dropdown ("2") and click on the pencil button to open the new tab's properties dialogue ("3").

The dialogue for the new tab looks like this:

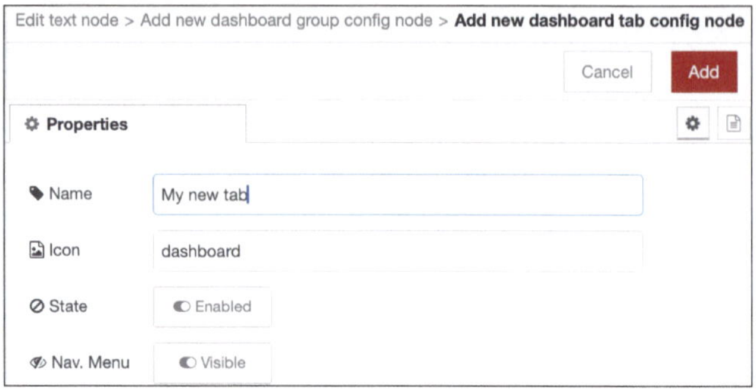

Figure 1.7.0.7: The properties window of the tab.

In the properties dialogue for the tab, add a name and leave the rest of the options as they are. Click "Add" to create this new tab. This will also add the group you created earlier to the new tab.

Click "Add" again to exit the group configuration dialogue, and notice that you are now back to the text node's properties dialogue, which now looks like this:

Figure 1.7.0.8: The updatred properties window of the text node.

In the Group field, notice that the newly created Tab ("1") and Group ("2") names appear. This indicates that this widget will appear in a tab named "My new tab" in the dashboard. Also, within the tab "My new tab", there is a group named "My new group", which is the container for the text widget.

Let's go to the dashboard to look at the result of all this work. Open the Dashboard in a new browser tab (go to http://<your_node_red_ip>:1880/ui or navigate using the right sidebar links as I show below).

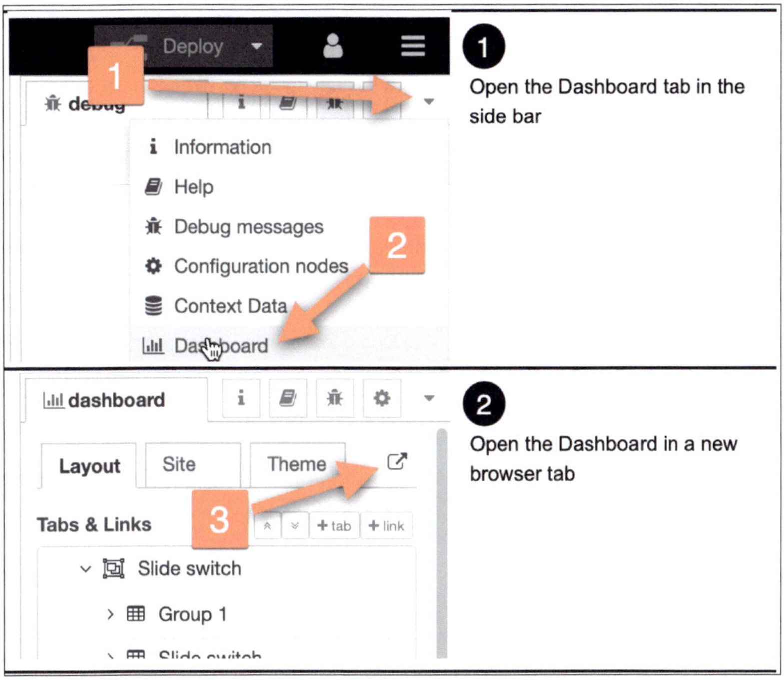

Figure 1.7.0.9: You can use the right sidebar to open the Dashboard in a new browser tab.

In the Dashboard, locate the new tab from the right navigation bar. Click it to show the newly created Group and its text widget.

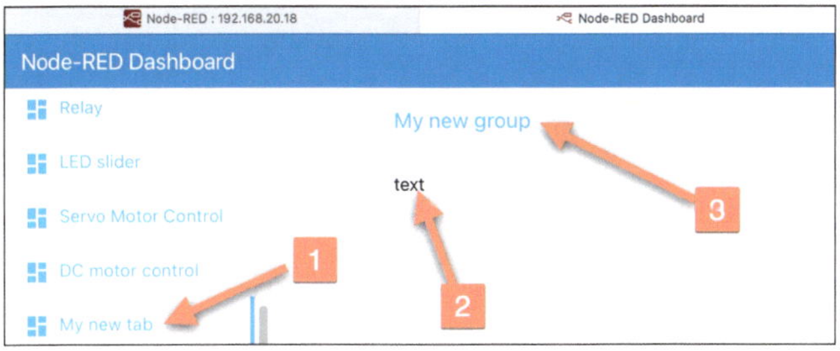

Figure 1.7.0.10: You will find the text widget within the newly created group and tab.

You will use the same process to create tabs and groups and assign widgets to groups in all experiments in this book involving the Dashboard.

The Node-RED Dashboard is a powerful tool for creating user interfaces for your Node-RED applications. It provides an easy and intuitive way to visualise data and interact with your flows. Whether you're building a home automation system, a data monitoring application, or an IoT project, the Node-RED Dashboard can be a valuable addition to your toolset.

7.1. Text Input and Output

One of the powerful features of Node-RED Dashboard is its ability to provide user interface elements, or widgets, that interact with your flows. Two of these widgets are the 'Text Input' and 'Text Output' nodes, which allow for the entry and display of text within your dashboard, respectively. This section introduces these nodes, detailing their configuration and use.

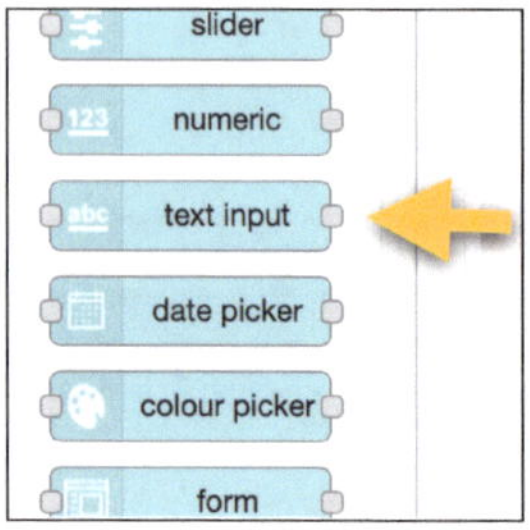

Figure 1.7.1.11: The Text Input node in the palette.

Text Input node

The 'Text Input' node provides an interface for users to enter data that can be used within your flows. It can receive simple text input but supports more complex types, such as a date or colour picker.

To configure a 'Text Input' node, follow these steps:

1. Drag and drop a 'Text Input' node from the dashboard category in the node palette to your workspace.

2. Double-click the node to open its configuration settings.

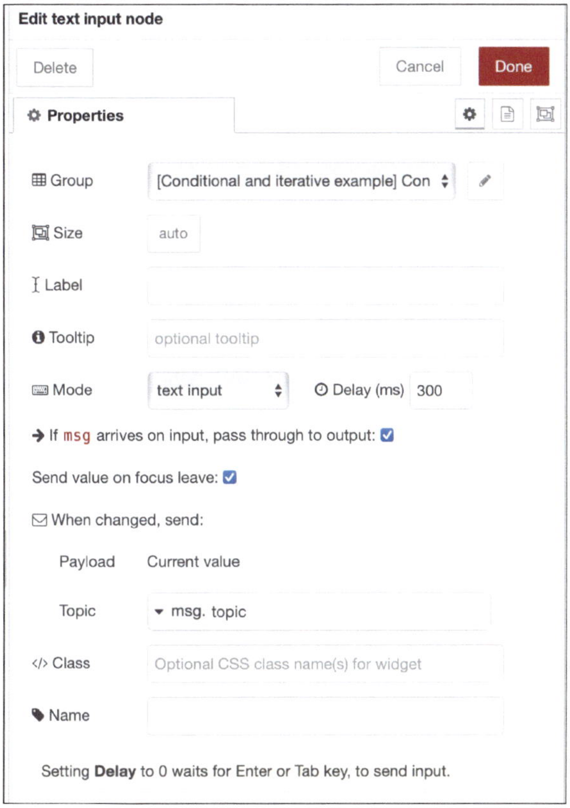

Figure 1.7.1.12: The Text Input node configuration options.

3. Set the 'Group' field to assign the node to a user interface group. You can select an existing group from the drop-down menu or click on the pen button to enter the Group dialogue box to create a new group and assign the group (and all its members) to a dashboard tab. The Dashboard may contain multiple tabs.

4. Under 'Mode', select the type of input you want (text, password, email address, number, date, time, colour, and many types of input text).

5. Optional fields such as 'Label' and 'Placeholder' can be filled to provide additional information for the user interface.

6. Click 'Done' to save your settings.

The 'Text Input' node sends a message every time the user interacts with the input field in the dashboard. The message's payload contains the user's input.

Text Output node

The 'Text Output' node provides a way to display text in the dashboard. This text can be static, but more commonly, it is dynamically updated based on the flow's operations.

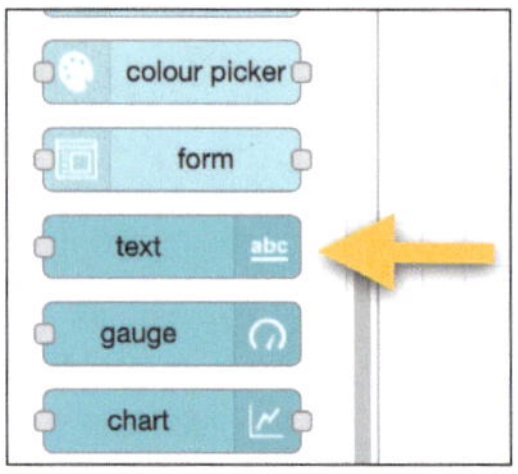

Figure 1.7.1.13: The Text Output node in the Palette.

To configure a 'Text Output' node, follow these steps:

1. Drag and drop a 'Text Output' node from the dashboard category in the node palette to your workspace.

2. Double-click the node to open its configuration settings.

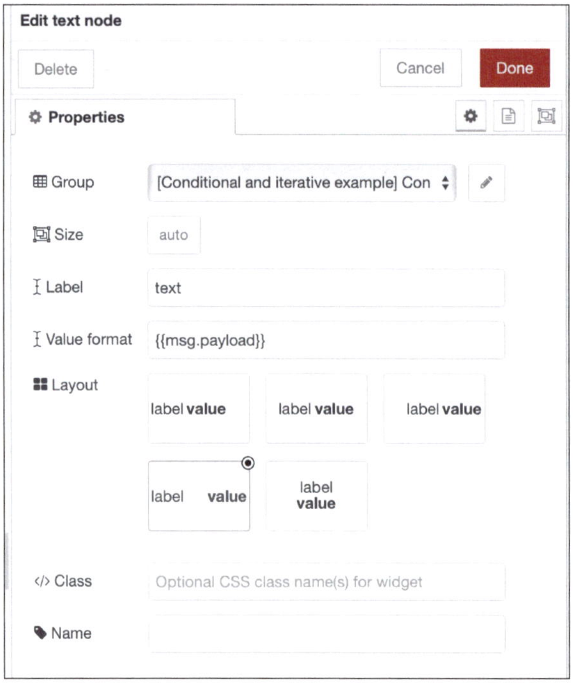

Figure 1.7.1.14: The Text node configuration options.

3. Set the 'Group' field to assign the node to a user interface group.

4. To format the displayed text, enter a Mustache template (more about Mustache in a moment) in the' Value Format' field. The default is "{{msg.payload}}", which will display the payload of the input message.

5. Other fields are optional.

6. Click 'Done' to save your settings.

The 'Text Output' node displays the input message's payload in the dashboard, formatted according to the 'Value Format' field.

About Mustache

"Mustache" is a simple web template system. It's often referred to as a "logic-less" system because it lacks the explicit control flow statements, like if and else conditionals or for loops, found in other templating engines. Instead, Mustache templates use tags, denoted by double curly braces (i.e., "{{" and "}}"), which get replaced with actual values at runtime.

The name "Mustache" comes from the heavy use of curly braces, which resemble a moustache when seen from the side. Mustache can be used for HTML, config files, and source code. In Node-RED dashboards, a Mustache template is often used to generate the HTML content displayed on the dashboard dynamically.

Here's an example of how you might use a Mustache template in Node-RED. Let's say you have a data object like this:

```
javascript
{
  "name": "John",
  "age": 30
}
```

You could create a Mustache template in Node-RED that looks like this:

```
<p>Hello, {{name}}. You are {{age}} years old.</p>
```

When the data object is passed into the Mustache template, it will replace the {{name}} and {{age}} tags with the corresponding values from the data object, resulting in the following HTML:

```
<p>Hello, John. You are 30 years old.</p>
```

This resulting HTML can then be displayed on your Node-RED dashboard.

In a more complex scenario, you might have an array of objects, and you could use Mustache to iterate over them and generate HTML for each one. For instance, consider the following data:

```
{
  "people": [
    {"name": "John", "age": 30},
    {"name": "Jane", "age": 28}
  ]
}
```

You could use the following Mustache template to generate HTML for each person:

```
<ul>
  {{#people}}
    <li>{{name}} is {{age}} years old.</li>
  {{/people}}
</ul>
```

This would result in the following HTML:

```
<ul>
  <li>John is 30 years old.</li>
  <li>Jane is 28 years old.</li>
</ul>
```

Node-RED uses Mustache syntax in the template node to access the properties of the incoming message object ("msg.payload", "msg.topic", etc.) as well. It is a vital component of the Node-RED ecosystem. You can learn more about the Mustache templating system on its website: https://mustache.github.io.

Using Text Input and Output nodes together

A simple use case for the 'Text Input' and 'Text Output' nodes is to create a flow where the user enters text, some processing occurs, and the result is displayed.

1. Connect a 'Text Input' node to a function node that modifies the input somehow.
2. Connect the function node to a 'Text Output' node.
3. Deploy the flow.

When a user enters text into the 'Text Input' field in the dashboard, the function node will process the text, and the result will be displayed in the 'Text Output' field.

The 'Text Input' and 'Text Output' nodes in the Node-RED Dashboard provide powerful and flexible ways to interact with your flows. Understanding and using these tools effectively can significantly enhance your Node-RED applications.

Example

Let's create a simple flow where we'll take text input from the user, process it, display it in the dashboard and debug it.

1. **Set up the Dashboard.** Open the Dashboard properties tab from the right side of the editor. You want to create a new tab and add a widget group to the tab. Click "+tab" to add a tab, and then "+group" to add a group.

Figure 1.7.1.15: Create a new Dashboard tab.

2. **Text Input Node**: First, drag and drop a 'Text Input' node onto your workspace. Double-click on the node to open its configuration settings. Set the 'Group' field to assign the node to a user interface group. Under 'Mode', select 'text input'. Leave other settings as default and click 'Done'. Setup the properties like in this example, including the Group (set to Group 1):

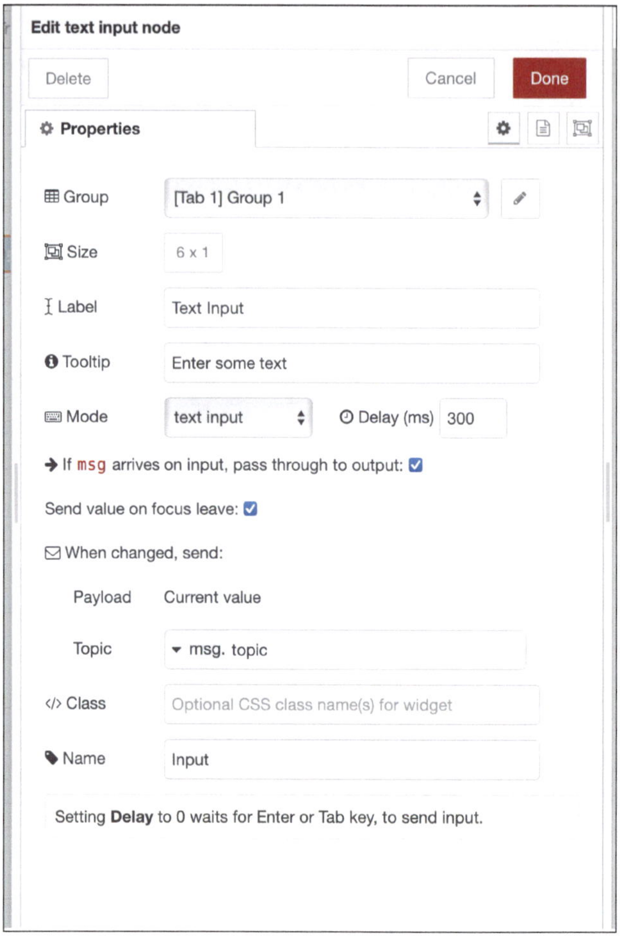

Figure 1.7.1.16: Create a new Dashboard tab.

3. **Function Node**: Drag and drop a 'Function' node onto your workspace—Double-click on it to open its configuration settings. Write a simple function to process the input in the' Function' field. For example, let's add a prefix to the input text:

```
msg.payload = "User said: " + msg.payload;
        return msg;
```

   Click 'Done'.

4. **Text Output Node**: Next, drag and drop a 'Text Output' node onto your workspace—Double-click on it to open its configuration settings. Set the 'Group' field to the same group as your 'Text Input' node. Leave the 'Value Format' field as default ("{{msg.payload}}") and click 'Done'. Set up this widget like this, including the Group:

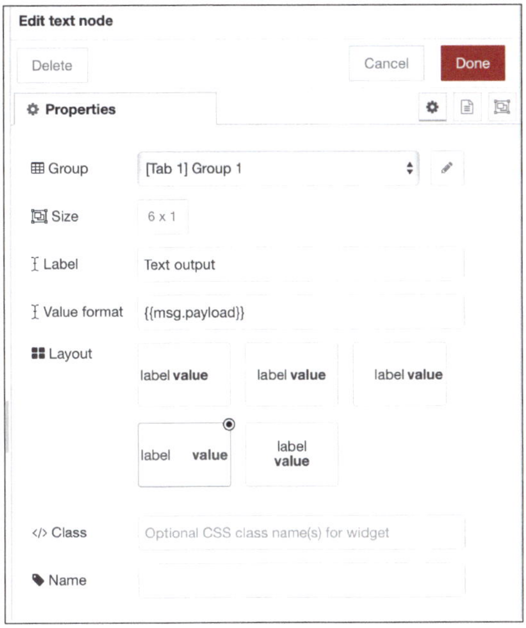

Figure 1.7.1.17: Setup the Text Output node.

5. Debug Node: Drag and drop a 'Debug' node onto your workspace. No configuration is needed for this node.

6. Now, connect the output port of the 'Text Input' node to the input port of the 'Function' node. Then, connect the output port of the 'Function' node to the input ports of both the 'Text Output' and 'Debug' nodes.

7. In the Dashboard layout, drag the widget icons in the list up or down to arrange the two widgets like this:

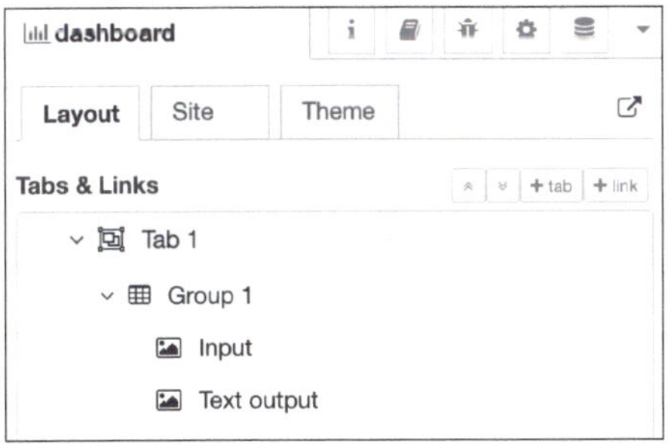

Figure 1.7.1.18: Open the dashboard.

8. Next, click 'Deploy' to deploy the flow.

9. Click the Open Dashboard button to open the Dashboard in a new browser window.

With this flow, when a user enters text into the 'Text Input' field in the dashboard, the function node will process the text, and the result will be displayed in the 'Text Output' field. The processed text will also be output to the debug tab. Here is a screenshot of this dashboard in the browser:

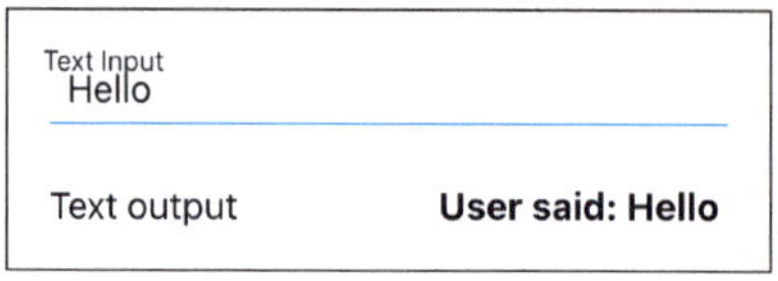

Figure 1.7.1.19: The dashboard we just created in the browser.

It is possible to define each browser's exact location and size in the dashboard by using the Size attribute in the dashboard node's properties window. For example, I have set the text output widget to occupy an area of six cells width and one cell height by defining this area in the Size field:

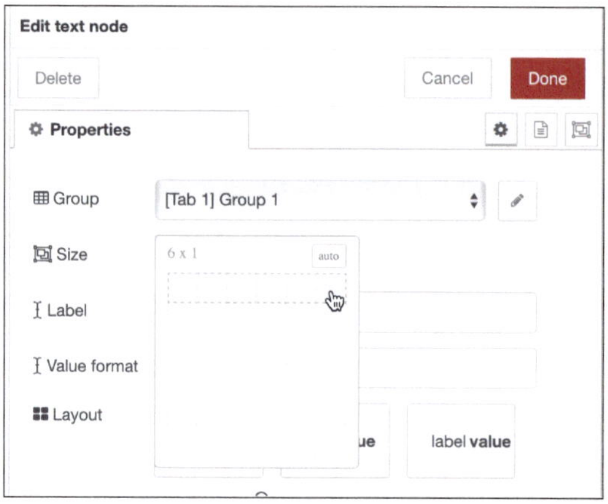

Figure 1.7.1.20: Define the size of the text widget.

You can further control the ordering of the widgets in the dashboard via the Tabs & Links box in the Layout tab of the Dashboard tab in the sidebar. Use the handles on the left of each widget to rearrange them:

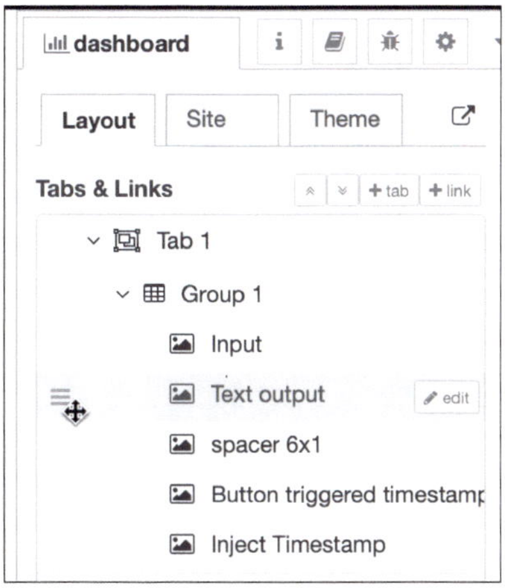

Figure 1.7.1.21: You can re-arrange the order of Dashboard widgets.

7.2. The Button

Button widgets are one of the most straightforward yet powerful tools in the Node-RED Dashboard. They allow users to interact with a flow clearly and intuitively. This section will introduce the button widget, including its configuration and use.

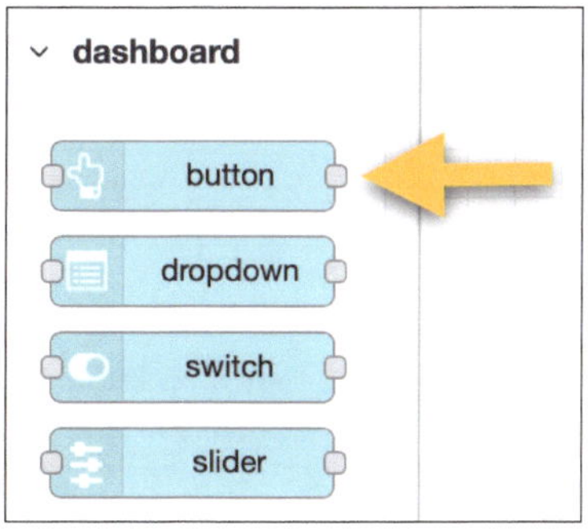

Figure 1.7.2.22: The Button node in the palette.

What is a button widget?

A button widget in Node-RED is a user interface element that sends a predefined message each time it is clicked or tapped. This makes it an excellent tool for triggering actions in your flow based on user interaction.

Setting Up a button widget

To use a button widget in Node-RED:

1. Drag and drop a "Button" node from the dashboard category in the node palette onto your workspace.

2. Double-click the node to open its configuration settings.

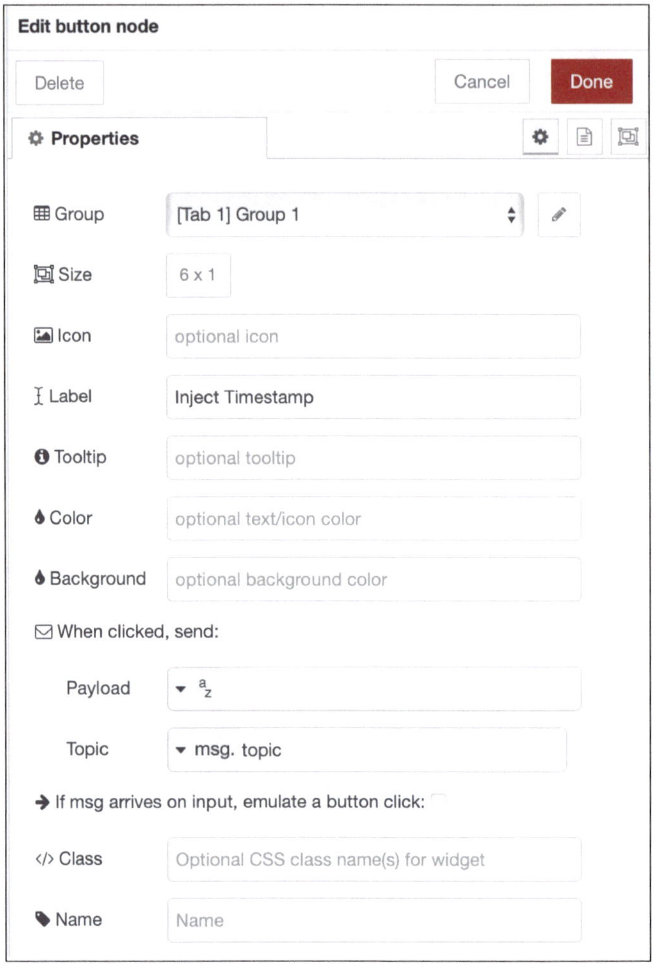

Figure 1.7.2.23: The Button node configuration options.

3. Set the "Group" field to assign the node to a user interface group.

4. In the "Label" field, provide a name for the button displayed on the dashboard.

5. In the "Payload" field, enter the message you want the button to send each time it is clicked. This can be a static or JavaScript expression that generates a dynamic value.

6. Set the "Payload Type" to the type of the payload value. This is usually "string" or "number" for static values. For JavaScript expressions, this should be set to "expression".

7. In the "Name" field, enter a name for the node. This name will be displayed in the workspace but not on the dashboard.

8. Click "Done" to save your settings.

Using a button widget

Once you've configured your button widget and deployed your flow, you can interact with it on your dashboard. Each time you click the button, it will send a message with its configured payload. You can use the button widget to trigger any action in your flow. This could be turning on a light, starting a timer, sending a request to a web server, or any other action that can be triggered by a message.

For example, you want to create a flow that turns on a lamp when clicking a button. You could configure a button node with a payload of "on", connect it to a function node that translates "on" to the appropriate command for your light, and then connect that to an output node that sends the command to your lamp.

Button widget demonstration

Let's create a simple flow using a button widget to trigger an action. In this case, we'll trigger the injection of a timestamp.

1. Drag and drop a "Button" node onto your workspace from the dashboard section—Double-click on it to open its configuration settings. Set the "Group" field to assign the node to a user interface group. In the "Label" field, provide a name for the button, say "Inject Timestamp". In the "Payload" field, set it to "String" and leave the value empty. You can experiment with other payload types, including the timestamp (use a debug node to see the value). Click "Done".

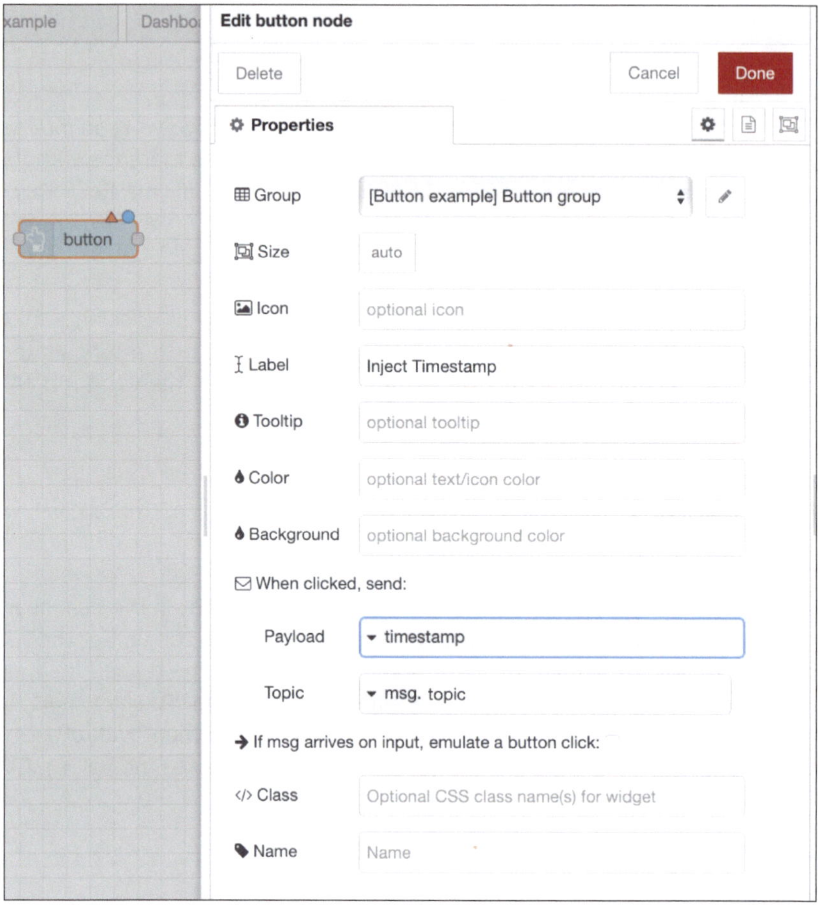

Figure 1.7.2.24: The example button node properties.

2. Drag and drop a "Function" node onto your workspace. Double-click on it to open its configuration settings. In the "Function" field, write a simple function to generate the current timestamp:

```
msg.payload = new Date().getTime();
return msg;
```

3. Click "Done".

4. Drag and drop a "Debug" node onto your workspace. No configuration is needed for this node.

5. Connect the output port of the "Button" node to the input port of the "Function" node. Then, connect the output port of the "Function" node to the input port of the "Debug" node.

6. Drag a text output widget and give it a suitable name, like "Button triggered timestamp". Add it to the button group.

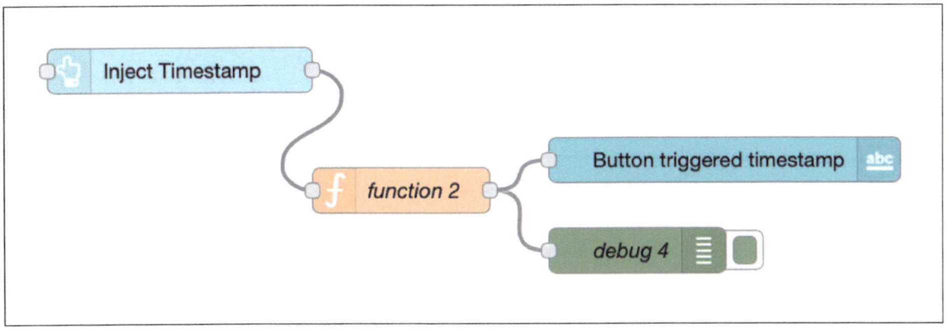

Figure 1.7.2.25: The button widget example flow.

7. Use the Dashboard layout editor to arrange the widgets. You can include a spacer to add a gap between the widgets.

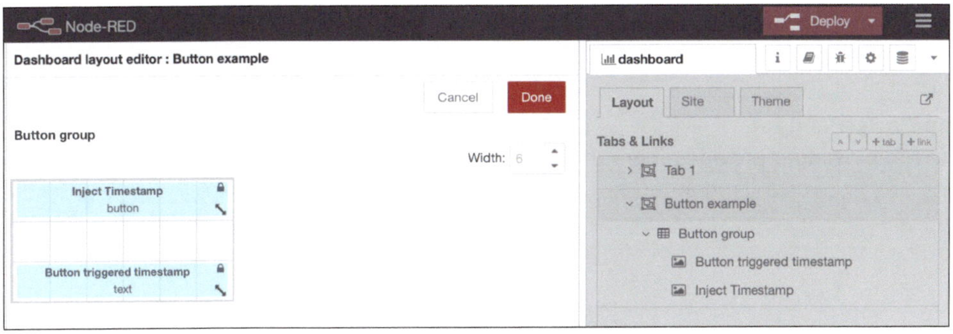

Figure 1.7.2.26: The button widget example dashboard layout.

8. Click "Deploy" to deploy the flow.

With this flow, when a user clicks the "Inject Timestamp" button in the dashboard, the current timestamp will be generated by the function node and output to the debug tab. You can see the dashboard for this flow below:

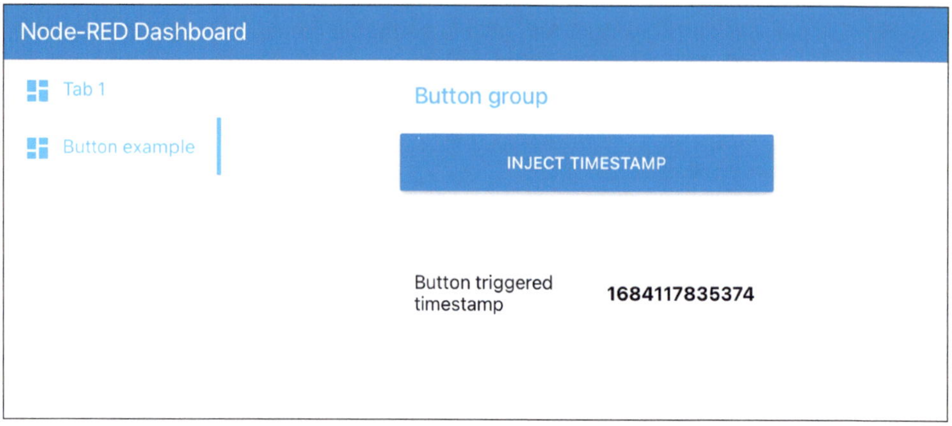

Figure 1.7.2.27: The button widget example dashboard in the browser.

7.3. The Gauge and Slider

The Node-RED Dashboard provides a set of nodes that make it easy to create interactive user interfaces. Among them, the Gauge and Slider widgets are versatile tools that display and control real-time data. This section will introduce these widgets, including their configuration and usage.

The Gauge widget

The Gauge widget provides a graphical representation of data, perfect for real-time sensor data or system status monitoring.

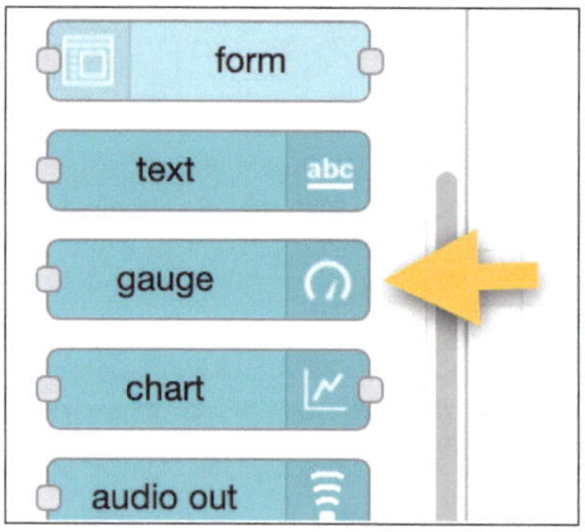

Figure 1.7.3.28: The Gauge node in the palette.

Setting up a Gauge widget

Here is how to set up a gauge widget in your Node-RED dashboard.

1. Drop a 'Gauge' node from the Dashboard category in the Palette onto your workspace.

2. Double-click the node to open its configuration settings.

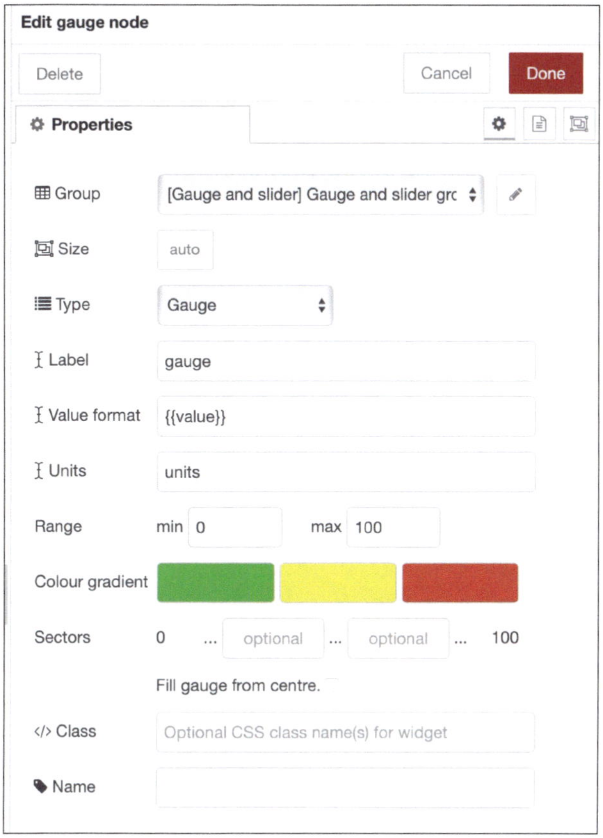

Figure 1.7.3.29: The Gauge widget configuration options.

3. Set the 'Group' field to assign the node to a user interface group.

4. In the 'Label' field, provide a name for the gauge displayed on the dashboard.

5. Set the minimum and maximum values that the gauge will represent.

6. Choose a type for the gauge (Gauge, Donut, Compass, or Wave).

7. Click 'Done' to save your settings.

Using a Gauge widget

The Gauge widget will display the value of the payload property of any message it receives. For example, if you have a flow that reads the temperature from a sensor and sends it as the payload of a message, connecting this flow to a gauge will display the temperature on your dashboard.

The Slider widget

The Slider widget allows users to input a numeric value by dragging a handle along a track. It's ideal for controlling devices that accept a range of numeric values, like dimming a lamp or setting a temperature.

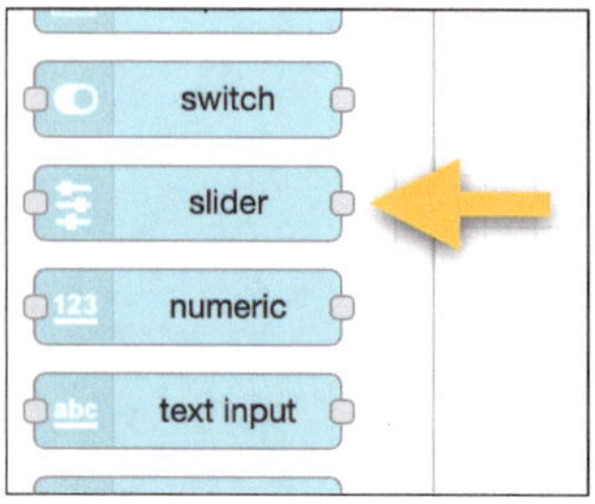

Figure 1.7.3.30: The Gauge node in the palette.

Setting up a Slider widget

Here is how to set up a gauge widget in your Node-RED dashboard.

1. Drag and drop a 'Slider' node from the Dashboard category onto your work-space.

2. Double-click the node to open its configuration settings.

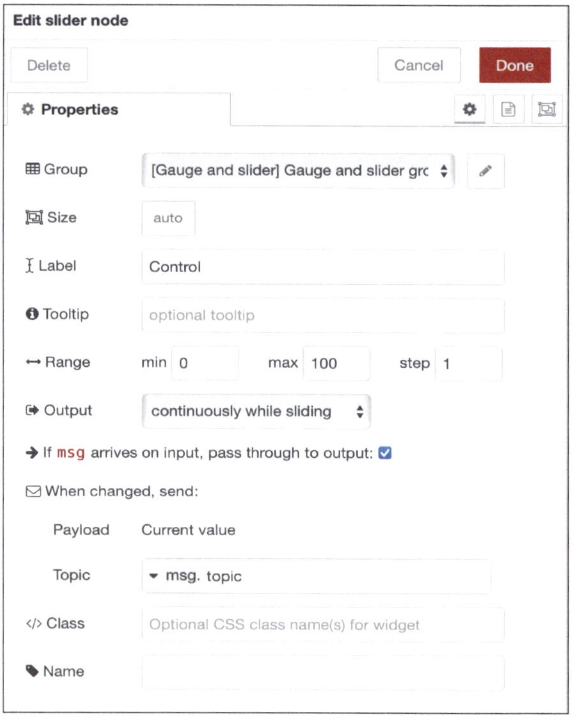

Figure 1.7.3.31: The Slider widget configuration options.

3. Set the 'Group' field to assign the node to a user interface group.

4. In the 'Label' field, provide a name for the slider displayed on the dashboard.

5. Set the minimum and maximum values that the slider will represent.

6. Click 'Done' to save your settings.

Using a Slider widget

The Slider widget will send a message with the current slider value as the payload each time the slider handle is moved. This makes it easy to incorporate the slider into your flows. For example, you could connect a slider to a function node that translates the slider value into a command for a device, like setting the brightness of a light or the temperature of a thermostat.

The Gauge and Slider widgets in the Node-RED Dashboard provide an easy way to interact with your flows. The Gauge widget lets you visualise data in real-time, while the Slider widget allows for interactive control of devices. Together, they can enhance your IoT projects and provide users with a rich, interactive experience.

A demo of the Gauge and Slider widgets

Let's create a simple flow that utilises a slider to control a value and a gauge to display that value.

1. Drag and drop a 'Slider' node from the dashboard section onto your workspace. Double-click the node to open its configuration settings. Set the 'Group' field to assign the node to a user interface group. In the 'Label' field, provide a name for the slider, say "Control". Set the minimum and maximum values per your requirement, say 0 to 100. Click 'Done'.

2. Drag and drop a 'Gauge' node onto your workspace—Double-click on it to open its configuration settings. Set the 'Group' field to assign the node to the same user interface group. In the 'Label' field, provide a name for the gauge, say "Display". Set the minimum and maximum values the same as the slider (0 to 100 in this case). Click 'Done'.

3. Connect the output port of the 'Slider' node to the input port of the 'Gauge' node. Click 'Deploy' to deploy the flow.

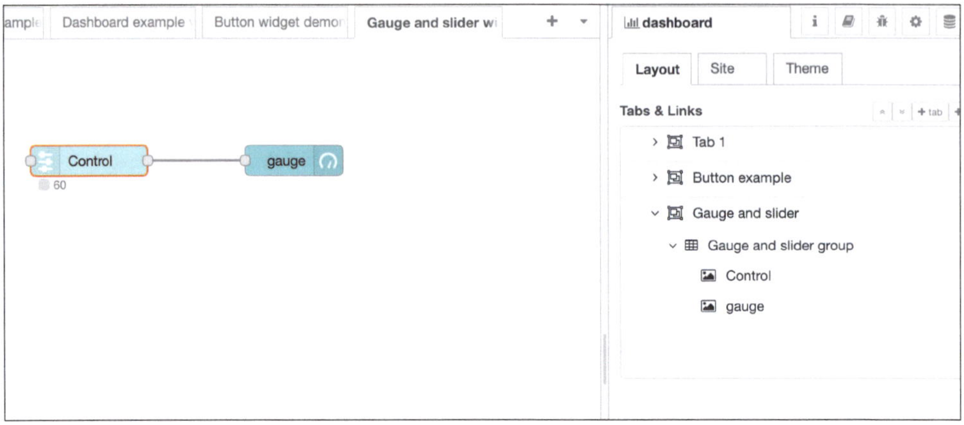

Figure 1.7.3.32: An example flow that combines Gauge and Slider widgets.

With this flow, when a user moves the slider in the dashboard, the gauge will display the slider's current value. It looks like this:

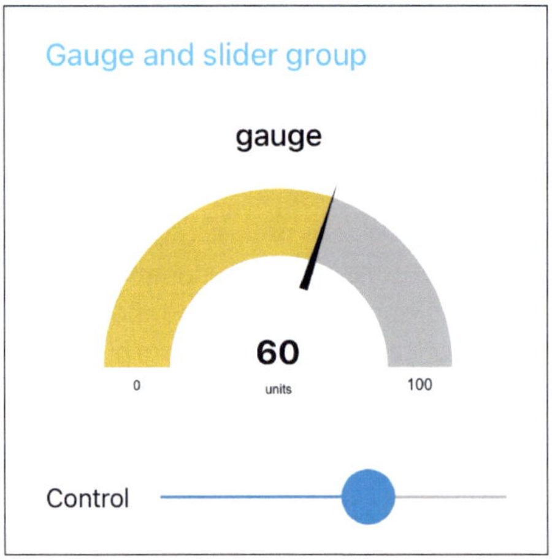

Figure 1.7.3.33: A dashboard that contains a gauge and a slider.

This simple example can be extended to control and monitor devices or systems. For instance, the slider could be used to manage the brightness of a light, with the gauge displaying the current brightness level.

7.4. The Switch

The Node-RED Dashboard provides a set of nodes that make it easy to create interactive user interfaces. Among these, the Switch widget is useful for controlling binary states, acting much like a physical switch. This section will introduce the Switch widget, including its configuration and usage.

The Switch widget in Node-RED is a user interface element that displays a switch on the dashboard. The user can toggle the switch on and off, allowing it to control your flows' binary states. For example, it could turn a light on and off or activate and deactivate a system or process.

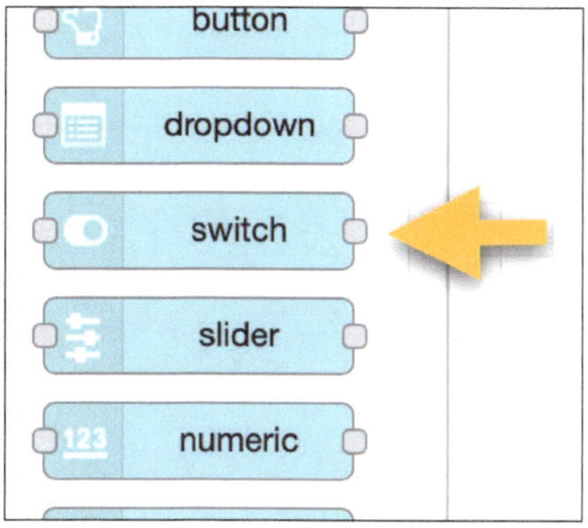

Figure 1.7.4.34: The Switch node in the palette.

Setting up a Switch widget

To use a Switch widget in Node-RED:

1. Drag and drop a 'Switch' node from the dashboard category in the node palette onto your workspace.

2. Double-click the node to open its configuration settings.

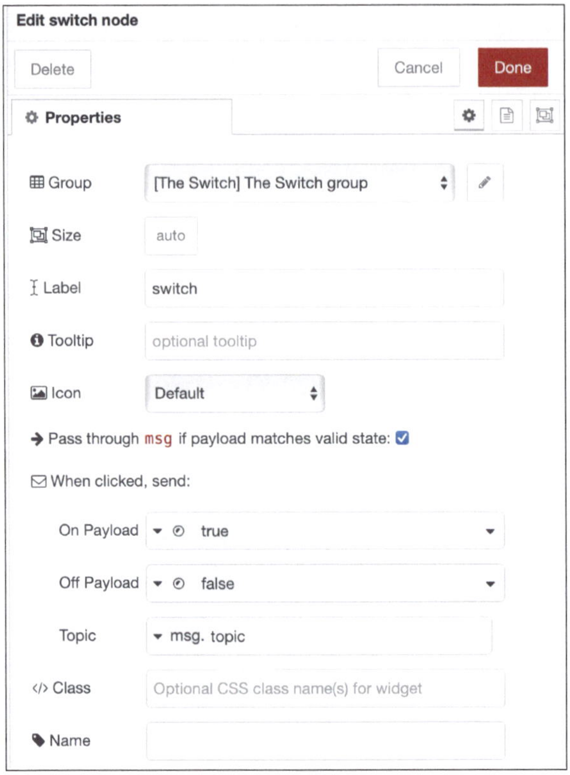

Figure 1.7.4.35: The Switch widget configuration options.

3. Set the 'Group' field to assign the node to a user interface group.

4. In the 'Label' field, provide a name for the switch displayed on the dashboard.

5. In the 'Payload' fields for 'On' and 'Off', enter the messages you want the switch to send when turned on and off, respectively. These messages can be static values or dynamic expressions.

6. Set the 'Payload Type' fields for 'On' and 'Off' to match the type of your payload values.

7. In the 'Name' field, enter a name for the node. This name will be displayed in the workspace but not on the dashboard.

8. Click 'Done' to save your settings.

Using a Switch widget
Once you've configured your Switch widget and deployed your flow, you can interact with it on your dashboard. When you toggle the switch on, it will send its 'On' payload, and when you toggle it off, it will send its 'Off' payload.

You can use these payloads to control any binary state in your flow. For example, you could connect the switch to an output node that controls a lamp. When the switch sends its 'On' payload, the light turns on, and when it sends its 'Off' payload, the light turns off.

The Switch widget in the Node-RED Dashboard is a powerful tool for interacting with binary states in your flows. By configuring a switch with appropriate 'On' and 'Off' payloads, you can give users control over these states straightforwardly and intuitively. Whether you're creating a home automation system, a control panel for a machine, or any other project that requires binary control, the Switch widget can be an invaluable tool.

An example flow with the Switch widget

Here's an example flow using the switch widget, inject, debug, and function nodes.

1. Drag and drop a 'Switch' node from the Dashboard category onto your workspace. Double-click the node to open its configuration settings. Set the 'Group' field to assign the node to a user interface group. In the 'Label' field, provide a name for the switch displayed on the dashboard, say "Light Control". Set the "On Payload" to "true" and "Off Payload" to "false". Click "Done".

2. Drag and drop a 'Function' node onto your workspace. In this function, we'll interpret the Boolean payload from the switch node and turn it into a string representing the light's state. Double-click the node to open its configuration settings. You might write a function like this:

```
if (msg.payload === true) {
       msg.payload = "The light is ON";
} else if (msg.payload === false) {
       msg.payload = "The light is OFF";
}
return msg;
```

   Click 'Done'.

3. Drag and drop a "Debug" node onto your workspace. This will allow you to view the output of the function node in the debug sidebar.

4. Add a text output widget to see the switch value in the dashboard.

5. Now connect the output of the "Switch" node to the input of the "Function" node and the output of the 'Function' node to the inputs of the 'Debug' and "Text" nodes.

6. Click 'Deploy' to deploy the flow. You can see the flow for this example below.

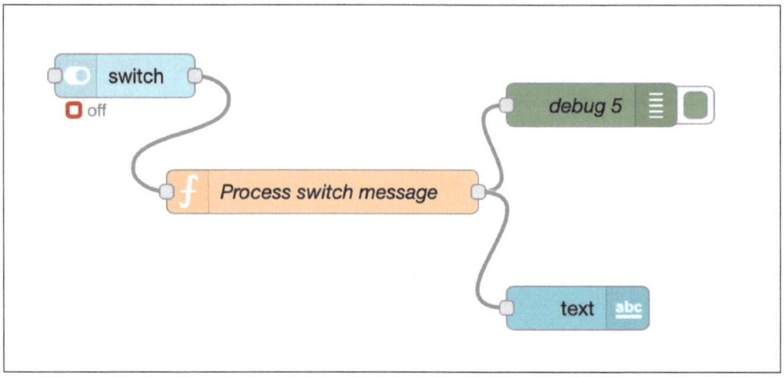

Figure 1.7.4.36: The Switch widget example flow.

You'll see a switch labelled "Light Control" when you go to your Node-RED dashboard. You can see the dashboard below.

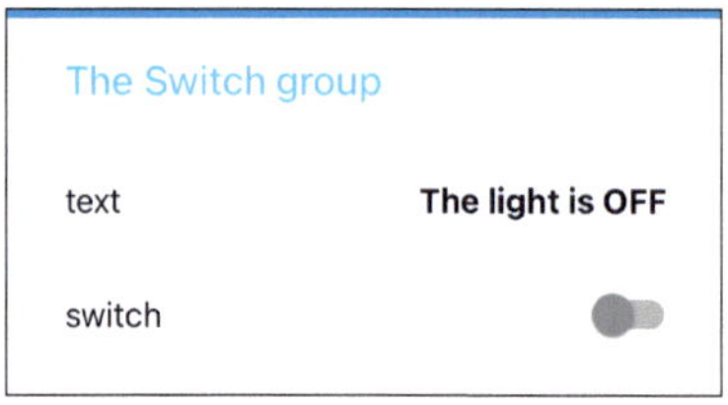

Figure 1.7.4.37: The Switch widget example dashboard.

Toggling this switch will send a message with a payload of "true" or "false" to the function node. The function node will translate this into "The light is ON" or "The light is OFF" and send this message to the debug node. The debug node will then print this message to the debug sidebar.

7.5. The Dropdown

The Node-RED Dashboard provides a set of nodes that make it easy to create interactive user interfaces. One such node is the Dropdown widget, which allows users to select from a list of options. This chapter will introduce the Dropdown widget, including its configuration and usage.

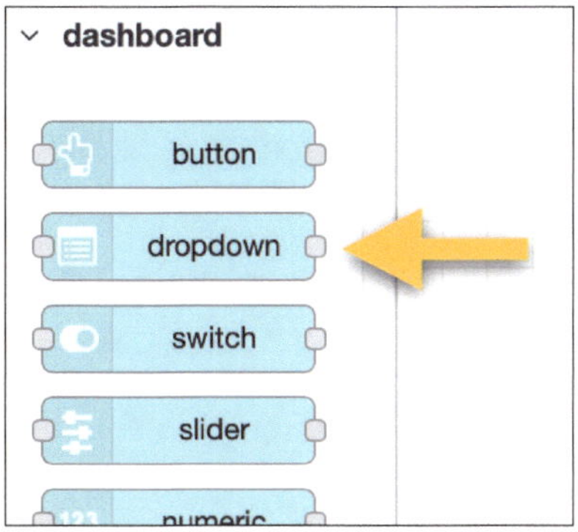

Figure 1.7.5.38: The Dropdown node in the palette.

The Dropdown widget in Node-RED is a user interface element that displays a dropdown list on the dashboard. The dropdown list can contain several options, of which the user can select one. This is useful when you want the user to be able to choose from a predefined set of options. For example, it could be used to select a mode of operation for a system, to choose a colour for a light, or to select a device to control.

Setting up a Dropdown widget

To use a Dropdown widget in Node-RED, follow these steps:

1. Drag and drop a 'Dropdown' node from the dashboard category in the node palette onto your workspace.

2. Double-click the node to open its configuration settings.

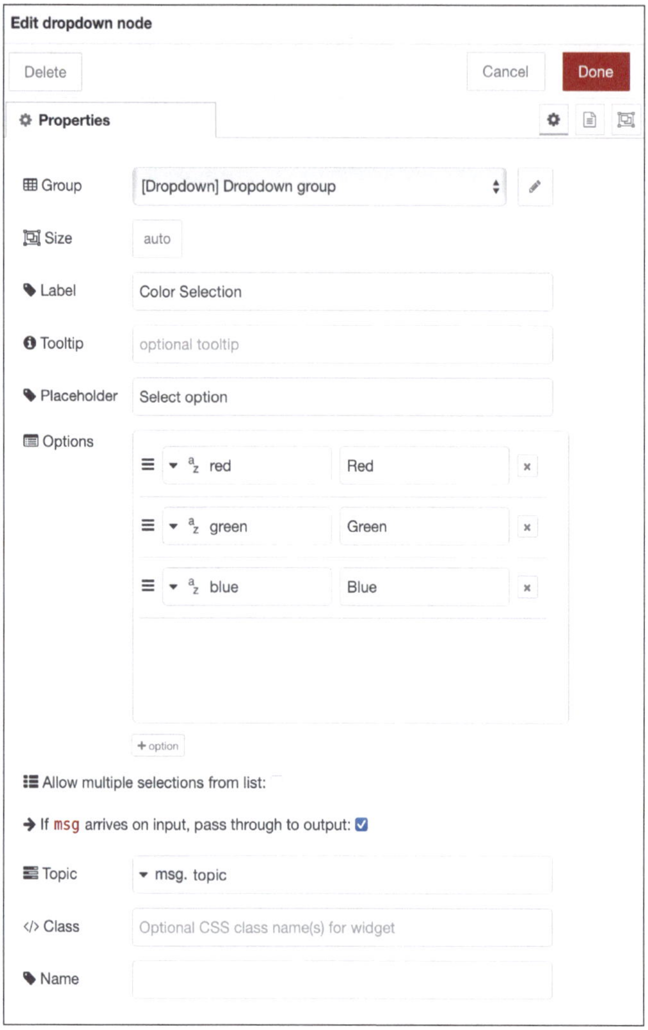

Figure 1.7.5.39: The Dropdown widget configuration.

3. Set the "Group" field to assign the node to a user interface group.

4. In the 'Label' field, provide a name for the dropdown displayed on the dashboard.

5. In the "Options" field, add the options that you want to appear in the dropdown list. Each option should have a "Label" (the text that will be displayed in the dropdown) and a "Value" (the payload that will be sent when this option is selected).

6. In the "Name" field, enter a name for the node. This name will be displayed in the workspace but not on the dashboard.

7. Click "Done" to save your settings.

Using a Dropdown widget

Once you've configured your Dropdown widget and deployed your flow, you can interact with it on your dashboard. Selecting an option from the dropdown will send a message with a payload equal to the 'Value' of the option chosen.

You can use these payloads to control different aspects of your flow. For example, if you connected the dropdown to an output node that contains a lamp, you could have other options for extra colours, and the light would change colour based on the selected option.

The Dropdown widget in the Node-RED Dashboard is a powerful tool for user interaction, allowing users to select from predefined options. You can create complex and interactive control systems by properly setting up the options and their associated payloads. Whether creating a home automation system, a control panel for a machine, or any other project requiring user input, the Dropdown widget is a versatile and valuable tool.

An example flow with the Dropdown widget

Here is an example flow using the dropdown widget, function, and debug nodes.

1. Drag and drop a "Dropdown" node from the Dashboard category onto your workspace. Double-click the node to open its configuration settings. Set the "Group" field to assign the node to a user interface group. In the "Label" field, provide a name for the dropdown displayed on the dashboard, say "Color Selection". In the "Options" field, add the options that you want to appear in the dropdown list. For instance, you might add options like:
   - Label: "Red", Value: "red"
   - Label: "Blue", Value: "blue"
   - Label: "Green", Value: "green"

   Click "Done".

2. Drag and drop a 'Function' node onto your workspace. This function will interpret the payload from the dropdown node and create a string message. Double-click the node to open its configuration settings. Here's an example function:

```
msg.payload = "The selected color is " + msg.payload;
return msg;
```

   Click 'Done'.

3. Drag and drop a Text widget onto your workspace. This will allow you to view the output of the function node in the Dashboard.

4. Now connect the output of the 'Dropdown' node to the input of the "Function" node and the output of the "Function" node to the input of the Text widget node.

5. Click 'Deploy' to deploy the flow. You can see the example flow in the editor below:

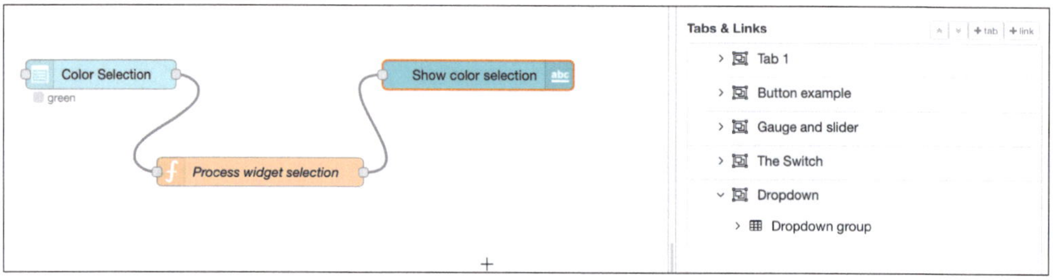

Figure 1.7.5.40: An example flow that contains a Dropdown widget.

You'll see a dropdown labelled "Color Selection" when you go to your Node-RED dashboard.

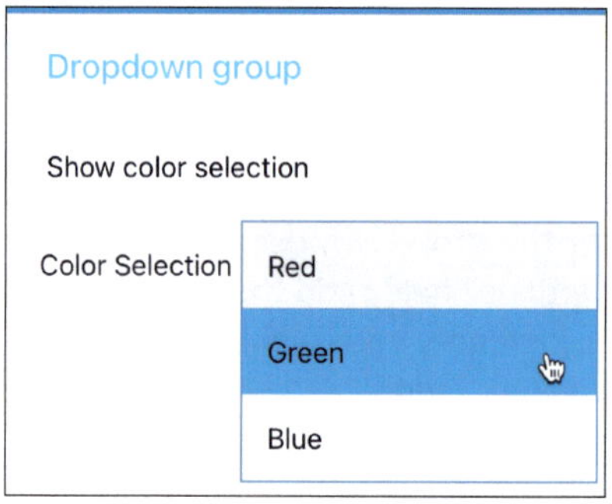

Figure 1.7.5.41: An example of a Dropdown widget in the dashboard.

Selecting a colour from this dropdown will send a message with a payload of the selected colour's value (for example, "red") to the function node. The function node will translate this into a message like "The selected colour is red" and send this message to the text widget. The debug node will then print this message to the dashboard.

7.6. The Form

Node-RED Dashboard provides an extensive set of nodes to create interactive user interfaces. One such node is the Form widget, which allows you to create and customise a form with multiple input fields. This section will introduce the Form widget, detailing its configuration and usage.

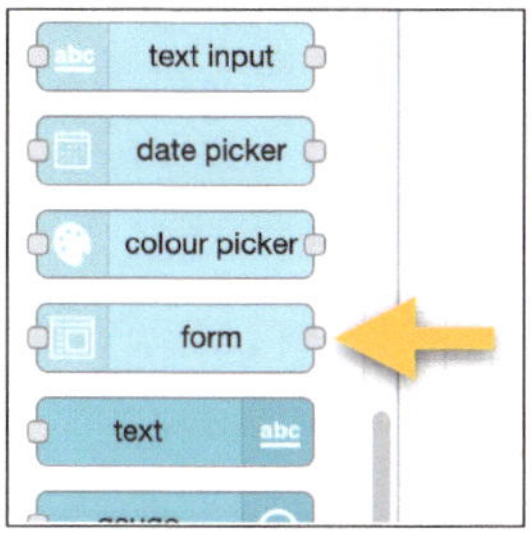

Figure 1.7.6.42: The Form node in the palette.

The Form widget in Node-RED is a user interface component that allows you to create a form with various input fields, such as text, number, date, and time. Users can enter information into the form and submit it. Once submitted, the form generates a message containing the input data as a payload, which can then be processed and used in your flow.

Forms are handy when collecting multiple pieces of information from the user, such as configuring a system, entering user credentials, or providing feedback.

Setting up a Form widget

To use a Form widget in Node-RED, follow this process:

1. Drag and drop a 'Form' node from the dashboard category in the node palette onto your workspace.

2. Double-click the node to open its configuration settings.

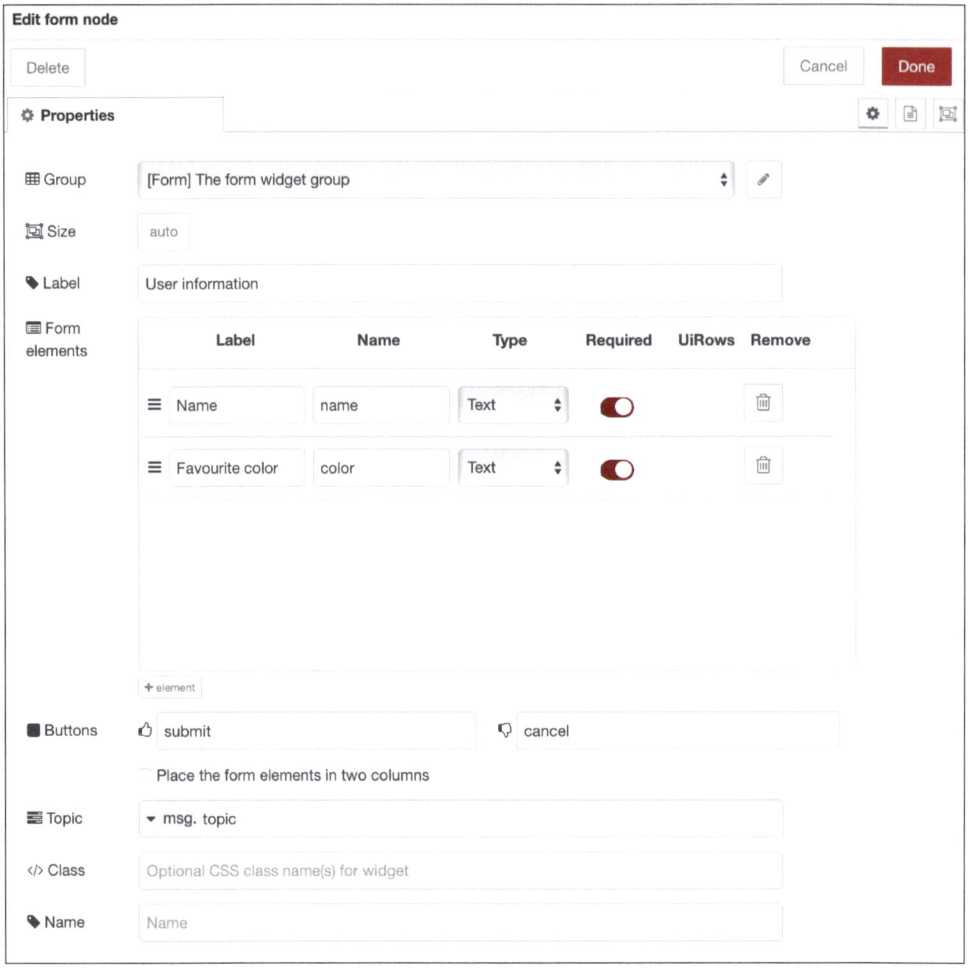

Figure 1.7.6.43: The "Form" node widget configuration.

3. Set the 'Group' field to assign the node to a user interface group.

4. In the 'Label' field, provide a name for the form displayed on the dashboard.

5. In the 'Form Fields' section, add the input fields you want to include in the form. Specify each input field's type, label, and options as required. You can add multiple input fields by clicking the "+" button.

6. In the 'Submit' field, provide the text to be displayed on the submit button.

7. In the 'Name' field, enter a name for the node. This name will be displayed in the workspace but not on the dashboard.

8. Click 'Done' to save your settings.

Using a Form widget

After configuring the Form widget and deploying your flow, you can interact with it on your dashboard. Users can enter information into the form's input fields and submit the form by clicking the submit button. A message containing the input data as a payload will be sent upon submission. The payload will be an object where each key corresponds to an input field's label, and the associated value is the user's input.

The Form widget in the Node-RED Dashboard is a powerful and versatile tool for user interaction, allowing the collection of multiple pieces of information through a single interface. By configuring the input fields and processing the submitted payloads, you can create complex and interactive control systems, gather user input, and drive your flows based on user-provided data. Whether you're building a configuration panel, a login system, or a feedback form, the Form widget is an invaluable tool for any Node-RED project.

An example flow with the Form widget

In this example, we'll create a form that asks for a user's name and favourite colour and then displays the submitted information on the dashboard using a 'Text' widget.

1. Drag and drop a 'Form' node from the Dashboard category onto your workspace. Double-click the node to open its configuration settings. Set the 'Group' field to assign the node to a user interface group. In the 'Label' field, provide a name for the form, say "User Information". In the 'Form Fields' section, add two input fields:
    - Label: "Name", Type: "text"
    - Label: "Favorite Color", Type: "text"

   In the 'Submit' field, provide the text to be displayed on the submit button, say "Submit".

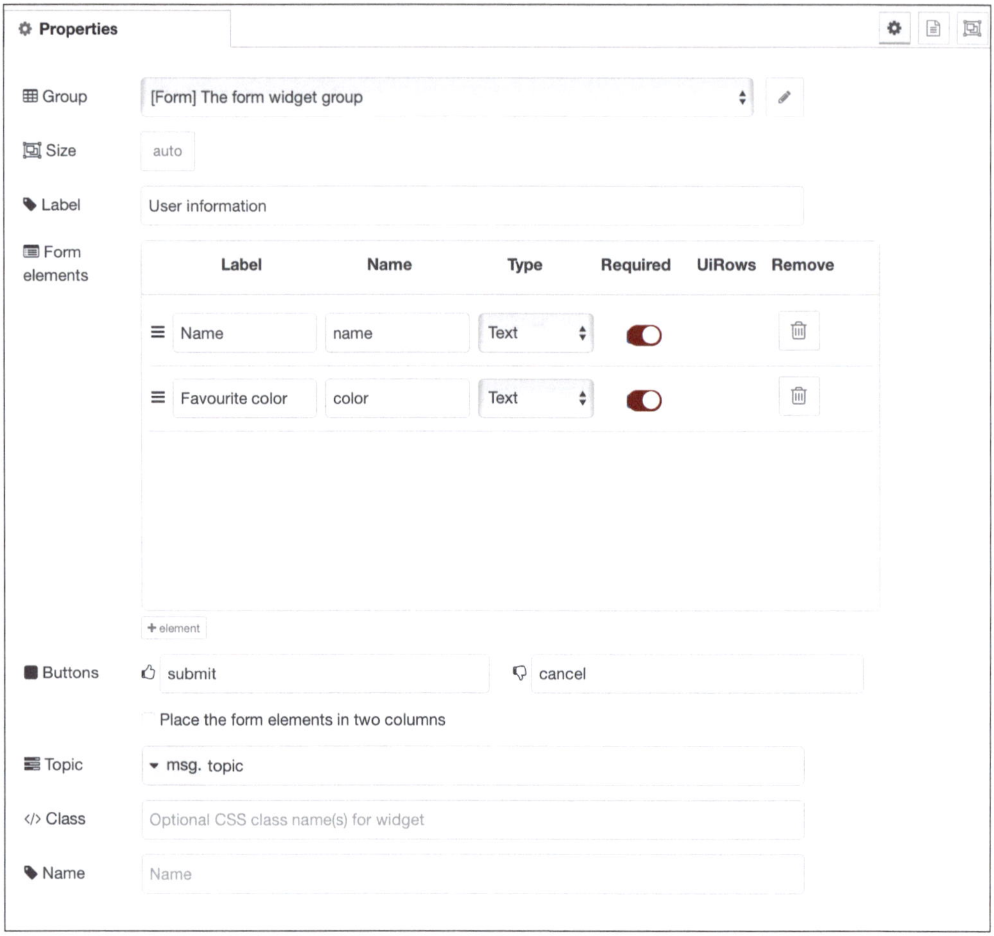

Figure 1.7.6.44: The example Form widget configuration.

Click 'Done'.

2. Drag and drop a 'Function' node onto your workspace. This function will interpret the payload from the form node and create a string message. Double-click the node to open its configuration settings. Here's an example function:

```
msg.payload = "Name: " + msg.payload.name + ", Favorite Color: " + msg.payload.
color;
return msg;
```

Click 'Done'.

3. Drag and drop a 'Text' node (from the Dashboard category) onto your workspace. This will allow you to display the output of the function node on the dashboard. Double-click the node to open its configuration settings. Set the

"Group" field to assign the node to a user interface group. In the "Label" field, provide a name for the text widget, say "User Info". Click "Done".

4. Now connect the output of the 'Form' node to the input of the "Function" node and the output of the 'Function' node to the input of the 'Text' node.

5. Click 'Deploy' to deploy the flow. Your flow should look like this:

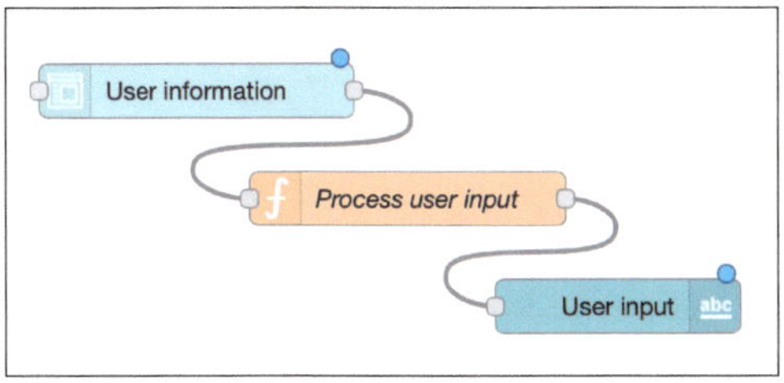

Figure 1.7.6.45: A flow with a Form widget.

When you go to your Node-RED dashboard, you'll see a form labelled "User Information" with fields for Name and Favorite Color. The form in the dashboard should look like this:

Figure 1.7.6.46: The example form in the dashboard.

After filling out the form and hitting "Submit", the submitted information will be displayed in the "User Info" section of the dashboard.

7.7. The UI Template

Node-RED Dashboard provides powerful nodes to create customisable and interactive user interfaces. Among these, the UI Template node is a versatile and flexible tool that allows you to define custom HTML, CSS, and JavaScript to create unique and personalised dashboard components. This seection will introduce the UI Template node, detailing its uses, configuration, and benefits.

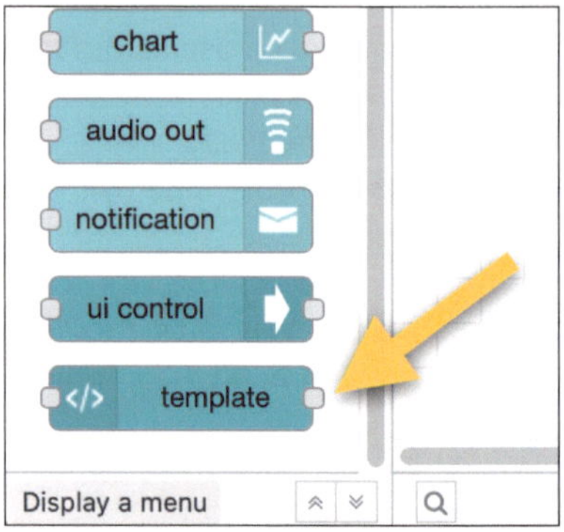

Figure 1.7.7.47: The UI Template node in the palette.

The UI Template node is a powerful node in the Node-RED Dashboard that allows you to create custom widgets using HTML, CSS, and JavaScript. Unlike other pre-built dashboard nodes, the UI Template node provides a blank canvas, allowing you to create complex, tailored widgets that perfectly fit your project's needs.

This node can display data in custom formats, create unique controls, integrate external libraries, or embed other websites or applications within your dashboard.

Setting up a UI Template node

To use a UI Template node in Node-RED, follow this process:

1. Drag and drop a 'ui_template' node from the Dashboard category in the node palette onto your workspace.

2. Double-click the node to open its configuration settings.

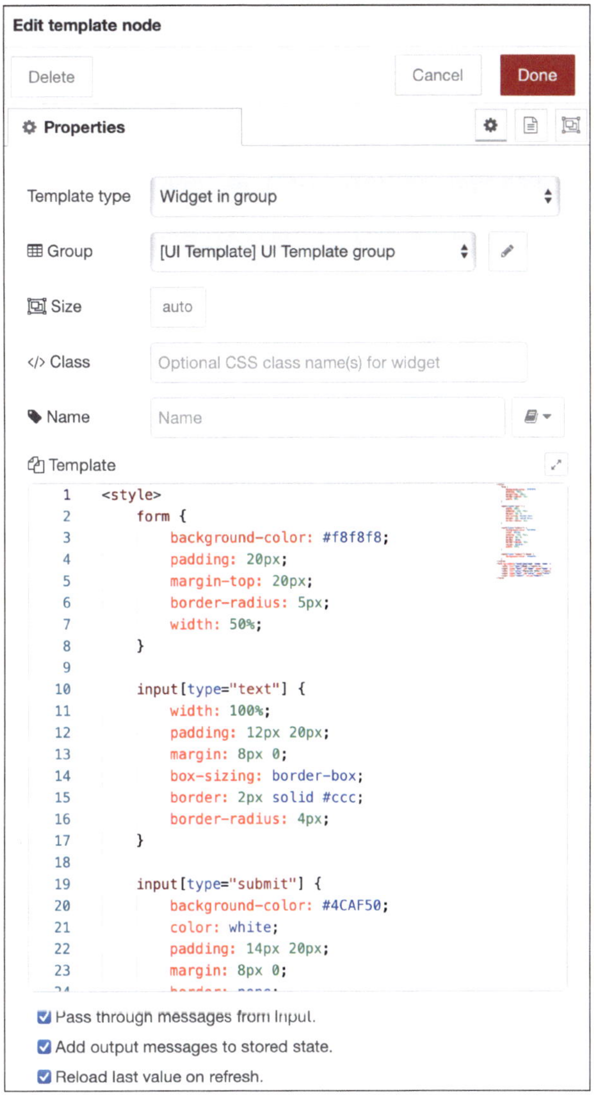

Figure 1.7.7.48: The Template node configuration window.

1. Set the 'Group' field to assign the node to a user interface group.

2. In the 'Template' field, you can write your custom HTML, CSS, and JavaScript code. You can also use AngularJS directives as the Node-RED Dashboard is built on AngularJS.

3. Optionally, in the 'Size' fields, you can specify the widget size on the dashboard.

4. Click 'Done' to save your settings.

Using a UI Template node

The UI Template node allows you to create fully customised widgets. However, it requires a good understanding of HTML, CSS, JavaScript, and AngularJS.

Once you've set up the node with your custom code and deployed your flow, your custom widget will appear on your dashboard. The UI Template node can receive input messages. When a message arrives, the payload can be accessed within the template using "msg. payload". This lets you dynamically update your custom widget with data from your flows.

The UI Template node in the Node-RED Dashboard opens up a new level of customisation and flexibility. It allows you to break free from the constraints of pre-built widgets and create unique and personalised components. Whether you're looking to create a specific control, display data in a unique way, or integrate external libraries, the UI Template node provides a powerful and flexible tool for your Node-RED project.

An example flow with the Form widget

Let's create a flow that uses the UI Template node to create a custom dashboard component. We'll build a simple custom form using HTML and AngularJS that takes the user's name and favourite colour and returns the information to them.

1. Drag and drop a 'ui_template' node from the Dashboard category onto your workspace. Double-click the node to open its configuration settings. Set the 'Group' field to assign the node to a user interface group. In the 'Template' field, enter the following HTML and CSS code (also see Figure 1.7.7.48 above):

```
<style>
        form {
        background-color: #f8f8f8;
        padding: 20px;
        margin-top: 20px;
        border-radius: 5px;
        width: 50%;
        }

        input[type="text"] {
        width: 100%;
        padding: 12px 20px;
        margin: 8px 0;
        box-sizing: border-box;
        border: 2px solid #ccc;
        border-radius: 4px;
        }

        input[type="submit"] {
        background-color: #4CAF50;
        color: white;
```

```
        padding: 14px 20px;
        margin: 8px 0;
        border: none;
        border-radius: 4px;
        cursor: pointer;
        width: 100%;
        }

        input[type="submit"]:hover {
        background-color: #45a049;
        }
</style>
<form ng-submit="send({payload: {name: name, color: color}})">
        <label for="name">Name:</label><br>
        <input type="text" id="name" name="name" ng-model="name"><br>
        <label for="color">Favorite Color:</label><br>
        <input type="text" id="color" name="color" ng-model="color"><br>
        <input type="submit" value="Submit">
</form>
```

This creates a form with two input fields: one for the user's name and another for the user's favourite colour. The "ng-submit" directive specifies what to do on form submission. In this case, it sends a message with a payload containing the user's name and favourite colour. The "ng-model" directive binds the input field's value to a variable (name or colour). Click 'Done'.

1. Drag and drop a 'function' node onto your workspace. This function will interpret the payload from the UI Template node and create a string message. Double-click the node to open its configuration settings. Here's an example function:

```
var name = { payload: msg.payload.name };
var color = { payload: msg.payload.color };
return [name, color];
```

   Configure this function node so that it has two outputs. The top will output the name value, and the bottom will output the colour value.

   Click 'Done'.

2. Create two text nodes:
   - "UI Text Node 1" will show the "name" value. It will be used to display the name. Connect the first output of the function node to this text widget.
   - "UI Text Node 2": Drag and drop a 'ui_text' node (from the Dashboard category) onto your workspace. This will be used to display the colour. Connect the second output of the function node to this text widget. Configure this

text widget in a similar way as the first one. You might want to give it a different 'Label', like "Favorite Color".

3. Now connect the output of the 'UI Template' node to the input of the "Function" node and the output of the 'Function' node to the input of the 'UI Text' node. Your flow should look like this:

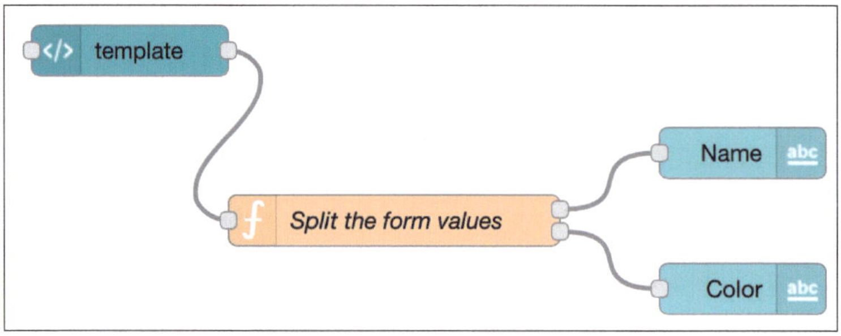

Figure 1.7.7.49: An example flow that contains a UI Template node.

4. Click "Deploy" to deploy the flow.

When you go to your Node-RED dashboard, you'll see a form labelled "User Info" with fields for Name and Favorite Color. After filling out the form and hitting "Submit", the submitted information will be displayed in the "User Info" section of the dashboard. You can see the resulting UI Template widget in the dashboard in the screenshot below:

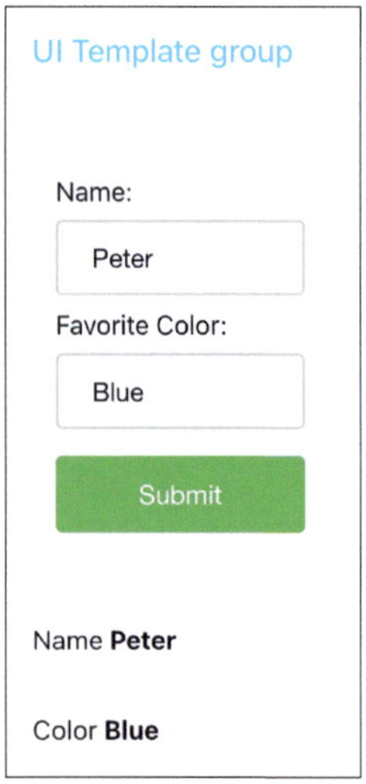

Figure 1.7.7.50: An example UI Template widget in the dashboard.

Chapter 1.8 • Node-RED and MQTT

In the introductory chapter of this book titled "Communication in Node-RED: Protocols and Methods", you learned that MQTT is one of the technologies used by Node-RED and its peripherals to communicate. MQTT (Message Queuing Telemetry Transport) is a lightweight publish-subscribe messaging protocol designed for constrained devices and low-bandwidth, high-latency, or unreliable networks. The protocol is ideal for Internet of Things (IoT) devices due to its minimal power usage, compact footprint, and secure data transmission. You can find details about MQTT, its technical specifications and example use cases on the MQTT website at https://mqtt.org/.

MQTT is used in various applications ranging from IoT and mobile applications to remote sensing and telemetry. It has gained widespread popularity in IoT due to its simplicity and efficiency, making it suitable for devices with limited processing capabilities and constrained network conditions. Some common use cases for MQTT include:

- IoT devices communicating environmental data like temperature or humidity
- Remote monitoring of systems or machinery
- Home automation systems
- Vehicle telematics and fleet management

Node-RED provides dedicated nodes for MQTT, making sending and receiving MQTT messages within a Node-RED flow easy. This aligns well with Node-RED's use as an IoT integration tool, as you can easily create flows that collect data from MQTT-enabled devices, process the data, and act based on it.

For example, you could have an "MQTT In" node that subscribes to an MQTT topic from a temperature sensor. When the temperature exceeds a certain threshold, your flow could send a warning via email, save the data to a database, or even trigger other IoT devices, like an air conditioning system.

MQTT has several notable advantages:

- Lightweight: MQTT messages are small, which benefits network-constrained devices.
- Quality of Service: MQTT supports three different levels of Quality of Service (QoS), allowing you to choose the right balance between message reliability and network bandwidth.
- Last Will and Testament: MQTT clients can specify a "last will and testament" message that is sent out if they disconnect unexpectedly, allowing other devices to know when a device goes offline.
- Retained Messages: MQTT brokers can keep the last message on a topic and send it to new subscribers, ensuring the latest state is always available.

However, MQTT has some potential disadvantages:

- Not fully HTTP compatible: MQTT uses its own protocol, which isn't directly compatible with HTTP. While MQTT is excellent for device-to-device communication, it may not work as smoothly when integrating with web services that use HTTP.
- Potentially insecure if not correctly configured: MQTT can be secure but requires proper configuration, including secure connections and authentication.

Before we proceed with any hands-on work with MQTT, let's take a few minutes to learn about MQTT's essential terms and concepts. Learning about these will make it easier for you to follow the remainder of this chapter.

MQTT Broker
The MQTT broker is the heart of any publish/subscribe protocol. As a server, it dispatches all messages between the sender and the rightful receiver. MQTT clients cannot interact with each other directly but have to communicate through the broker. This central hub receives all messages, filters them, decides who interests them, and then sends them to all subscribed clients.

MQTT Client
An MQTT client is any device (from a microcontroller to a full-fledged server) running an MQTT library and connecting to an MQTT broker over a network. The client can publish information to the broker or subscribe to certain topics to receive specific information.

Publish/Subscribe
MQTT uses a publish/subscribe model, an alternative to the traditional client-server model. In this model, a client (publisher) creates and sends messages to a broker, delivering those messages to any clients (subscribers) that have subscribed to the relevant topic. This allows for the decoupling of the publisher and subscriber, as they do not need to know about each other's existence, just about the topic they are interested in.

Topic
A topic in MQTT is a UTF-8 string that the broker uses to filter messages for each connected client. Topics are hierarchical and are defined using a slash (/) as a separator. For example, "home/bedroom/temperature" is a topic associated with a temperature sensor in a bedroom. A client can subscribe to a specific topic, a hierarchy of topics, or all topics.

Publisher
In MQTT terminology, a publisher is an MQTT client that produces and sends messages to an MQTT broker. This message contains the data the publisher wants to share. The type of data sent can vary widely: the temperature reading from a sensor, a status update from a device, or a message generated by a user or an application. The publisher sends this data wrapped in an MQTT message, including the topic to which it should be assigned. This topic serves as the routing information for the broker.

The publisher's role is to produce and send data without knowing who (if anyone) will receive or process it. This is the fundamental principle of decoupling in the MQTT protocol.

Subscriber

Conversely, a subscriber in MQTT is also an MQTT client, but its role is to receive messages from the MQTT broker. It does this by declaring to the broker which topics it is interested in, in a process called 'subscribing'. Once subscribed, the MQTT broker will forward any message that arrives on that topic to the subscriber.

The subscriber can subscribe to one or many topics depending on its requirements. Like the publisher, the subscriber also does not need to know anything about the publishers on the topic, further underlining the decoupling principle of MQTT.

These roles are not exclusive. A single MQTT client can be both a publisher and a subscriber, often on different topics but potentially even on the same topic if that fits the use case. This flexibility is one of the strengths of MQTT and allows it to be applied to various scenarios in the IoT space.

Quality of Service (QoS)

QoS in MQTT specifies how hard the broker/client will try to ensure a message is received. It has three levels:

- QoS 0 (At most once): Messages are delivered according to the best efforts of the operating environment. Message loss can occur.
- QoS 1 (At least once): Messages are assured to arrive, but duplicates can occur.
- QoS 2 (Exactly once): Messages are assured to arrive exactly once. This is the safest but slowest method.

MQTT is an excellent protocol for IoT and other applications requiring lightweight, efficient messaging. Node-RED's MQTT nodes make it easy to use MQTT in your flows, allowing you to leverage this robust protocol for your IoT projects. In this chapter, you will learn how to install an MQTT broker (i.e. an MQTT server) on your virtual machine and use it with your Node-RED instance. You will use this MQTT setup extensively in the various experiments in this book.

8.1. Installing MQTT Mosquitto on Ubuntu Server 22.04

Node-RED can work with any MQTT broker that implements the standard MQTT protocol. In this book, we will work with Eclipse Mosquitto (https://mosquitto.org/), one of the most popular MQTT brokers. It is lightweight, supports secure connections, and is suitable for small and large deployments.

In this section, you will learn how to install Mosquitto on Ubuntu Server 22.04. Please remember that although we have installed Node-RED as a Docker container, we will install the Mosquito MQTT broker at the operating system level for simplicity. This information will be necessary when we work to access the new MQTT broker from Node-RED.

Update System Packages

Before installing any new package on your Ubuntu system, ensure that all the existing packages are up to date. You can do this by running the following commands:

```
$ sudo apt update
$ sudo apt upgrade
```

Install Mosquitto

Once the system packages are up to date, you can install Mosquitto using the package manager. Run the following command to install Mosquitto:

```
$ sudo apt install mosquitto mosquitto-clients
```

The "mosquitto-clients" package will install additional command-line clients that can be used to send and receive MQTT messages, which is helpful for testing and debugging.

Verify Installation

After installing Mosquitto, it should start running automatically. You can verify this by checking its status with the following command:

```
$ sudo systemctl status mosquitto
```

If Mosquitto is running correctly, you should see an output indicating that the Mosquitto service is active (running).

At this point, you have installed Mosquitto on your Ubuntu Server 22.04. You can now start creating MQTT clients to publish and subscribe to topics on your MQTT broker. Note that, by default, Mosquitto listens on port 1883. If you want to secure your MQTT broker with SSL/TLS, you will need to configure additional settings, which is beyond the scope of this chapter.

8.2. Test the MQTT Service on the Command Line

Testing the Mosquitto MQTT broker using the command line is straightforward. You will need two terminal windows. The first one will play the role of the "publisher", and the second one will be the "subscriber". You will use the MQTT client utilities "mosquitto_pub" (for publishing) and "mosquitto_sub" (for subscribing) that come with the "mosquitto-clients" package.

Let's start. The following steps take place on the Ubuntu server's command line. To login to the server, use SSH as I do in the example below (replace the IP address with the one that belongs to your Ubuntu server):

```
$ ssh peter@192.168.20.18
```

1. Open a new terminal window, log in to the Ubuntu server with SSH, and subscribe to a topic. Let's start by subscribing to a topic named "test/topic" using "mosquitto_sub":

```
$ mosquitto_sub -h localhost -t test/topic
```

   - The "-h" option specifies the host where the Mosquitto broker is running (in this case, it's running on the same machine, so we use 'localhost'). You can also use an IP address or a domain name. The "-t" option specifies the topic. The topic can be any valid alphanumeric string, including forward slash and underscore characters. See the relevant section where I provide more information about topic names in MQTT. Also, note that we are subscribing to a topic and a broker without authentication in this example. While this is not "best practice" due to its obvious security implications, it is acceptable for now as we test our new MQTT broker. Later in this chapter, you will learn how to set up authentication.

   This command will start a subscriber client that listens for messages on the "test/topic" topic.

2. In a second terminal window, log in to the Ubuntu server with SSH and publish a message to the "test/topic" topic using "mosquitto_pub":

```
$ mosquitto_pub -h localhost -t test/topic -m "Hello, MQTT!"
```

   The "-m" option specifies the message to publish.

3. Return to the first terminal window (where "mosquitto_sub" is running). You should see the message "Hello, MQTT!" that you published from the second terminal. See my session below:

Figure 1.8.2.1: MQTT Mosquitto broker test.

   You can re-run the "mosquitto_pub" command multiple times, and each time, you will see the message appearing under the "mosquitto_sub" command.

That's it! This simple test verifies that your Mosquitto broker receives and routes MQTT messages correctly.

About topic names

MQTT topic names are case-sensitive and must contain one or more topic levels, separated by the forward slash (/) character. A topic level is a string that can contain any character, but it must not include the null character (U+0000). It's important to note that leading or trailing "/" creates a distinct topic level: a blank string.

Regarding valid characters, the MQTT specification does not explicitly prohibit any characters in the topic string apart from the null character. You can use any characters you wish in a topic string, including spaces and special characters. However, using some characters could cause compatibility issues with MQTT libraries or tools that don't correctly handle unusual characters.

Conventionally, MQTT topics are used to structure and filter messages meaningfully. As such, topic naming usually reflects the type and source of data in a hierarchical format, such as "device_type/device_id/sensor_type".

For example, the temperature of a particular thermostat could be published on the topic "thermostat/001/temperature". This convention makes it easy to subscribe to all temperature updates from all thermostats (thermostat/+/temperature) or all updates from a specific thermostat (thermostat/001/+), for example.

Knowing the difference between topic names and filters is also important. Topic filters introduce two wildcard characters:

- '+' can be used as a wildcard for a single hierarchy level.
- '#' can represent all remaining levels of the hierarchy.

So a topic filter of +/+/temperature would match any topic with three levels, where the last is "temperature", and a filter of sensors/# would match any topic starting with "sensors/". However, these wildcards are not valid in topic names when publishing a message. They can only be used when subscribing.

8.3. Using Authenticated Sub and Pub

A newly installed Mosquitto MQTT broker will accept unauthenticated subscribers and publishers by default. This is not best practice; the broker should use authentication as a minimum security precaution. In this section, you will learn how to set up authentication for your MQTT Mosquitto broker. This will ensure that your MQTT broker is secure and can prevent unauthorised access to your IoT network.

We will use a username and password-based authentication mechanism built into Mosquitto.

Create a password file

To begin, create a password file. This file will store the username and password used to authenticate MQTT clients. To make the password file, use the "mosquitto_passwd" utility with the Mosquitto broker.

Login to the server, and create a user named "mqtt_user":

```
$ sudo mosquitto_passwd -c /etc/mosquitto/passwd mqtt_user
```

You will be prompted to enter the password for this user. Make sure to choose a strong password. The -c option creates a new password file. If you wish to add more users, omit the "-c" option:

```
$ sudo mosquitto_passwd /etc/mosquitto/passwd another_user
```

Edit the Mosquitto configuration file

Next, modify the Mosquitto configuration file to use the password file. The configuration file is typically located at "/etc/mosquitto/mosquitto.conf". Open the configuration file with your preferred text editor. For example, you can use nano:

```
$ sudo nano /etc/mosquitto/mosquitto.conf
```

Add the following lines to the file:

```
allow_anonymous false
password_file /etc/mosquitto/passwd
listener 1883
```

Let's look at each line and the effect it has on the broker's configuration:

- allow_anonymous false: This line specifies the authentication setting for Mosquitto clients. When set to false, it means that anonymous clients (clients that do not provide a username and password) are not allowed to connect to the broker.
- password_file /etc/mosquitto/passwd: This line points to the file that contains the usernames and passwords for client authentication. The file /etc/mosquitto/passwd will have the list of users and their respective hashed passwords. When a client tries to connect, Mosquitto will check this file to authenticate the provided credentials.
- listener 1883: This line configures the port on which the Mosquitto broker will listen for incoming MQTT client connections. The number 1883 is the standard port for MQTT connections. You can change this number if you want to set up Mosquitto to listen on a different port.

Once you're finished editing, save and close the file. If you used nano, you can do this by pressing Ctrl + X, Y, and Enter.

Restart the Broker

After updating the configuration file, you'll need to restart the Mosquitto broker for changes to take effect:

```
$ sudo systemctl restart mosquitto
```

Test the configuration

Now, let's test the new configuration. Open up two terminal windows. In the first terminal window, subscribe to a topic using mosquitto_sub with the username and password:

```
$ mosquitto_sub -h localhost -t "test/topic" -u "mqtt_user" -P "password"
```

Replace "password" with the password you chose earlier. Double quotes are optional if your password and username do not contain white spaces. The "-h" option specifies the MQTT broker hostname, "-t" specifies the topic, "-u" provides the username, and "-P" provides the password.

In the second terminal window, publish a message to the same topic using mosquitto_pub:

```
$ mosquitto_pub -h localhost -t "test/topic" -m "Hello, MQTT!" -u "mqtt_user" -P
"password"
```

Again, replace "password" with the password you chose earlier. The -m option specifies the message to publish. If everything works correctly, "Hello, MQTT!" appears in the first terminal window. The MQTT client could authenticate with the broker and receive the published message. Here is an example screenshot from my test session:

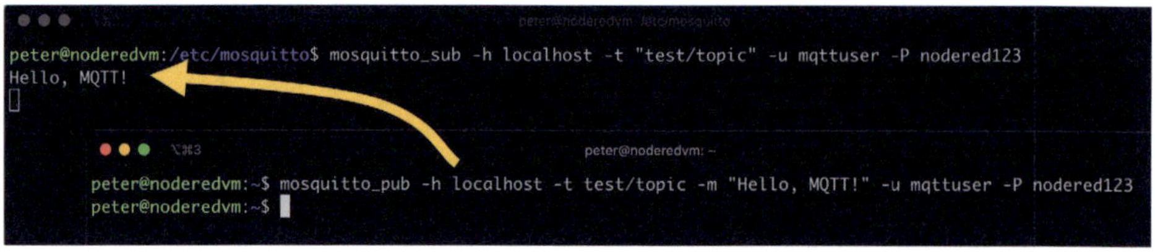

Figure 1.8.3.2: Authenticated MQTT Mosquitto broker test.

Your Mosquitto MQTT broker is now configured to require authenticated subscribers and publishers and will reject non-authenticated connections. Always secure your MQTT broker, especially if it's accessible online.

8.4. Test MQTT in Node-RED

Let's turn our attention to Node-RED and set up a simple flow that uses the built-in "MQTT in" and "MQTT out" nodes to send and receive messages via the Mosquitto broker. If you have not yet set up the broker on the Ubuntu server or are not set up for authenticated communications, complete the relevant sections earlier in this chapter before you continue.

MQTT nodes in Node-RED

Node-RED comes with MQTT nodes pre-installed. You will find the MQTT nodes in the Node-RED palette on the left side under the "network" category.

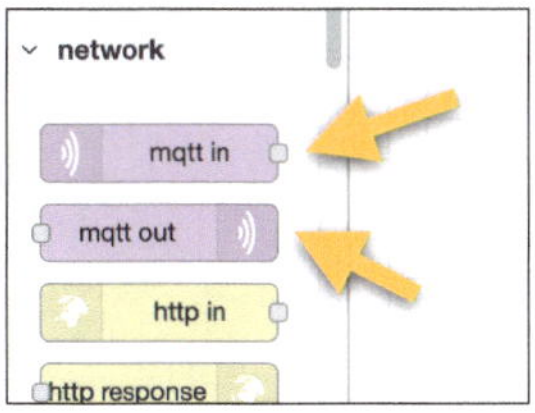

Figure 1.8.4.3: The MQTT-in and MQTT-out nodes are available by default.

The MQTT In and MQTT Out nodes are essential components in Node-RED for building IoT applications that leverage the MQTT protocol. These nodes enable your flows to interact with MQTT brokers and connected MQTT devices.

The MQTT In node subscribes to a topic from an MQTT broker. It acts as a subscriber in the publish-subscribe pattern of MQTT. Whenever a message is published to the topic the MQTT In node is subscribed to, it receives and passes it along the flow in Node-RED. To use it, you must drag the MQTT In node into your flow and then configure it with the details of the MQTT broker, including its URL and the topic you want to subscribe to.

The MQTT Out node, on the other hand, is used to publish a message to a topic on an MQTT broker. It takes the «msg.payload" from the incoming message and publishes it to the specified topic on the broker. Similarly to the MQTT In node, you'd drag the MQTT Out node into your flow and configure it with your MQTT broker's details and the topic you want to publish to.

The MQTT In and MQTT Out nodes provide a straightforward way to implement the publish-subscribe messaging pattern in your IoT applications. This makes it possible to create flows that can receive real-time data from multiple devices, process it, and then send commands back to those or other devices.

Configure MQTT nodes in Node-RED
Let's start by configuring MQTT nodes for our Mosquitto broker.

MQTT In (Subscribe) Node
Let's kick off with the MQTT-in node. This node will receive notifications from the broker when a message is received in the specified topic.

1. Drag and drop the "mqtt in" node (it's labelled as "mqtt" in the palette) onto a new canvas.

2. Double-click on it to open the node configuration panel.

3. Click the pencil icon next to "Add new mqtt-broker..." in the "Server" field.

4. In the "Connection" tab of the "Edit mqtt-broker node" dialogue, fill in the details:

- **Broker**: 192.168.20.18 (the IP address of your MQTT broker)
- **Port**: 1883 (default MQTT port; adjust if your broker is set up differently)
- Leave the "Client ID" blank to allow the broker to assign one automatically or fill in a specific ID if desired.

5. Click on the "Security" tab. Enter the Username and Password ("mqtt_user" and "nodered123" respectively in this case). Take care to enter the correct credentials.

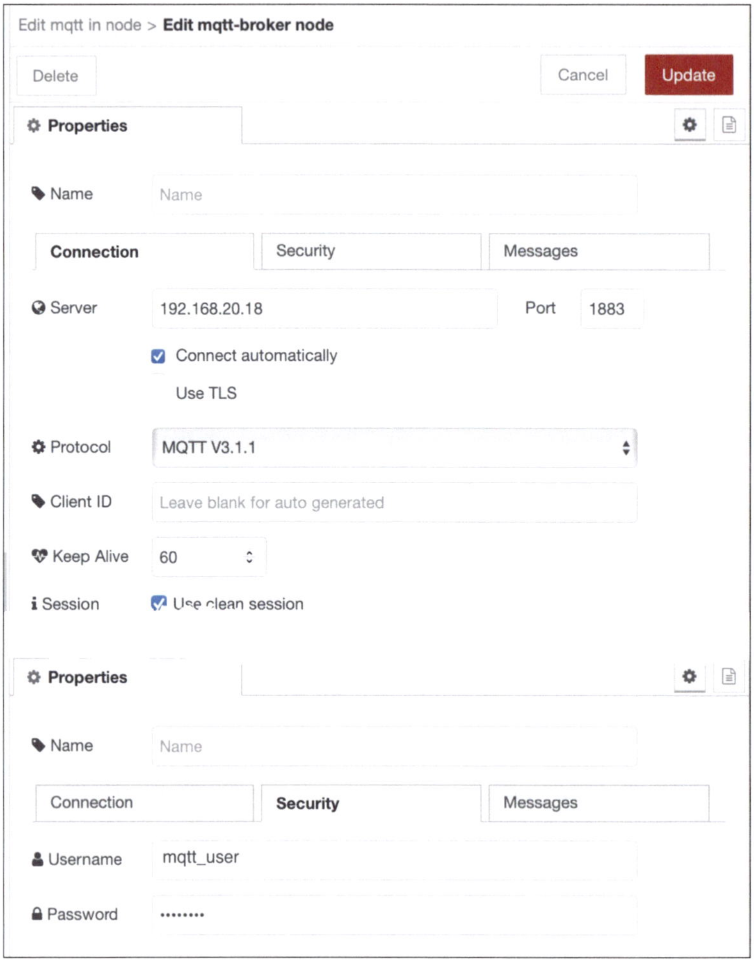

Figure 1.8.4.4: The MQTT broker configuration, howing the Connection and Security tabs.

6. Click "Update".

7. Back in the "Edit mqtt in node" dialog, fill in the "Topic" field with "test/topic".

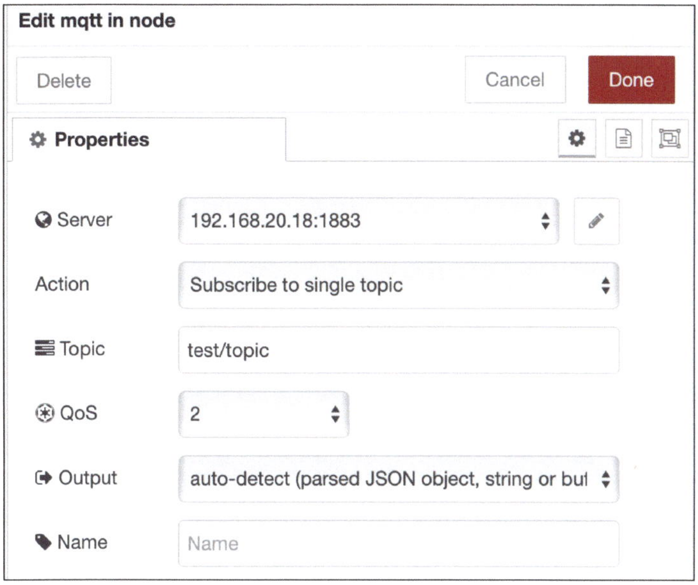

Figure 1.8.4.5: The MQTT-in node configuration.

8. Click "Done".

MQTT Out (Publish) node

Follow these steps:

1. Drag and drop the "mqtt out" node (it's labelled as "mqtt" in the palette) onto the canvas.
2. Double-click on it to open the node configuration panel.
3. Follow the same steps (3-8) as in the MQTT In node configuration above, using the same broker IP, port, username, and password, and the topic "test/topic".

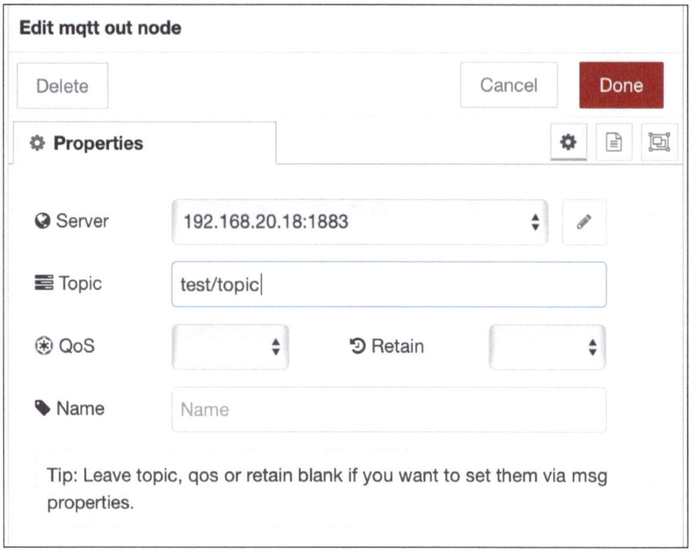

Figure 1.8.4.6: The MQTT-out node configuration.

At this point, you've configured your MQTT In and Out nodes to communicate with your broker.

Create a Subscription/Publication flow

Let's create a simple flow demonstrating subscribing and publishing to an MQTT topic. We'll start with the Subscription.

1. Drag and drop an "inject" node onto the canvas. This will simulate an event, such as a button press.

2. Connect the output of the "inject" node to the input of the MQTT Out node.

3. Double-click on the "inject" node and set "Payload" to the string "Hello, MQTT!" and click "Done".

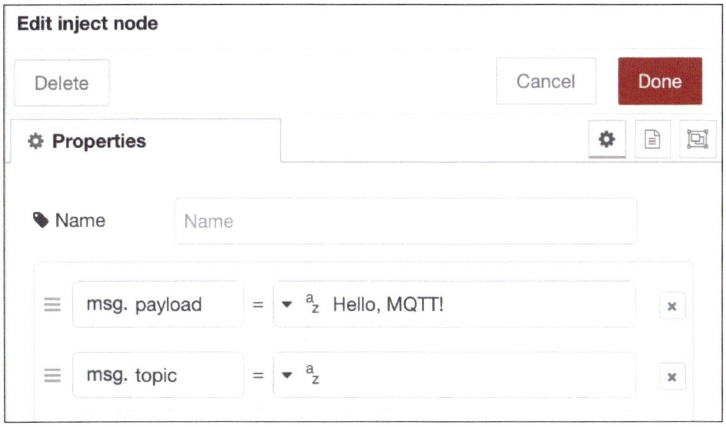

Figure 1.8.4.7: The Inject node configuration.

4. Connect the output of the MQTT In node to a "debug" node (found under the "output" category in the palette). This node will display the messages received on the topic. Your flow should look like this:

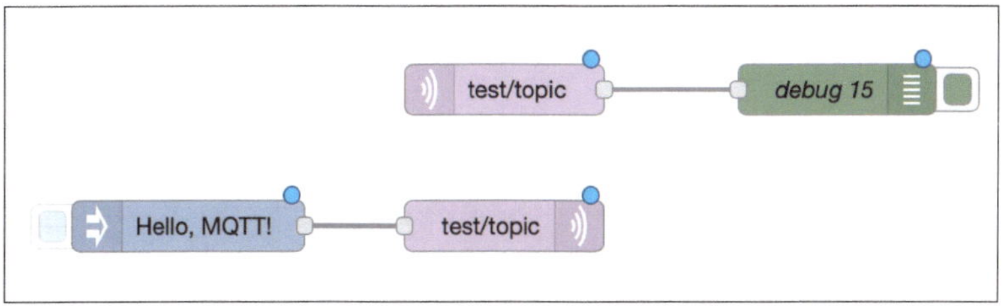

Figure 1.8.4.8: The MQTT test flow node configuration.

The MQTT Out node connected to the Inject node will serve as our publishing flow. When the inject node is triggered, it will send the message to the MQTT broker, publishing it to the "test/topic".

Deploy and test

Follow these steps to test our new flow and verify it can communicate with the MQTT broker.

1. Click on the "Deploy" button in the top-right corner of the Node-RED interface. This deploys your flow onto the Node-RED server.

2. Click on the button on the left side of the "inject" node to trigger the event. This will send a "Hello, MQTT!" message to the MQTT broker.

3. You will see the published message appearing in the debug tab on the right side of the Node-RED interface. This shows that the MQTT In node has received the published message. You can see my experiment in the screenshot below:

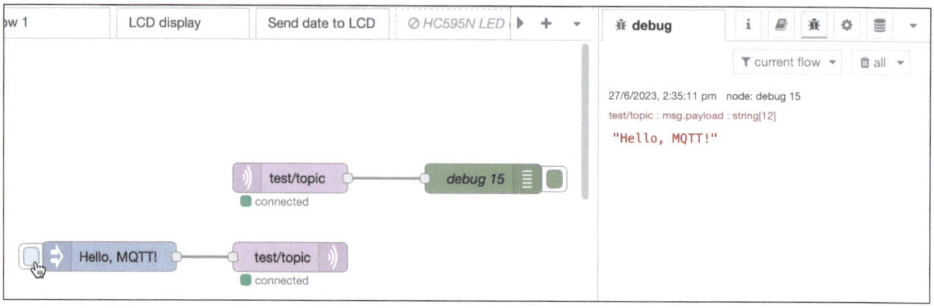

Figure 1.8.4.9: The MQTT test flow in operation.

If your "mosquitto_sub" utility is still running in the terminal as per the previous section in this chapter, it will also display the "Hello, MQTT!" message in the console as it is also subscribed to the same topic. It looks like this (showing a second message):

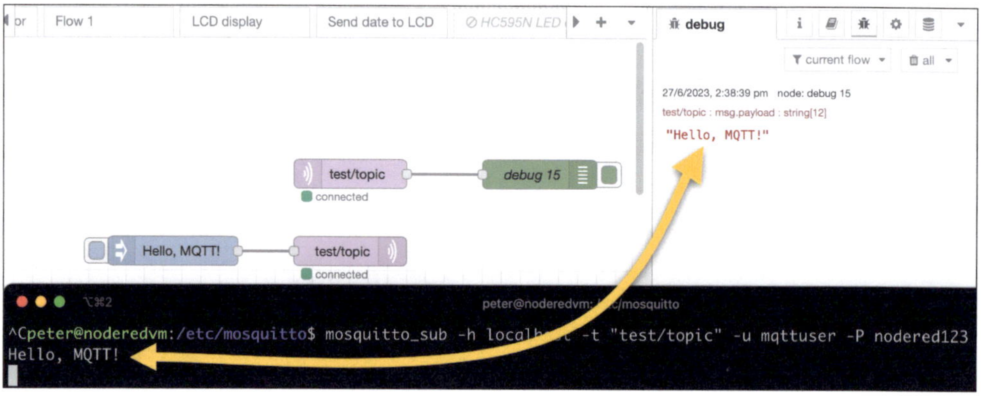

Figure 1.8.4.10: All subscribers to the same topic will receive the message.

Congratulations, you have successfully set up MQTT in Node-RED and tested a simple flow; you will use this setup extensively in the experiments in this book.

8.5. MQTT with Raspberry Pi Pico

In this section, you will learn how to use MQTT to send and receive simple messages to your Node-RED instance with a Raspberry Pi Pico W. To keep this section concise, I have assumed that you know the basics of the Raspberry Pi Pico W. Specifically, I assume that you know how to use the Thonny editor to write and upload MicroPython scripts to the Raspberry Pi Pico and Pico W and that you are familiar with MicroPython code that relates to Wifi connectivity.

If you are unfamiliar with these topics, please review Part 3 of this book, "Raspberry Pi Pico and MicroPython Primer". The chapters in that part will teach you everything you need to know to follow the instructions in this section and beyond.

To use MQTT with the Raspberry Pi Pico W, we will be using the "micropython-umqtt.simple" MicroPython module.

The "micropython-umqtt.simple" module is a simple MQTT client explicitly written for the MicroPython language. MicroPython is a lean and efficient Python 3 programming language implementation that includes a small subset of the Python standard library and is optimised to run on microcontrollers and in constrained environments.

The primary purpose of the micropython-umqtt.simple module enables MicroPython applications to communicate with other devices across a network using the MQTT protocol. As the name suggests, the design goal of "umqtt.simple" is to keep things as simple as possible, which aligns well with the general philosophy of MicroPython. In terms of its notable characteristics:

1. **Lightweight:** "micropython-umqtt.simple" is designed to be lightweight and suitable for resource-constrained environments. It's perfectly tailored for the needs of microcontrollers, which have limited processing power and memory.

2. **MQTT Compatibility:** The module is compatible with MQTT 3.1 and 3.1.1 versions. It supports the basic MQTT commands, including CONNECT, PUBLISH, SUBSCRIBE, and PING.

3. **QoS Support:** "micropython-umqtt.simple" supports Quality of Service level 0 (deliver the message at most once) and level 1 (deliver the message at least once). This provides flexibility in managing the trade-off between network bandwidth and message reliability.

4. **Non-blocking:** Operations in "umqtt.simple" do not block other processes. You can perform different tasks concurrently without the MQTT client causing delays.

5. **Easy to Use:** With a focus on simplicity, the module provides an easy-to-use API that can be quickly learned by developers familiar with Python.

While these are some of the major points, it's essential to mention that "micropython um qtt.simple" doesn't support some advanced MQTT features, such as Automatic Reconnection and QoS level 2. If you need those features, you may need to handle them in your application code or look for other libraries offering them.

It's worth noting that as of the time of this writing, the micropython-umqtt.simple module can be found on the Python Package Index (PyPi). The pypi (Python Package Index) page is here: https://pypi.org/project/micropython-umqtt.simple/. You can also access the "umqtt.simple" project homepage and documentation from there.

Install umqtt.simple in Thonny

As in this book, we use the Thonny IDE for all our MicroPython programming projects. Let's use Thonny to install the "umqtt.simple" module to use it with the Raspberry Pi Pico.

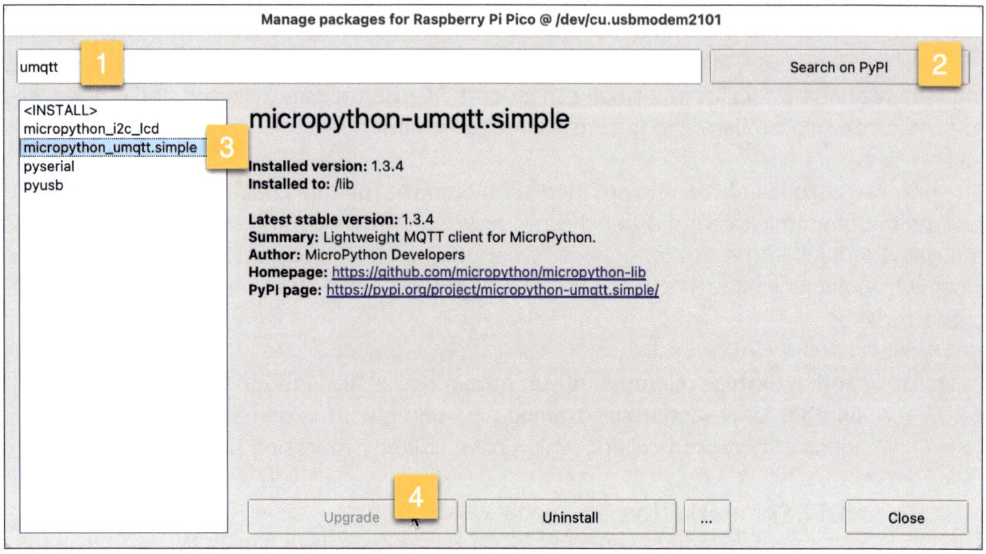

Figure 1.8.5.11: The MQTT test flow node configuration.

Concerning the screenshot in the image above, here is the process:

1. Open Thonny IDE on your computer.
2. Connect your Raspberry Pi Pico W to your computer using a USB cable.
3. In Thonny IDE, click "Tools" in the menu bar and select "Manage Packages".
4. In the Package Manager window, type "umqtt" in the search bar and click "Search on PyPi" ("1" and "2" in the image).
5. Select the "micropython_umqtt.simple" package from the list of results and click on the "Install" button ("3" and "4" in the image). The button with the label "4" is titled "Upgrade" in the screenshot because this module is already installed on my computer.
6. Wait for the installation and close the package manager.

You've successfully installed the "umqtt.simple" MicroPython module to your Thonny IDE. Now, you can use it in your Raspberry Pi Pico W scripts.

To use the "umqtt.simple" module in your script, import it at the beginning of your code like this:

```
import umqtt.simple as mqtt
```

You can then use the various functions provided by the module to connect to an MQTT broker and publish or subscribe to messages. In the following sections, you will learn how to publish and subscribe to an MQTT topic on your Node-RED instance.

8.6. MQTT Pub Example

In this section, you will try a simple example of connecting a Raspberry Pi Pico W to a Node-RED flow using MQTT protocol. The Raspberry Pi Pico W, powered by MicroPython, will emulate a sensor producing temperature and humidity readings. Node-RED will be used to process and print this data to the debug window.

Let's divide the process into MicroPython script creation for Raspberry Pi Pico W and the Node-RED flow setup.

Beware! For this to work, the Pico and the MQTT broker must be on the same network, or their networks must be able to communicate.

MicroPython script for Raspberry Pi Pico W

First, create a MicroPython script that emulates temperature and humidity readings. These values will be randomly generated and published to a specific MQTT topic.

We'll use the "micropython-umqtt.simple" module for this script. You should have installed this module in the previous section.

In Thonny, create a new file on the Raspberry Pi, and copy this code:

```
from machine import Pin
from time import sleep
from umqtt.simple import MQTTClient
import network
from random import *

# Configure your Wi-Fi credentials
SSID     = '<your_ssid>'
PASSWORD = '<your_password>'

wlan = network.WLAN(network.STA_IF)
wlan.active(True)
wlan.connect(SSID,PASSWORD)

while not wlan.isconnected():
    pass

print('Connected to Wi-Fi')
print(wlan.ifconfig())
```

```
# Setup MQTT
SERVER      = '192.168.20.16'  # Replace with your MQTT broker IP
CLIENT_ID   = b'PICO'
TOPIC_TEMP = b'sensor_temperature'
TOPIC_HUMD = b'sensor_humidity'
USER        = b'mqttuser'      # Replace with your MQTT username
PASSWORD    = b'<password>'    # Replace with your MQTT password

def mqtt_connect():
   client = MQTTClient(CLIENT_ID, SERVER, user=USER, password=PASSWORD,
keepalive=3600)
   client.connect()
   print('Connected to %s MQTT Broker'%(SERVER))
   return client

def reconnect():
   print('Failed to connect to the MQTT Broker. Reconnecting...')
   sleep(5)
   machine.reset()

try:
   client = mqtt_connect()
except OSError as e:
   reconnect()

while True:
    temp = randint(10, 30)
    humd = randint(50, 90)
    client.publish(TOPIC_TEMP, str(temp))
    client.publish(TOPIC_HUMD, str(humd))
    print("Published temp = ", temp, ", humd = ", humd)
    sleep(5.0)
```

Below is my Thonny IDE with the script titled "mqtt_pub_test.py". I have obscured my WiFi network SSID, password, and MQTT user's password.

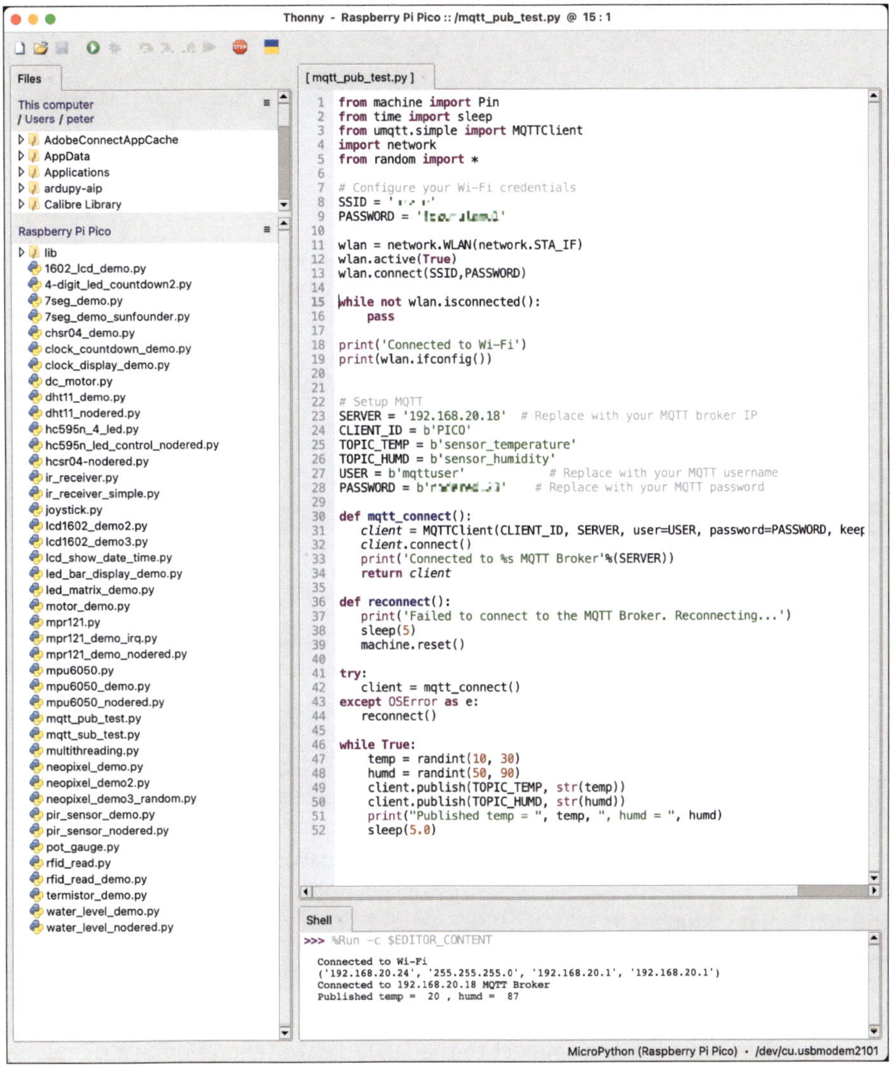

Figure 1.8.6.12: The MQTT test MicroPython script that sends simulated temperature and humidity values to the broker.

This script generates random values for temperature (between 10 and 30) and humidity (between 50 and 90) every 5 seconds and publishes them to the "sensor_temperature" and "sensor_humidity" topics on the MQTT broker.

Replace the IP address in the "SERVER" variable with the address of your actual MQTT broker.

Script line-by-line review

Here's a line-by-line breakdown of how the script works:

1-3: Importing required modules.

- machine: to control the hardware components.
- time: to control timing, delays and sleeps.
- umqtt.simple: to use the MQTT protocol.
- network: to manage the Wi-Fi connection.
- random: to generate random numbers.

5-6: Your Wi-Fi credentials are defined here. Replace <your_ssid> and <your_password> with your Wi-Fi SSID and password.

8-9: A station interface (STA_IF) is created and activated for the device to connect to a Wi-Fi network.

10: The script attempts to connect to the defined Wi-Fi network using the provided credentials.

12-14: A loop continues until the device is connected to the Wi-Fi network.

16-18: It prints a success message and the network configuration once connected.

21-26: These lines configure MQTT broker details.

- SERVER: the IP address of your MQTT broker.
- CLIENT_ID: the unique identifier for the MQTT client. Here, it is 'PICO'.
- TOPIC_TEMP and TOPIC_HUMD: the MQTT topics where the temperature and humidity data will be published.
- USER and PASSWORD: the username and password for the MQTT broker.

28-33: mqtt_connect function:

- This function creates an MQTTClient object with the previously defined CLIENT_ID, SERVER, USER, and PASSWORD.
- The client then tries to connect to the MQTT broker and prints a success message if it is connected.

35-38: reconnect function:

- If the MQTT client fails to connect to the MQTT broker, this function will print a message, wait for 5 seconds and then reset the machine.

40-44: This try-except block calls the mqtt_connect function to connect the MQTT client to the MQTT broker.

- If the connection is unsuccessful, it calls the reconnect function to attempt reconnection.

46-53: Main loop of the script:

- It generates random temperature and humidity values in 10-30 and 50-90, respectively.
- These values are then published to the respective MQTT topics.
- After publishing the values, it prints them and pauses for 5 seconds before the next iteration.

This script continuously generates random temperature and humidity values, simulating a sensor, and publishes them to the specified MQTT topics via your MQTT broker.

Node-RED flow setup

Let's continue with Node-RED. Create a new flow and drop two "mqtt_in" and two "debug" nodes. We'll set up a Node-RED flow to subscribe to the "sensor_temperature" and "sensor_humidity" topics on the MQTT broker and print the received values.

Wire the nodes as in the screenshot below:

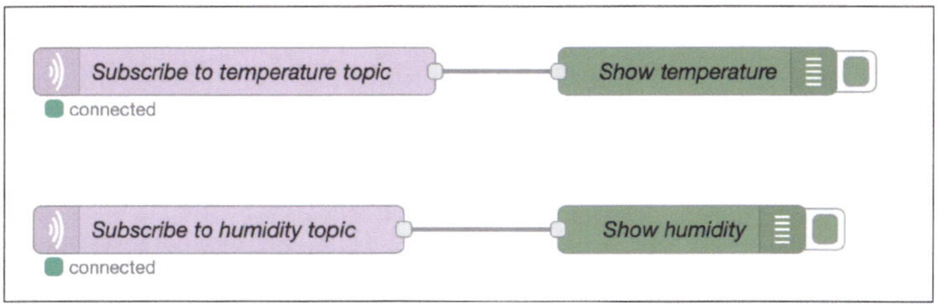

Figure 1.8.6.13: The MQTT test flow node configuration.

Double-click on each of the MQTT-in nodes to configure them. Ensure that you have selected the correct MQTT server (broker). The "Action" should be "Subscribe to single topic", and the "Output" should be "auto-detect". One MQTT-in node should subscribe to the "sensor_temperature" topic and the other to "sensor_humidity". Below is the configuration for the MQTT-in node that gets the temperature. You can use this example to configure the second MQTT-in node to subscribe to the "sensor_humidity" topic.

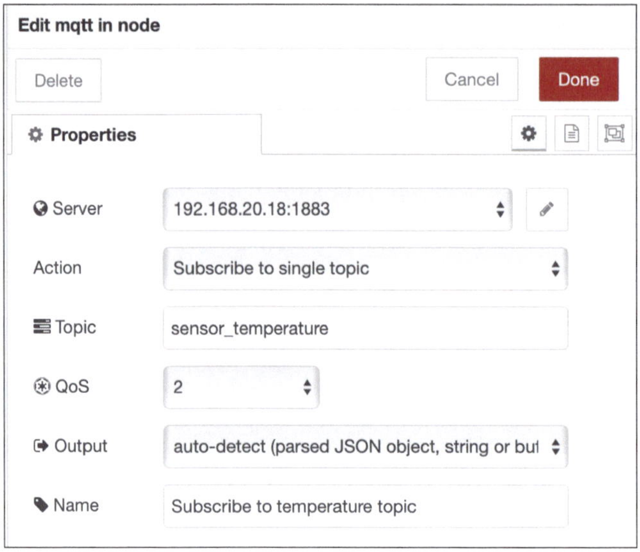

Figure 1.8.6.14: The configuration for the first MQTT-in node.

Deploy the flow and watch the temperature and humidity (simulated) values appear in the debug tab (see my screenshot below).

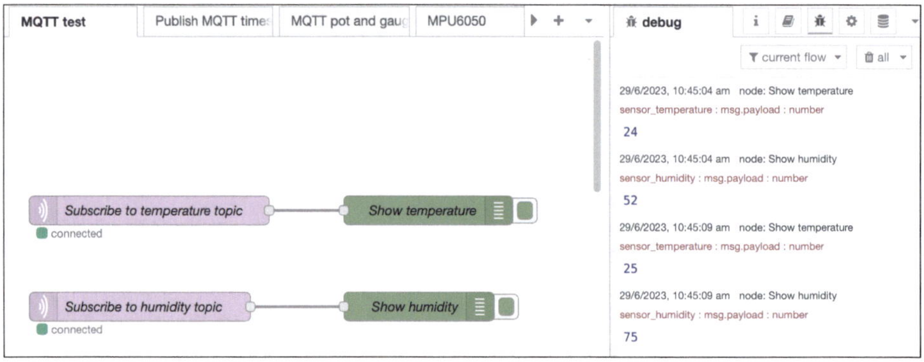

Figure 1.8.6.15: The flow receives simulated temperature and humidity values from the Raspberry Pi Pico W and prints them to the debug tab.

Let's look at another example: the Raspberry Pi Pico subscribes to a topic to which a Node-RED flow publishes data. Combining these examples allows you to create flows that allow two-way communication between Node-RED and a Raspberry Pi Pico.

I provide the complete JSON code for this flow here for your reference:

```
[
    {
        "id": "76b35a127cfa010e",
        "type": "tab",
        "label": "MQTT test",
        "disabled": false,
        "info": "",
        "env": []
    },
    {
        "id": "8be6a3ef.7638d",
        "type": "debug",
        "z": "76b35a127cfa010e",
        "name": "Show temperature",
        "active": true,
        "tosidebar": true,
        "console": false,
        "tostatus": false,
        "complete": "payload",
        "targetType": "msg",
        "statusVal": "",
        "statusType": "auto",
        "x": 530,
        "y": 180,
        "wires": []
    },
    {
        "id": "4ee4c43816604b28",
        "type": "mqtt in",
        "z": "76b35a127cfa010e",
        "name": "Subscribe to temperature topic",
        "topic": "sensor_temperature",
        "qos": "2",
        "datatype": "auto-detect",
        "broker": "694bd92dc7019404",
        "nl": false,
        "rap": true,
        "rh": 0,
        "inputs": 0,
        "x": 230,
        "y": 180,
        "wires": [
            [
                "8be6a3ef.7638d"
            ]
        ]
```

```
    },
    {
        "id": "3a04496a6b7887aa",
        "type": "debug",
        "z": "76b35a127cfa010e",
        "name": "Show humidity",
        "active": true,
        "tosidebar": true,
        "console": false,
        "tostatus": false,
        "complete": "payload",
        "targetType": "msg",
        "statusVal": "",
        "statusType": "auto",
        "x": 520,
        "y": 280,
        "wires": []
    },
    {
        "id": "1871816e7c287390",
        "type": "mqtt in",
        "z": "76b35a127cfa010e",
        "name": "Subscribe to humidity topic",
        "topic": "sensor_humidity",
        "qos": "2",
        "datatype": "auto-detect",
        "broker": "694bd92dc7019404",
        "nl": false,
        "rap": true,
        «rh»: 0,
        «inputs»: 0,
        «x»: 220,
        «y»: 280,
        «wires»: [
            [
                "3a04496a6b7887aa"
            ]
        ]
    },
    {
        "id": "694bd92dc7019404",
        "type": "mqtt-broker",
        "name": "",
        "broker": "192.168.20.18",
        "port": "1883",
        "clientid": "",
```

```
        "autoConnect": true,
        "usetls": false,
        "protocolVersion": "4",
        "keepalive": "60",
        "cleansession": true,
        "birthTopic": "test_topic",
        "birthQos": "0",
        "birthPayload": "",
        "birthMsg": {},
        "closeTopic": "",
        "closeQos": "0",
        "closePayload": "",
        "closeMsg": {},
        "willTopic": "",
        "willQos": "0",
        "willPayload": "",
        "willMsg": {},
        "userProps": "",
        "sessionExpiry": ""
    }
]
```

8.7. MQTT Sub Example

Let's look at a new example where the Raspberry Pi Pico W subscribes to an MQTT topic to which a Node-RED flow publishes data. This time, we'll start with the Node-RED flow and finish with the MicroPython script that will run on the Raspberry Pi Pico. This example aims to learn how the Raspberry Pi Pico W can receive data from Node-RED using the MQTT broker as an intermediary.

Node-RED Flow Setup

In Node-RED, create a new flow that contains a trigger, debug, and MQTT-out nodes. Do the wiring as in the screenshot below:

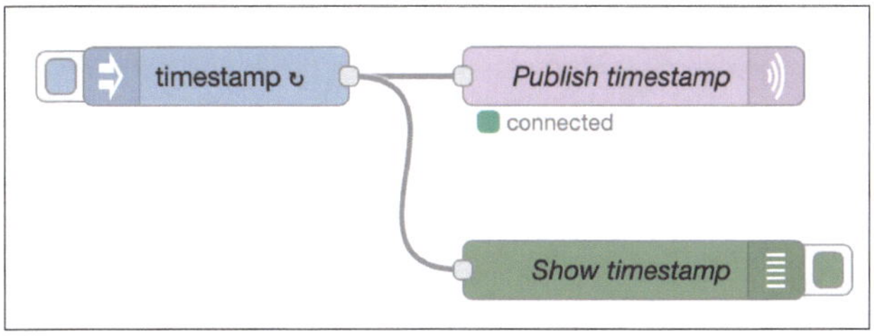

Figure 1.8.7.16: The flow publishes a timestamp every second to an MQTT topic from where the Raspberry Pi can read it.

For the trigger node, use these settings (leave default settings if I don't mention them below):

- msg.payload = timestamp
- Repeat: interval
  - every 1 seconds

For the MQTT-out node, use these settings:

- Server: Set your actual MQTT server.
- Topic: nodered_timestamp

Here is the setup of the MQTT-out node in my flow:

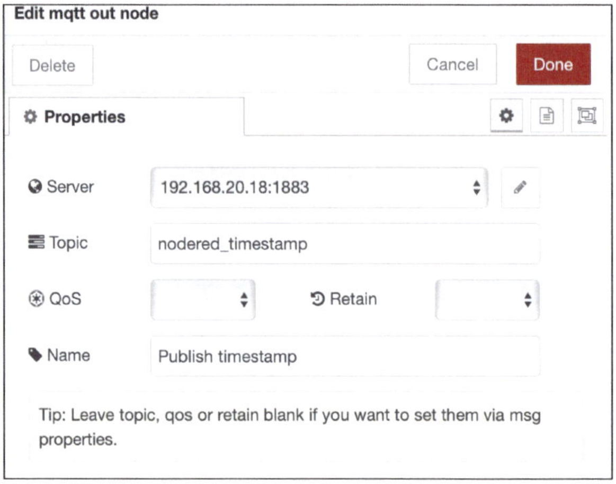

Figure 1.8.7.17: The configuration for the MQTT-out node.

Deploy this flow, and notice that the timestamp values start to appear in the debug tab, as in the example below:

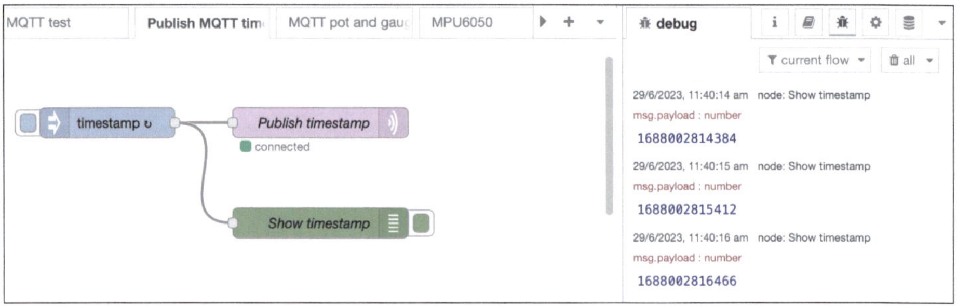

Figure 1.8.7.18: The MQTT-out example flow in operation.

MicroPython script for Raspberry Pi Pico W

On the Raspberry Pi Pico W side, we'll use a MicroPython script to subscribe to the "nodered_timestamp" topic, read the newly-published value as soon as it is published, and print the timestamp value to the console. As with our first example, we'll use the "micropython-umqtt.simple" module. Most of the MicroPython script you see below is identical to the one from the first example. In particular, all code that relates to the Wi-Fi connectivity is unchanged. Most of the code that relates to the MQTT broker is mostly intact. The only significant difference is using the "subscribe" function of the MQTT client object. Here is the script:

```
from machine import Pin
from time import sleep
from umqtt.simple import MQTTClient
import network
from random import *

# Configure your Wi-Fi credentials
SSID     = '<your_ssid>'
PASSWORD = '<your_password>'

wlan = network.WLAN(network.STA_IF)
wlan.active(True)
wlan.connect(SSID,PASSWORD)

while not wlan.isconnected():
    pass

print('Connected to Wi-Fi')
print(wlan.ifconfig())

# Setup MQTT
SERVER          = '192.168.20.18'  # Replace with your MQTT broker IP
CLIENT_ID       = b'PICO'
TOPIC_TIMESTAMP = b'nodered_timestamp'

USER      = b'mqttuser'            # Replace with your MQTT username
PASSWORD  = b'<password>'    # Replace with your MQTT password

def mqtt_connect():
   client = MQTTClient(CLIENT_ID, SERVER, user=USER, password=PASSWORD,
   keepalive=3600)
   client.connect()
   print('Connected to %s MQTT Broker'%(SERVER))
   return client

def reconnect():
   print('Failed to connect to the MQTT Broker. Reconnecting...')
   sleep(5)
   machine.reset()

def on_message(topic, msg):
    print("Received message:", msg.decode())
```

```
def main(server, topic, username, passw):
    client = MQTTClient(CLIENT_ID, server, user=username, password=passw,
    keepalive=3600)
    client.set_callback(on_message)
    client.connect()
    client.subscribe(topic)
    print("Connected to %s, subscribed to %s topic" % (server, topic))

    try:
        while True:
            client.check_msg()
            sleep(1)

    finally:
        client.disconnect()

main(SERVER, TOPIC_TIMESTAMP, USER, PASSWORD)
```

Upload this script to your Raspberry Pi Pico W. If your Node-RED flow is operating, timestamps appear in the Thonny editor's shell window. Here is a screenshot of my Thonny editor showing this script in operation.

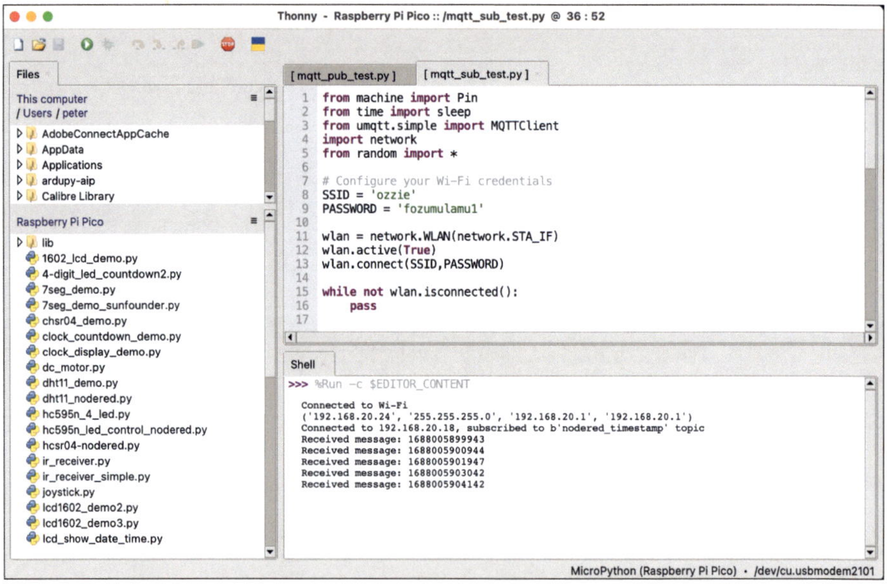

Figure 1.8.7.19: The example MicroPython script in operation. Timestamp values published by the Node-RED flow are printed in the Shell.

Script line-by-line review

Let's break this down the line by line:

1-5: Importing necessary modules:

- machine: to control the hardware components.
- time: to manage timing and delays.
- umqtt.simple: to use the MQTT protocol.
- network: to manage the Wi-Fi connection.
- random: to generate random numbers.

7-9: Define your Wi-Fi credentials. Replace <your_ssid> and <your_password> with your Wi-Fi SSID and password.

11-13: A station interface (STA_IF) is created and activated for the device to connect to a Wi-Fi network.

14: The script attempts to connect to the defined Wi-Fi network using the provided credentials.

16-18: A loop continues until the device is connected to the Wi-Fi network.

20-22: It prints a success message and the network configuration once connected.

25-30: These lines configure MQTT broker details.

- SERVER: the IP address of your MQTT broker.
- CLIENT_ID: the unique identifier for the MQTT client. Here, it is 'PICO'.
- TOPIC_TIMESTAMP: the MQTT topic where the timestamp data will be received.
- USER and PASSWORD: the username and password for the MQTT broker.

32-37: mqtt_connect function:

- This function creates an MQTTClient object with the previously defined CLIENT_ID, SERVER, USER, and PASSWORD.
- The client then tries to connect to the MQTT broker and prints a success message if it is connected.

39-42: reconnect function:

- If the MQTT client fails to connect to the MQTT broker, this function will print a message, wait for 5 seconds and then reset the machine.

44-45: on_message function:
This function is a callback that prints any received message from the broker.

47-61: main function:

- An MQTT client is created with a callback function (on_message).
- The client connects to the broker and subscribes to the TOPIC_TIMESTAMP.
- A loop is started to check for incoming messages continuously. If a message is received, the on_message function is called.
- If an error occurs or the script ends, the client disconnects from the broker.

63: The main function is called with the defined MQTT broker details.

This script connects to a Wi-Fi network and an MQTT broker, subscribes to a timestamp topic, and prints any messages it receives on that topic.

I provide the complete JSON code for this example's Node-RED flow here for your reference:


```
[
    {
        "id": "24c36904ca68652a",
        "type": "tab",
        "label": "Publish MQTT timestamp",
        "disabled": false,
        "info": "",
        "env": []
    },
    {
        "id": "52ade0556fb90c0d",
        "type": "inject",
        "z": "24c36904ca68652a",
        "name": "",
        "props": [
            {
                "p": "payload"
            },
            {
                "p": "topic",
                "vt": "str"
            }
        ],
        "repeat": "1",
        "crontab": "",
        "once": false,
        "onceDelay": 0.1,
        "topic": "",
        "payload": "",
        "payloadType": "date",
        «x»: 170,
        «y»: 160,
        «wires»: [
            [
                «ebc538601927dca8»,
                «db1fad03141a75d2»
            ]
        ]
```

```
    },
    {
        "id": "ebc538601927dca8",
        "type": "mqtt out",
        "z": "24c36904ca68652a",
        "name": "Publish timestamp",
        "topic": "nodered_timestamp",
        "qos": "",
        "retain": "",
        "respTopic": "",
        "contentType": "",
        "userProps": "",
        "correl": "",
        "expiry": "",
        "broker": "694bd92dc7019404",
        "x": 390,
        "y": 160,
        "wires": []
    },
    {
        "id": "db1fad03141a75d2",
        "type": "debug",
        "z": "24c36904ca68652a",
        "name": "Show timestamp",
        "active": true,
        "tosidebar": true,
        "console": false,
        "tostatus": false,
        "complete": "payload",
        "targetType": "msg",
        "statusVal": "",
        "statusType": "auto",
        "x": 390,
        "y": 260,
        "wires": []
    },
    {
        "id": "694bd92dc7019404",
        "type": "mqtt-broker",
        "name": "",
        "broker": "192.168.20.18",
        "port": "1883",
        "clientid": "",
        "autoConnect": true,
        "usetls": false,
        "protocolVersion": "4",
```

```
        "keepalive": "60",
        "cleansession": true,
        "birthTopic": "test_topic",
        "birthQos": "0",
        "birthPayload": "",
        "birthMsg": {},
        "closeTopic": "",
        "closeQos": "0",
        "closePayload": "",
        "closeMsg": {},
        "willTopic": "",
        "willQos": "0",
        "willPayload": "",
        "willMsg": {},
        "userProps": "",
        "sessionExpiry": ""
    }
]
```

Part 2: Node-RED & Raspberry Pi Pico Experiments

Chapter 2.1 • Frequently Used Patterns

In the experiments I have included in this part of the book, there are several programming components and patterns that I will include and use frequently.

These components and patterns are:

1. Wi-Fi: This provides network connectivity so your Raspberry Pi Pico W can connect to the Node-RED instance.

2. MQTT: All data interchange between the Raspberry Pi Pico W and the Node-RED instance will be implemented with the MQTT broker as the intermediary.

3. Node-RED: Any data in a flow can be presented in the dashboard with one of the available widgets. Data can also be passed on to other nodes or external services (such as a Google Sheet or other Cloud API).

A node that I use everywhere in my flows is "Debug". In all experiments in this book, the Debug node is used extensively to make it easy to inspect incoming, outgoing, and calculated or converted data. For example, we'll use Debug to print the data that a flow has received from MQTT or the data that the flow sends to the Raspberry Pi Pico W. You have already learned how to use the Debug node, so I will not repeat this content in this chapter.

I will discuss the three patterns I listed above in detail in the first few experiments. However, in the interest of brevity, in most of the experiments, I assume that you understand how they work and focus on what is new in each experiment. If you need a revision, you can return to this chapter. Also, for the same reason, I will only use the Dashboard to display flow data in some experiments but not all. For any experiment in this book without a Dashboard, you can easily add one by applying the knowledge you acquired from the experiments that use the Dashboard.

Let's look at these patterns in detail, including their code and instructions.

1.1. Wi-Fi

The Raspberry Pi Pico W has integrated Wi-Fi. In MicroPython, we can use the "network" module to access the Wi-Fi functionality and connect the board to the local Wi-Fi network. Unfortunately, the MicroPython documentation website does not (yet) contain information on how to use the Network module with the Raspberry Pi Pico; however, you can use the WLAN information provided for the ESP32. You will find the relevant documentation at this location: https://docs.micropython.org/en/latest/esp32/quickref.html#wlan

In the experiments in this book, the code that relates to the Wifi connectivity looks like this:

```
import network

# Configure your Wi-Fi credentials
SSID     = '<your_wifi_network_SSID>'
PASSWORD = '<your_wifi_network_password>'

wlan = network.WLAN(network.STA_IF)
wlan.active(True)
wlan.connect(SSID,PASSWORD)

while not wlan.isconnected():
    pass

print('Connected to Wi-Fi')
print(wlan.ifconfig())

# the board is now connected to Wi-Fi…
```

Let's look at this code in detail.

1. "import network": This line imports the "network" module, part of the Micro-Python standard library. This module provides functionality to configure and interact with different types of networks, including Wi-Fi.

2. "SSID = '<your_wifi_network_SSID>'": Here you're defining a variable called SSID (Service Set Identifier). This is the name of the Wi-Fi network you want your device to connect to. Replace <your_wifi_network_SSID> with the name of your Wi-Fi network.

3. "PASSWORD = '<your_wifi_network_password>'": Similarly, you're defining a variable called PASSWORD to store the password for the Wi-Fi network. Replace <your_wifi_network_password> with the actual password for your Wi-Fi network.

4. "wlan = network.WLAN(network.STA_IF)": This line creates a WLAN (Wireless Local Area Network) object in station mode. "network.STA_IF" represents the station interface, which is used when connecting to a Wi-Fi network.

5. "wlan.active(True)": This line activates the WLAN interface. Without activating the interface, you can't connect to a network.

6. "wlan.connect(SSID,PASSWORD)": This line initiates a connection to the Wi-Fi network using the SSID and password you specified earlier. The connect method takes the SSID and password as arguments.

7. "while not wlan.isconnected(): pass": This is a busy loop that will continue to run as long as the device is not connected to the Wi-Fi network. "wlan.isconnected()" checks the connection status of the WLAN interface. The pass statement is a placeholder and does nothing; it simply prevents a syntax error in a case where no action is intended.

8. "print('Connected to Wi-Fi')": Once the device is connected to the Wi-Fi network, the loop will exit, and this line will execute, printing the string 'Connected to Wi-Fi' to the console.

9. "print(wlan.ifconfig())": The "ifconfig" method returns a tuple containing the IP address, subnet mask, gateway, and DNS server. When you print this, you can see the network configuration details for your device. This can be useful for debugging network issues or finding the device's IP address.

With the Raspberry Pi Pico connected to the WiFi network (the same WiFi network that the Node-RED instance and MQTT broker belong to), your MicroPython script can use the MQTT module. Let's see how that works.

1.2. MQTT Sub and Pub

As you learned in Part 1 of this book, we will use the "micropython-umqtt.simple" module (or simply "mqtt.simple") to communicate with the MQTT broker. Below, we will use the related code in all experiments where MQTT functionality is needed.

I have removed the Wifi code to keep this example simple. Here is the code — this example shows how to publish a string of text to a topic:

```
from umqtt.simple import MQTTClient
from time import sleep
import network

# WIFI related code
# WIFI will be connected before continuing

# Configure MQTT
SERVER    = '<your_MQTT_broker_IP_address>'
CLIENT_ID = b'<your_device_client_ID>'
TOPIC     = b'<the_topic_to_subscribe_to>'
USER      = b'<your_MQTT_username>'
PASSWORD  = b'<your_MQTT_username>'

def mqtt_connect():
   client = MQTTClient(CLIENT_ID, SERVER, user=USER, password=PASSWORD,
   keepalive=3600)
   client.connect()
   print('Connected to %s MQTT Broker'%(SERVER))
```

```
    return client

def reconnect():
    print('Failed to connect to the MQTT Broker. Reconnecting...')
    sleep(5)
    machine.reset()

try:
    client = mqtt_connect()
except OSError as e:
    reconnect()


while True:
     value = "Hello world"
     client.publish(TOPIC, value)
     sleep(1.0)
```

Here's a detailed explanation of how this code works:

1. "from umqtt.simple import MQTTClient": This line imports the MQTTClient class from the umqtt.simple module. This class provides methods for creating a client that can connect to an MQTT (Message Queue Telemetry Transport) server, send messages (publish), and wait for messages (subscribe).

2. "from time import sleep": This imports the sleep function from the time module, which can be used to create delays in your code execution. We'll use these delays to throttle the messages from the Raspberry Pi Pico to the MQTT broker.

3. "import network": This imports the network module, which provides the capabilities to interact with various network types, including Wi-Fi.

4. The comment "# WIFI related code" suggests that the code for connecting to Wi-Fi should be placed here. This comment serves as a placeholder and doesn't perform any operation.

5. The comment "# WIFI will be connected before continuing" is an instruction or reminder that the code after this point assumes that a Wi-Fi connection has been established.

6. "SERVER = '<your_MQTT_broker_IP_address>'": This defines the IP address of the MQTT broker/server you want to connect to. Replace <your_MQTT_broker_IP_address> with the actual IP address.

7. "CLIENT_ID = b'<your_device_client_ID>'": Here you're defining a variable called CLIENT_ID which identifies your device to the MQTT broker. Replace <your_device_client_ID> with a unique identifier for your device.

8. "TOPIC = b'<the_topic_to_subscribe_to>'": This is the MQTT topic on which your client will publish messages. Replace <the_topic_to_subscribe_to> with the actual topic you're using.

9. "USER = b'<your_MQTT_username>'" and "PASSWORD = b'<your_MQTT_password>'": These lines define the username and password used to authenticate with the MQTT server. Replace <your_MQTT_username> and <your_MQTT_password> with the actual credentials.

10. "def mqtt_connect():": This line begins the definition of a function called mqtt_connect, which will be used to establish a connection to the MQTT broker.

11. "client = MQTTClient(CLIENT_ID, SERVER, user=USER, password=PASSWORD, keepalive=3600)": Inside the mqtt_connect function, this line creates an MQTTClient object with the specified client ID, server IP, user, password, and keepalive time.

12. "client.connect()": This line initiates the connection to the MQTT server using the details provided when the MQTTClient object was created.

13. "print('Connected to %s MQTT Broker'%(SERVER))": If the connection is successful, this line will print a success message, indicating that the client has connected to the MQTT broker.

14. "return client": This line returns the client object to the function's caller. This allows the caller to use the client for further MQTT operations like "publish" or "subscribe".

15. "def reconnect():": This begins the definition of a function called reconnect, which will be used if an error occurs when connecting to the MQTT broker.

16. "print('Failed to connect to the MQTT Broker. Reconnecting...')": Inside the reconnect function, this line prints an error message indicating a failed connection.

17. "sleep(5)": This line delays 5 seconds before attempting to reconnect.

18. "machine.reset()": This line resets the machine, similar to pushing a reset button on a physical machine. It's a way of starting fresh if the connection attempt fails.

19. "try: client = mqtt_connect()": This is the beginning of a try/except block. It attempts to connect to the MQTT broker by calling the mqtt_connect function.

20. "except OSError as e: reconnect()": If an OSError occurs during the connection attempt (for example, if the broker is unreachable), the code in the except block runs, which calls the reconnect function to reset the device and attempt a fresh connection.

21. "while True:": This begins an infinite loop. The code inside this loop will continue running indefinitely.

22. "value = "Hello world"": Inside the loop, this line defines a variable value and sets it to the string "Hello world".

23. "client.publish(TOPIC, value)": This line uses the "publish" method of the MQTTClient object to send the value "Hello world" to the MQTT broker on the specified topic.

24. "sleep(1.0)": This line pauses the execution of the loop for 1 second before it starts over. This effectively means the client will publish the message "Hello world" to the MQTT broker every second.

Now that you understand how Wi-Fi and MQTT work on the Raspberry Pi Pico W let's combine the two code examples and see the whole picture. Here is the combined script:

```
from umqtt.simple import MQTTClient
from time import sleep
import network

# Configure your Wi-Fi credentials
SSID     = '<your_wifi_network_SSID>'
PASSWORD = '<your_wifi_network_password>'

wlan     = network.WLAN(network.STA_IF)
wlan.active(True)
wlan.connect(SSID,PASSWORD)

while not wlan.isconnected():
    pass

print('Connected to Wi-Fi')
print(wlan.ifconfig())

# Configure MQTT
SERVER    = '<your_MQTT_broker_IP_address>'
CLIENT_ID = b'<your_device_client_ID>'
```

```
TOPIC     = b'<the_topic_to_subscribe_to>'
USER      = b'<your_MQTT_username>'
PASSWORD  = b'<your_MQTT_password>'

def mqtt_connect():
   client = MQTTClient(CLIENT_ID, SERVER, user=USER, password=PASSWORD,
   keepalive=3600)
   client.connect()
   print('Connected to %s MQTT Broker'%(SERVER))
   return client

def reconnect():
   print('Failed to connect to the MQTT Broker. Reconnecting...')
   sleep(5)
   machine.reset()

try:
   client = mqtt_connect()
except OSError as e:
   reconnect()

while True:
    value = "Hello World!"
    client.publish(TOPIC, value)
    sleep(1.0)
```

You will see this code in almost all experiments in this part of the book. Now, let's look at the Dashboard patterns.

1.3. Node-RED

As you know, Node-RED makes it easy to present any data in a visual Dashboard. There are widgets that you can use to give numerical data as numbers (I.e. with the "numeric" widget) or a range (i.e. with the "gauge" widget). You can show text strings with the "text" widget. You can also take user input using widgets such as the "slider", the "switch", the "dropdown" and the "text input".

The same input and output patterns can be applied regardless of which widget you use in a Dashboard. To simplify learning, I have divided these patterns into two main categories: Raw I/O and Processed I/O. Let's look at each one.

Raw I/O

In Raw Input/Output, the data is passed without modifying the dashboard widget or published to an MQTT topic. For example, consider the flow below:

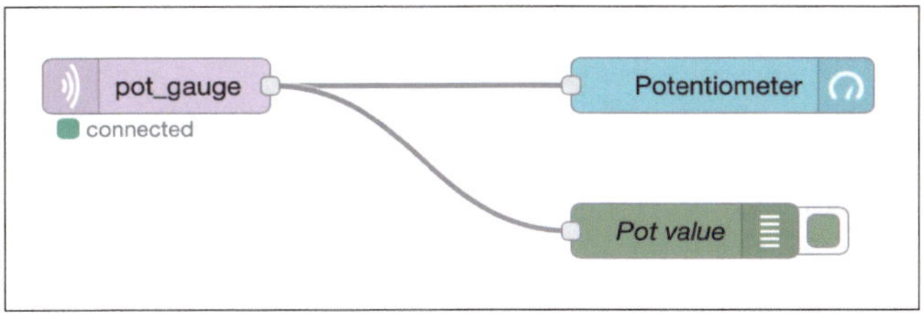

Figure 2.1.3.1: The value from the pot_gauge topic is passed unmodified to the Potentiometer gauge.

This flow starts with an MQTT_in node. The node gets values from the "pot_gauge" topic and passes them to the Potentiometer dashboard gauge without modification.

Here is another example:

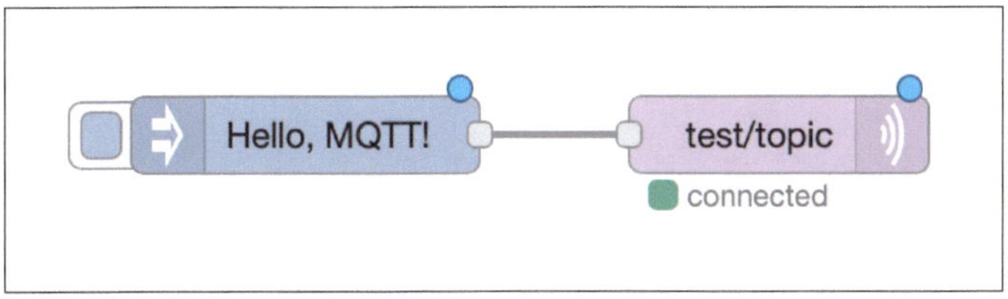

Figure 2.1.3.2: The value from the inject node is passed unmodified to the MQTT-out node.

In this example, the Inject node sends a text string to a MQTT-out node, which will send it to the "test/topic" topic of the MQTT broker. The text string is not modified in any way during its journey.

Processed I/O

Passing raw input/output data through a flow is sufficient in many use cases, but not all. Often, data will need to be processed in some way before it can continue its journey through the flow. For example, consider the dashboard below:

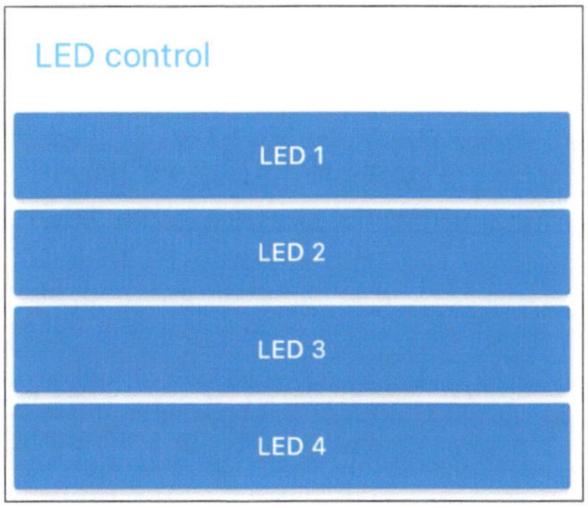

Figure 2.1.3.3: A button press will update the value in a single MQTT topic.

When you press one of these buttons, an LED (out of an array of four connected to a Raspberry Pi Pico W) will turn on, and the rest will turn off. You use a single MQTT topic containing the integer representing the button pressed to accomplish this.

You could design a flow with one MQTT-out node connected to each button; this would be four MQTT-out nodes. This is a waste because it requires a lot of duplicate code, which is frowned upon in software development circles. It is better to handle this case with a flow like in this example:

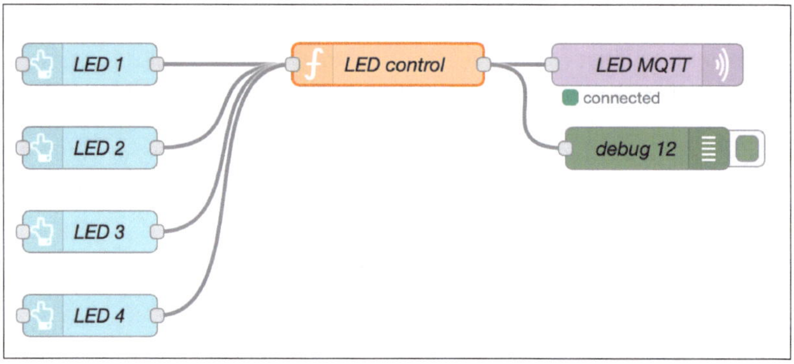

Figure 2.1.3.4: The Function node eliminates duplication and increases versatility.

In this example flow, the four button widgets send their value (when pressed) to a Function node. The Function node will process the incoming data and send an appropriate value to the MQTT topic via a single MQTT-out node. The code inside the Function node is this:

```
var intToSend = 0;
if (msg.payload == 1) intToSend = 1;
if (msg.payload == 2) intToSend = 2;
if (msg.payload == 3) intToSend = 3;
if (msg.payload == 4) intToSend = 4;
return { payload: intToSend};
```

This is an example of data being processed before it progresses in the flow or (in this case) being sent out of the flow to an external service.

The opposite can also happen. It is possible to receive data from an external service and process it before allowing it to continue its journey in the flow.

Here is a straightforward example:

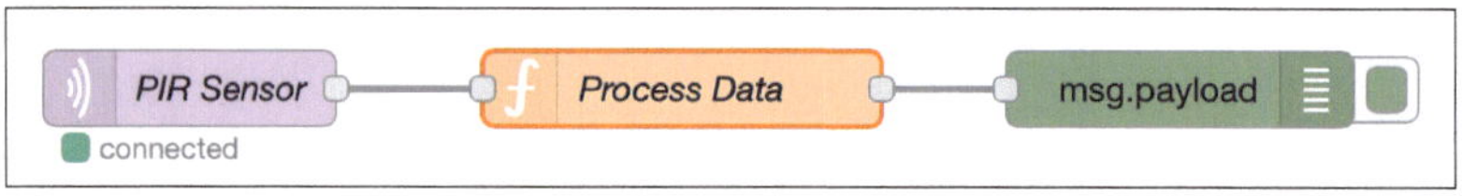

Figure 2.1.3.5: Data incoming from the MQTT-in node is processed in the Function node.

In this example, the data that is entering the flow via the MQTT-in node is processed by the Function node. In this simple case, the function Javascript is this:

```
if (msg.payload === 'Motion detected!') {
    return { payload: 'ALARM: Intruder detected!' };
}
return { payload: 'All quiet.' };
```

As you can see, it simply sends an appropriate message to the debug node. But it can be more complicated, as you can see in the code below, where a humidity value from a soil sensor is used to determine if an irrigation pump should be turned on:

```
var pump_state = '0';

var device_string       = msg.payload.split(",");
var raw_humidity_value  = device_string[0];
var device_name         = device_string[1];

flow.set("raw_humidity_value",raw_humidity_value);
flow.set("device_name",device_name);

if (raw_humidity_value > flow.get("soil_humidity_threshold"))
{
    pump_state = '1';
    flow.set("pump_state","On");
} else
{
    pump_state = '0';
    flow.set("pump_state","Off");
}

return {payload: pump_state};
```

Here is one last example of processed I/O. See the flow below:

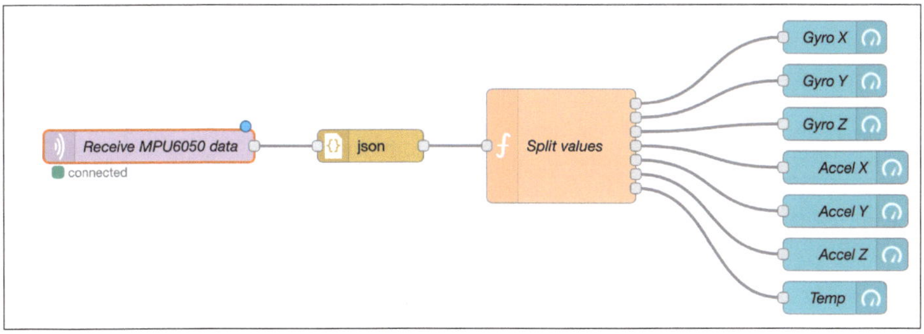

Figure 2.1.3.6: Processing Json data from a sensor.

In this example, we have an MPU (motion sensor) with 6 degrees of freedom. This means this sensor outputs six values (three for the gyro and three for the accelerometer) plus the current temperature. While the Raspberry Pi could send each value separately into a dedicated MQTT topic, again, it would be wasteful in resources. Instead, the Raspberry Pi bundles the values into a single JSON document and publishes it to a single topic. The Node-RED flow will get this JSON document and use some Javascript code to split it to its values. Here is the script that does that:

```
var gx = msg.payload.gyro_x;
var gy = msg.payload.gyro_y;
var gz = msg.payload.gyro_z;
var ax = msg.payload.accel_x;
var ay = msg.payload.accel_y;
var az = msg.payload.accel_z;
var temp = msg.payload.temp;

return [
    {payload: gx},
    {payload: gy},
    {payload: gz},
    {payload: ax},
    {payload: ay},
    {payload: az},
    {payload: temp}
];
```

As you can see, using Javascript embedded in Function nodes makes it possible for us to design very complex yet efficient flows. In the following experiments, we'll use the patterns you learned in this chapter.

Chapter 2.2 • Warm Up

In this chapter, you'll work on a series of "warming up" experiments. These are simple experiments that use simple hardware, such as buttons and LEDs. The objective of these experiments is to apply the programming patterns you learned about in the previous chapter.

Let's begin with a potentiometer.

2.1. Gauge and Potentiometer

In this experiment, you will learn how to take position readings from a potentiometer connected to your Raspberry Pi Pico W and depict the potentiometer value using a gauge in a Node-RED dashboard. At the end of this experiment, the dashboard will look like this:

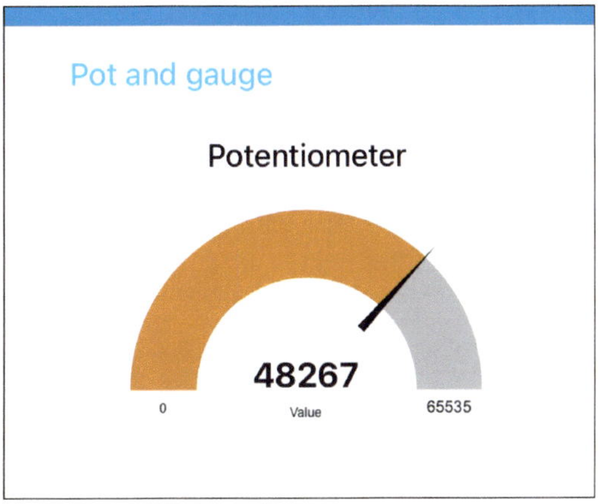

Figure 2.2.1.1: The potentiometer position value depicted in a gauge widget in Node-RED.

Setup

For this experiment, you will need these materials.

1. A Raspberry Pi Pico W
2. A 10-kΩ potentiometer
3. Breadboard and jumper wires
4. The Node-RED instance you set up earlier in this book.
5. The Thonny editor

To wire the hardware, follow these instructions:

1. Connect one of the outer pins of the potentiometer to the 3V3 pin on the Raspberry Pi Pico.

2. Connect the other outer pin of the potentiometer to a GND pin.

3. Connect the middle pin of the potentiometer to GP26 (ADC0) or GP27 (ADC1) pin on the Raspberry Pi Pico.

When you have finished with the wiring, your circuit should look like this:

Figure 2.2.1.2: The circuit of this experiment.

You can see the wiring diagram below (made with Fritzing):

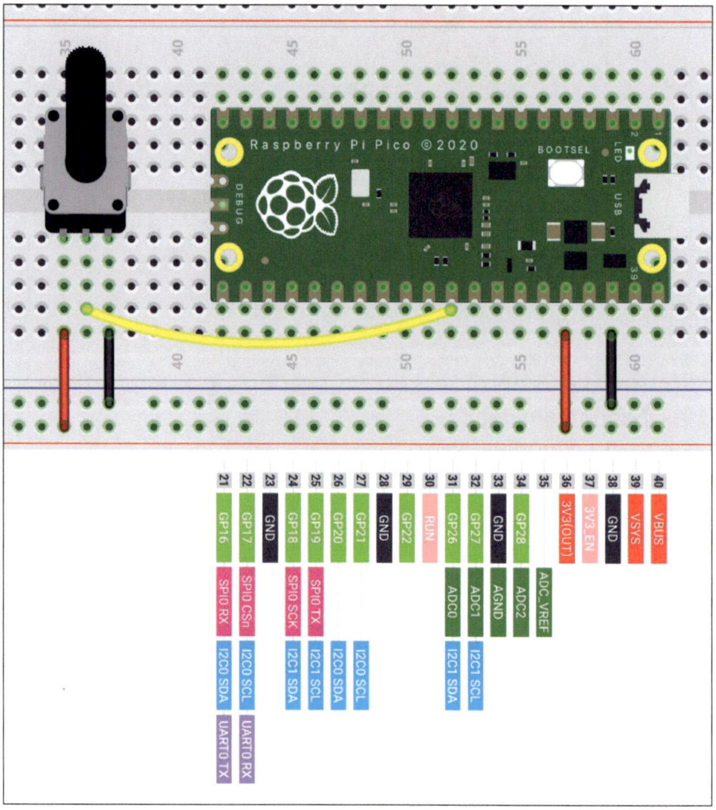

Figure 2.2.1.3: The wiring diagram for the potentiometer.

Raspberry Pi Pico Code

Connect your Raspberry Pi Pico W to your computer and open the Thonny editor. In Thonny, create a new document with the file name "pot_gauge.py" and save it in the root of the Raspberry Pi Pico file system. MicroPython code that you see below reads the analog value from the potentiometer and publishes it to the MQTT topic every second:

```
from machine import Pin, ADC
from umqtt.simple import MQTTClient
from time import sleep
import network

# Configure the potentiometer
pot = ADC(26) # if connected to GP26 which is ADC0

# Configure your Wi-Fi credentials
SSID     = '<your_wifi_network_SSID>'
PASSWORD = '<your_wifi_network_password>'

wlan = network.WLAN(network.STA_IF)
wlan.active(True)
wlan.connect(SSID,PASSWORD)

while not wlan.isconnected():
       pass

print('Connected to Wi-Fi')
print(wlan.ifconfig())

# Configure MQTT
SERVER    = '<your_MQTT_broker_IP_address>'
CLIENT_ID = b'<your_device_client_ID>'
TOPIC     = b'<the_topic_to_subscribe_to>'
USER      = b'<your_MQTT_username>'
PASSWORD  = b'<your_MQTT_password>'

def mqtt_connect():
   client = MQTTClient( CLIENT_ID,
                        SERVER,
                        user=USER,
                        password=PASSWORD,
                        keepalive=3600)
   client.connect()
   print('Connected to %s MQTT Broker'%(SERVER))
   return client

def reconnect():
```

```
    print('Failed to connect to the MQTT Broker. Reconnecting...')
    sleep(5)
    machine.reset()

try:
    client = mqtt_connect()
except OSError as e:
    reconnect()

while True:
        value = pot.read_u16()
        client.publish(TOPIC, str(value))
        print("Pot value = ", value)
        sleep(1.0)
```

To make this code work, you must replace the Wi-Fi SSID and password, MQTT server IP and user information with your own. Please note the following:

1. Python is sensitive to indentation. Use spaces (not tabs!) to indent code inside loops and functions.

2. It is OK to break a long line into multiple lines and to use spaces to make them more readable. I have done this with the first line of the "mqtt_connect()" function. In that line, I broke the parameters into multiple lines and used spaces to align them so they were easier to read.

3. Ensure that the single-quote characters that enclose strings, such as the SERVER IP address and the TOPIC variables, is the one that is assigned to the key next to the return key. Do not use the backtick character set to the top-left key on your keyboard next to the numeral "1" (the key assignment in the US keyboard layout). We want this character: "'" not this "`".

Go ahead and upload the script now. Once the script starts running on the Raspberry Pi Pico W, it will print the pot value in the shell and send the same value to the MQTT broker's "pot_gauge" topic, ready for Node-RED to read.

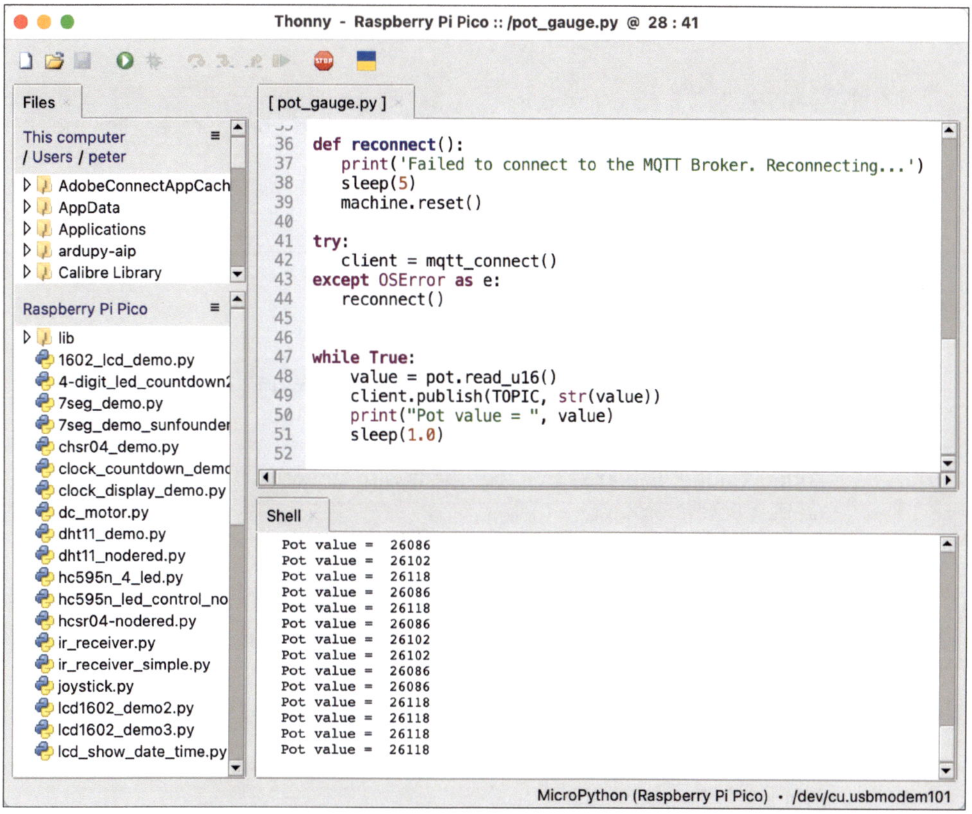

Figure 2.2.1.4: Thonny has uploaded the script to the Raspberry Pi Pico and shows output in the Shell.

Node-RED flow

The flow for this experiment is effortless. It consists of an MQTT-in node and a gauge widget node. As usual, we'll add a debug node to see the incoming data in the debug tab.

The flow looks like this:

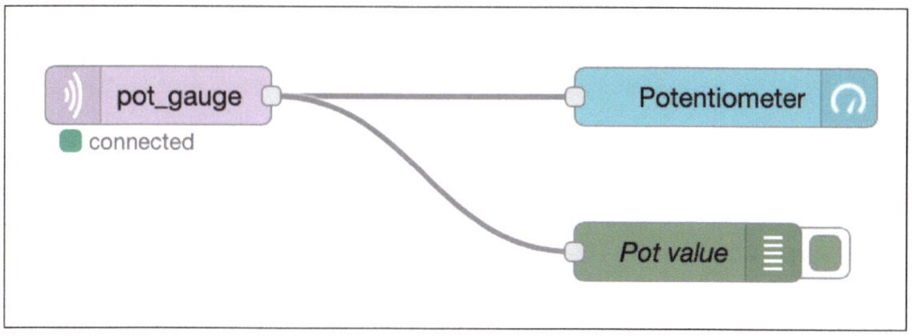

Figure 2.2.1.5: The flow of this experiment.

Setup each node like this:

- MQTT-in node
    - Server: your local MQTT server IP address and port number. For me, it is 192.168.20.18:1883
    - Action: Subscribe to a single topic.
    - Topic: pot_gauge
    - QoS: 2
    - Output: auto-detect (string or buffer)
    - Name: blank, or a name you prefer.
- Gauge node
    - Group: Create a new group named "Pot and gauge". Assign this group to a new Dashboard tab named "Pot Gauge".
    - Size: auto or set your size.
    - Type: Gauge. Feel free to experiment with the other types.
    - Label: Potentiometer.
    - Value format: {{value}}
    - Units: Leave blank
    - Range: Min and max depend on the values coming from the source. In the case of the Raspberry Pi Pico, the ADC converter is 16 bits and produces values from zero to 65535.
        - min: 0
        - max: 65535
    - Colour gradient and sectors: Leave defaults or experiment with customer values.
    - Class: blank.
    - Name: blank, or a name you prefer.

Go ahead and deploy this flow. In the Debug tab, you will see the potentiometer position values, as in the screenshot below:

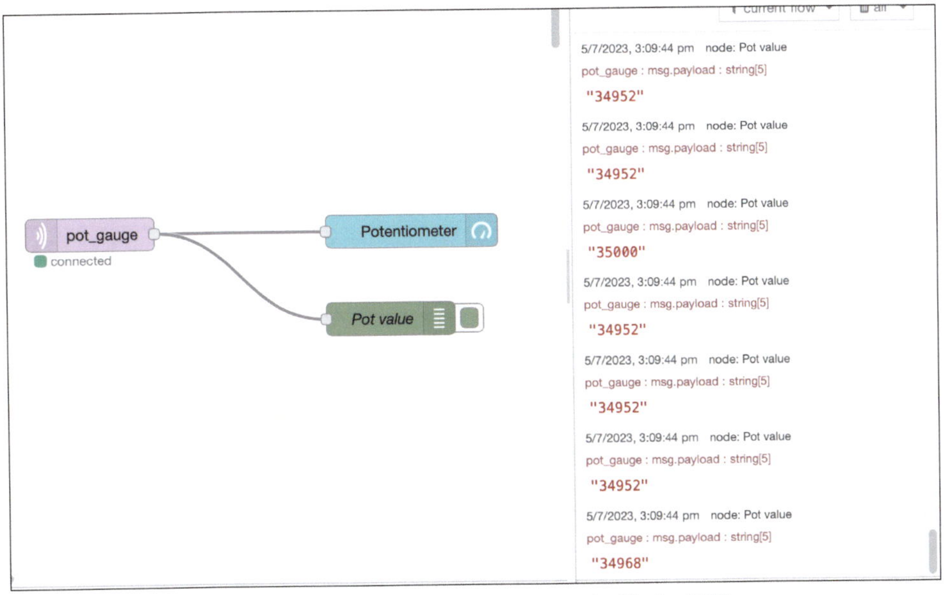

Figure 2.2.1.6: The debug output in Node-RED.

Open the Dashboard and click on the Pot Gauge tab. When you change the pot's position, the gauge needle and colour will change with a second or so, as in the screenshot at the start of this experiment.

How to modify the gauge range

In this iteration of the experiment, the dashboard gauge shows the raw values of the potentiometer as they are sampled by the Raspberry Pi Pico on the analog input pin and sent to the Node-Red flow. In most realistic projects, you will want to map the raw input value to a more suitable range of numbers. In this experiment, let's modify the flow so that the input values are mapped from their original range of 0 to 65535 to a more useful range, such as 0 to 100. With this modification, we can refer to the position of the potentiometer in terms of a percentage, which is more intuitive.

To do this, we can add the "range" node to our flow and wire it to receive input from the MQTT-in node and send its output to the gauge node. You can see the re-wired flow below:

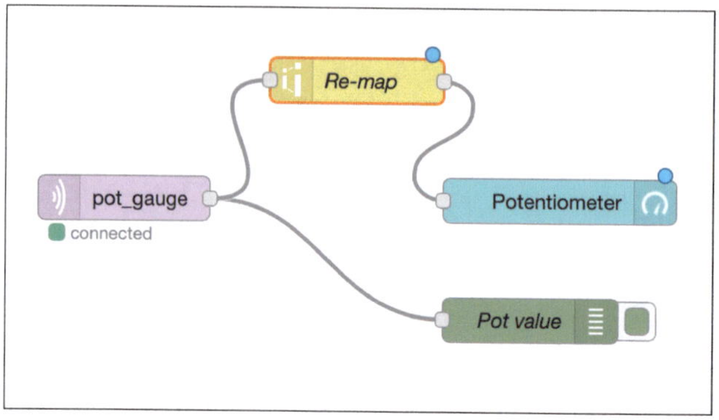

Figure 2.2.1.7: The "range" node is named "Re-map".

Let's configure the new "range" node. Double-click the node to open its properties, and set it up like this:

- Property: msg.payload
- Action: Scale the message property
- Map the input range:
  - From: 0
  - To: 65535
- ... to the target range:
  - from: 0
  - to: 100
  - Check the "Round result to nearest integer" box.
- Name: Re-map.

You can see the properties form of the "range" node below:

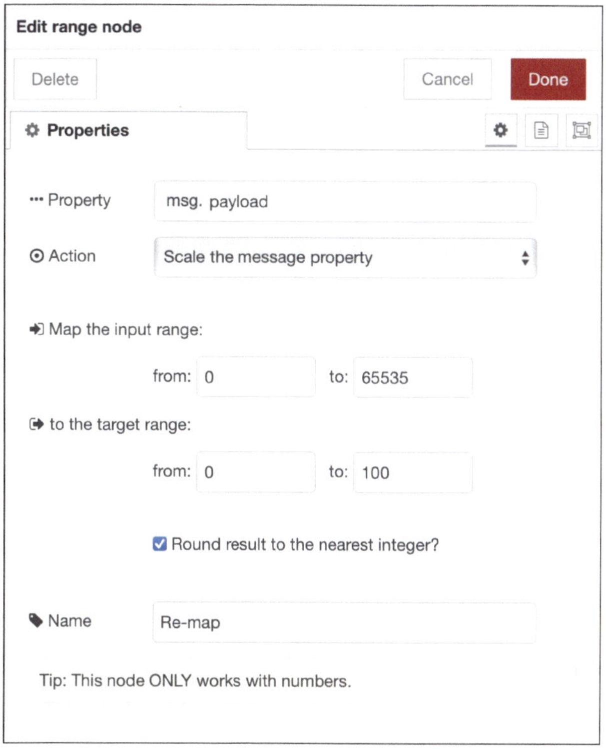

Figure 2.2.1.8: The properties of the "range" node.

With this change, the gauge will receive numbers ranging from 0 to 100. Therefore, you must change the gauge node to reflect the new number range. Double-click the gauge node to open its properties window and change the Range min and max values to 0 and 100, respectively.

Deploy the changed flow and use your browser to see the Dashboard for the flow. It should look like this:

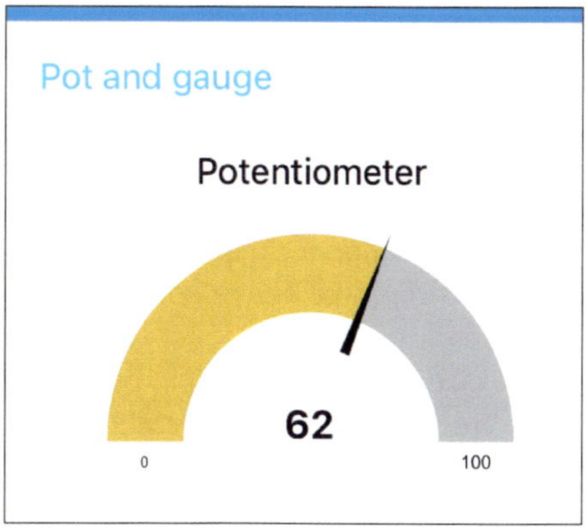

Figure 2.2.1.9: The gauge now uses the new number range.

The gauge now displays the potentiometer position in the percentage range 0 to 100.

2.2. Button

In this section, you will embark on a stimulating project connecting the Raspberry Pi Pico, a physical button, and the Node-RED Dashboard. This project aims to use a physical button wired to a Raspberry Pi Pico to control the status message on a Node-RED dashboard widget. The message "Pressed" will be displayed when the button is pressed. The message "Unpressed" will appear when the button is not pressed. This experiment is a practical lesson in device-to-dashboard communications and the fundamentals of wiring and coding in Node-RED and MicroPython.

Wiring instructions

To accomplish this, you'll need the following materials:

- A Raspberry Pi Pico
- A pushbutton
- A 10-kΩ pull-down resistor
- Jumper wires
- Breadboard

Here are the steps to wire the push button to your Raspberry Pi Pico:

1. Place your Raspberry Pi Pico W on the edge of the breadboard with the USB connector facing outwards (this will make it easier to attach the USB cable).

2. Use a short jumper to connect the GND pin of the Pico (any GND pin will do; I use pin 38) to the blue power rail of the breadboard.

3. Use a short jumper wire to connect the 3V3 (pin 36) pin of the Pico to the red power rail of the breadboard.

4. Place the button on the breadboard so that its two sets of pins (top and bottom row) are connected across the breadboard's two sides (see the photograph below).

5. Connect the pull-down resistor to connect one button pin to the blue (GND) power rail.

6. Use a short jumper wire to connect the other pin of the button (on the same side as the resistor) to the red power rail of the breadboard.

7. Use a longer jumper wire to connect the pin on the opposite side of the resistor to GP14 of the Pico. We will use GP14 to read the state of the button.

Your circuit should look like this:

Figure 2.2.2.10: A momentary button connected to the Raspberry Pi Pico W.

You can see the wiring diagram below (made with Fritzing):

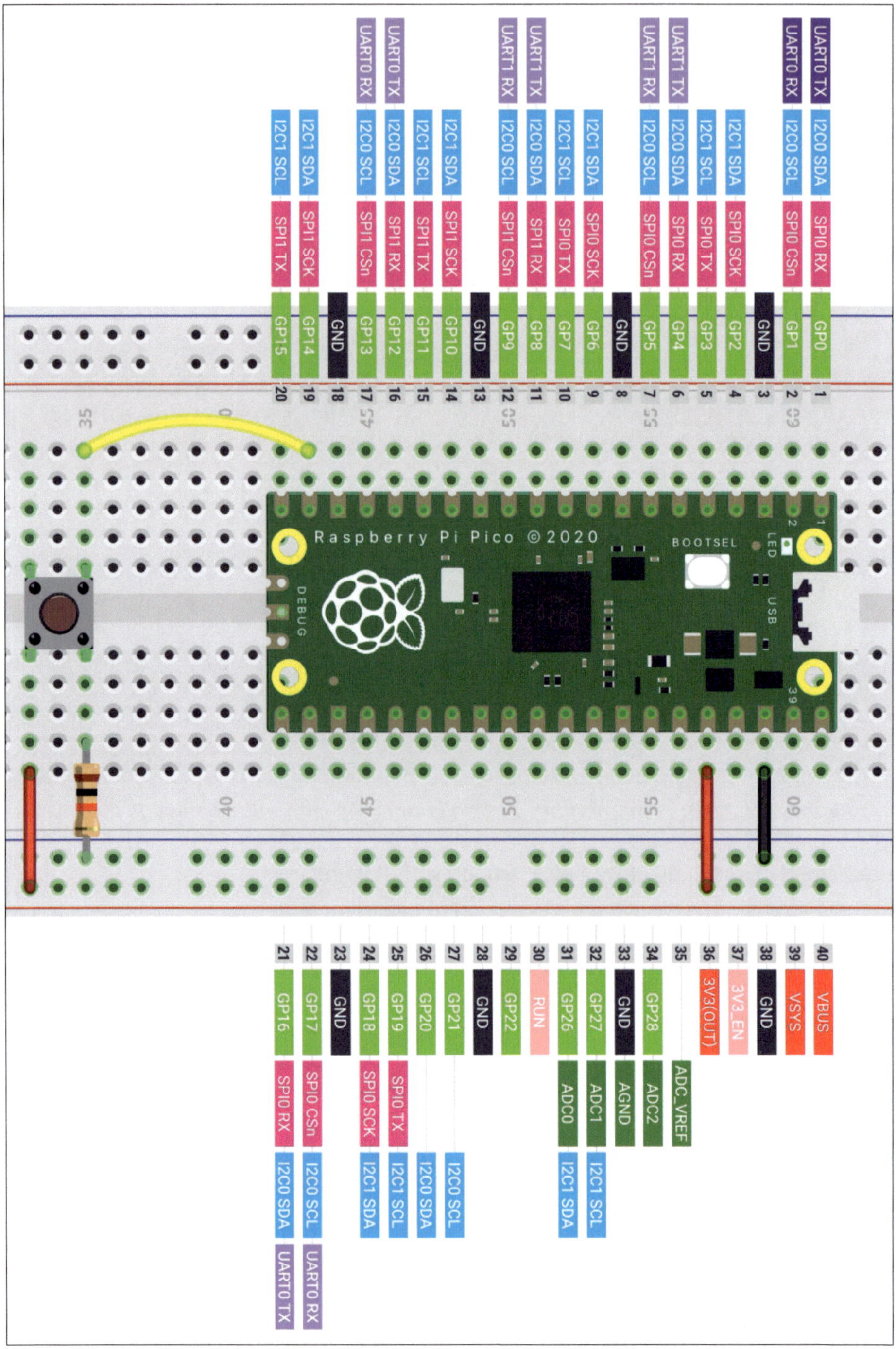

Figure 2.2.2.11: The wiring diagram for the momentary button.

Raspberry Pi Pico MicroPython Code

After the wiring, the next step is to program the Raspberry Pi Pico using MicroPython. We will use Thonny, a Python IDE perfect for running MicroPython on the Pico.

The basic MicroPython code that will read the button status and print the status to the console is this:

```
from machine import Pin
import time
button = Pin(14, Pin.IN, Pin.PULL_UP)

while True:
    if button.value():
        print('Pressed')
    else:
        print('Unpressed')
    time.sleep(0.1)
```

This script sets up a loop that checks the button's value every 100 milliseconds and prints "Unpressed" or "Pressed" accordingly.

We aim to transmit the button status to the Node-RED flow (which we have not constructed yet). Therefore, we will add the necessary communications code that allows the Pico to connect to the local Wi-Fi network and then publish button status values to the MQTT broker. You may recall that this scenario is similar to what you did in the previous chapter. Let's re-use the relevant MicroPython code from the potentiometer example. Here's the new code:

```
from machine import Pin, ADC
from umqtt.simple import MQTTClient
from time import sleep
import network

button = Pin(14, Pin.IN, Pin.PULL_UP)

# Configure your Wi-Fi credentials
   SSID     = '<your_wifi_network_SSID>'
   PASSWORD = '<your_wifi_network_password>'

wlan = network.WLAN(network.STA_IF)
wlan.active(True)
wlan.connect(SSID,PASSWORD)

while not wlan.isconnected():
    pass

print('Connected to Wi-Fi')
```

```
print(wlan.ifconfig())

# Configure MQTT
   SERVER      = '<your_MQTT_broker_IP_address>'
   CLIENT_ID  = b'<your_device_client_ID>'
   TOPIC       = b'button_state'
   USER        = b'<your_MQTT_username>'
   PASSWORD    = b'<your_MQTT_password>'

def mqtt_connect():
   client = MQTTClient( CLIENT_ID,
                        SERVER,
                        user=USER,
                        password=PASSWORD,
                        keepalive=3600
                      )
   client.connect()
   print('Connected to %s MQTT Broker'%(SERVER))
   return client

def reconnect():
   print('Failed to connect to the MQTT Broker. Reconnecting...')
   sleep(5)
   machine.reset()

try:
   client = mqtt_connect()
except OSError as e:
   reconnect()

while True:
    if button.value():
        print('Pressed')
        client.publish(TOPIC, "1")
    else:
        print('Unpressed')
        client.publish(TOPIC, "0")
    sleep(1)
```

The new code version will sample the button every 1 second and send "1" to the broker if it is pressed and "0" if it is not. Notice that the messages to the MQTT broker are text strings, not integers.

Also, notice that the 1-second sampling period is not the best way to implement the button functionality. As you can imagine, if I press the button between the moments when the script takes a sample, the Pico will not notice the press. For the Pico to see a button press,

I have to press the button (and keep it pressed) when a sample is taken. We will improve on this later, but the 1-second sampling period is sufficient.

Upload the script to your Pico and continue with the Node-RED flow.

Node-RED flow

Let's will create the Node-RED flow to listen for the Raspberry Pi Pico's output and update the dashboard widget accordingly. Start Node-RED, navigate your workspace and open a new flow editor.

Construct the flow like this:

1. From the nodes palette on the left side, drag an 'mqtt in' node (found under the input section) onto the workspace.

2. Double-click on the 'mqtt in' node to configure it:
   - For 'Server,' click the edit button and add the details of your MQTT broker.
   - Set the 'Topic' to "button_state". This should match the topic string you used in the MicroPython script.
   - For the name, type "Button status".
   - Click 'Done' to save the settings.

3. From the "Function" section in the nodes palette, drag a "function" node onto the workspace. This node will convert the button status values from "0" and "1" to "Not Pressed" and "Pressed".

4. Double-click on the 'function in' node to configure it:

   In the "On Message" tab, copy this Javascript code:

```
if (msg.payload == "1") {
msg.payload = "Pressed";
} else if (msg.payload == "0") {
msg.payload = "Not Pressed";
}
return msg;
```

1. Under the 'dashboard' section in the nodes palette, drag a 'text' node onto the workspace.
   - Double-click the 'text' node to configure it:
     - Set the 'Group' by clicking the edit button and then 'Add new ui_group'. Name the group "[Button]" and the tab "Button".
     - Set 'Label' as 'Button State'.
     - Click 'Done' to save the settings.

2. Add a debug node to see the button state in the debug tab.
   - Connect the 'mqtt in' node to the 'function' node and the function node to the text node. Also, connect the 'mqtt in' node to the debug node.
   - Click the red 'Deploy' button in the top right corner to save and deploy the flow.

Your flow should look like this:

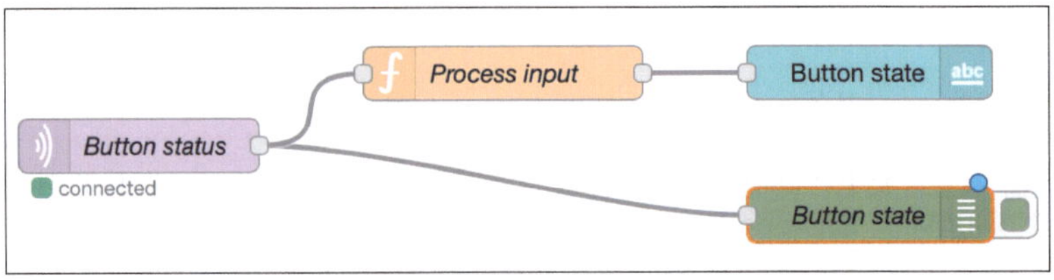

Figure 2.2.2.12: The button flow.

After you deploy this flow, Node-RED will listen to the output of the Raspberry Pi Pico. The Pico will send "Pressed" or "Unpressed" to the Node-RED server whenever the button is pressed or unpressed. This server will then update the widget on the dashboard accordingly.

In the Dashboard, the text widget looks like this when the button is not pressed:

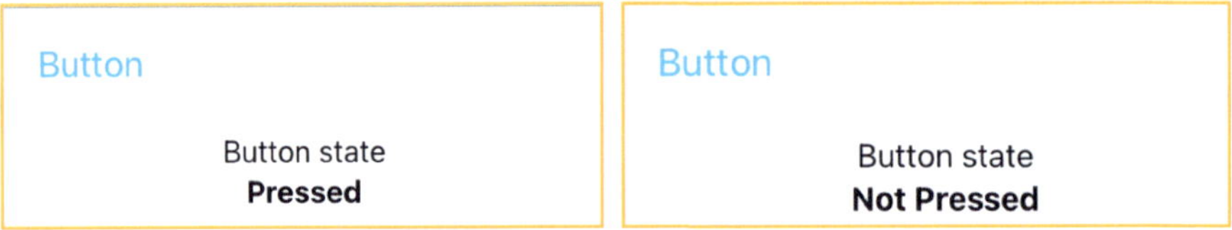

Figure 2.2.2.13: The button flow.

As you press and de-press the button, notice how its status updates in the dashboard. Because of the 1-second sampling rate, the updates are delayed up to one second. As I mentioned earlier, even though the system works, this design is not optimal. In the next section, we'll slightly improve the MicroPython code, significantly improving the system's overall performance.

2.3. Sample Button with Interrupts

Interrupts are a great way to make your script more responsive and efficient. Rather than continually polling the button status in a loop, you can use an interrupt to trigger a function whenever the button is pressed or released.

To use interrupts with MicroPython on the Raspberry Pi Pico, you should use the "irq" method of the Pin object to attach an interrupt service routine (ISR) to the pin. Here's how you could modify the script to do this (we will ignore the MQTT code for this example):

```
from machine import Pin

def button_callback(pin):
    if pin.value():
        print('Pressed')
    else:
        print('Unpressed')

button = Pin(14, Pin.IN, Pin.PULL_UP)
button.irq(trigger=Pin.IRQ_FALLING | Pin.IRQ_RISING, handler=button_callback)

while True:
    pass
```

In this script, button_callback is an ISR called every time the button's state changes. If the button is pressed (causing a falling edge as the button is pulled down), the function prints "Pressed" to the console. If the button is released (causing a rising edge as the pull-up resistor pulls the pin back to high), it prints "Unpressed".

Note that the while True: pass at the end is needed to prevent the script from ending, as the main thread of execution does nothing but wait for button presses.

Now that you understand how the interrupt method works, you can use it to replace the infinite loop with the 1-second delay. Here is the entire MicroPython script using the interrupt method:

```
from machine import Pin, ADC
from umqtt.simple import MQTTClient
from time import sleep
import network

def button_callback(pin):
    if button.value():
        print('Pressed')
        client.publish(TOPIC, "1")
    else:
        print('Unpressed')
        client.publish(TOPIC, "0")

button = Pin(14, Pin.IN, Pin.PULL_UP)
button.irq(trigger=Pin.IRQ_FALLING | Pin.IRQ_RISING, handler=button_callback)

# Configure your Wi-Fi credentials
   SSID     = '<your_wifi_network_SSID>'
   PASSWORD = '<your_wifi_network_password>'
```

```
wlan = network.WLAN(network.STA_IF)
wlan.active(True)
wlan.connect(SSID,PASSWORD)

while not wlan.isconnected():
    pass

print('Connected to Wi-Fi')
print(wlan.ifconfig())

# Configure MQTT
   SERVER     = '<your_MQTT_broker_IP_address>'
   CLIENT_ID  = b'<your_device_client_ID>'
   TOPIC      = b'button_state'
   USER       = b'<your_MQTT_username>'
   PASSWORD   = b'<your_MQTT_password>'

def mqtt_connect():
   client = MQTTClient( CLIENT_ID,
                        SERVER,
                        user=USER,
                        password=PASSWORD,
                        keepalive=3600
                       )
   client.connect()
   print('Connected to %s MQTT Broker'%(SERVER))
   return client

def reconnect():
   print('Failed to connect to the MQTT Broker. Reconnecting...')
   sleep(5)
   machine.reset()

try:
   client = mqtt_connect()
except OSError as e:
   reconnect()

while True:
    pass
```

Upload the script to the Pico. The Node-RED flow does not require any modifications. Once the script is uploaded, press and release the button a few times. Notice how the status changes are transmitted to the Node-RED flow and appear in the dashboard almost instantaneously.

Also, notice that the interrupt service function "button_callback" is defined at the top of the script before any line that may be called. This order of definition is essential. In MicroPython, a function has to be (at least) declared before it is named for the first time.

In the future, we will use both methods to sample buttons, potentiometers, or sensors on the Pico side. However, remember that in almost every practical case, the use of interrupts and interrupt service functions is the preferred method.

2.4. LED Control

In this section, you will work on a project involving Node-RED, a Raspberry Pi Pico, and an LED. This project aims to create a Node-RED dashboard with two buttons that control the state of an LED connected to the Raspberry Pi Pico. When you click one button, the LED lights up; when you click the other button, the LED will turn off.

This project demonstrates how you can interact with physical hardware, such as a Raspberry Pi Pico and LED, using Node-RED's user-friendly dashboard.

Wiring

In the previous chapter, you wired a button to your Pico. Even though you will not use the button in this experiment, leave it on as you will use it in the next chapter. In the wiring instructions below, however, I am not mentioning the button's existence for simplicity.

You will need:

1. Raspberry Pi Pico W
2. Breadboard
3. A red LED
4. A 330 Ω resistor
5. Jumper wires

Here is how to connect the LED:

1. Insert the LED into your breadboard. Note that the longer pin of the LED is the positive side (anode), and the shorter pin is the negative side (cathode).

2. Connect one end of the resistor to the same row as the anode of the LED on your breadboard. Connect the other end of the resistor to the GP15 of the Raspberry Pi Pico.

3. Connect a jumper wire from a GND pin on the Raspberry Pi Pico to the cathode of the LED.

The circuit should look like this (reminder: ignore the button and its pull-down resistor for this project).

Figure 2.2.4.14: The LED circuit (we'll use the button in the next chapter).

You can see the wiring diagram below (made with Fritzing):

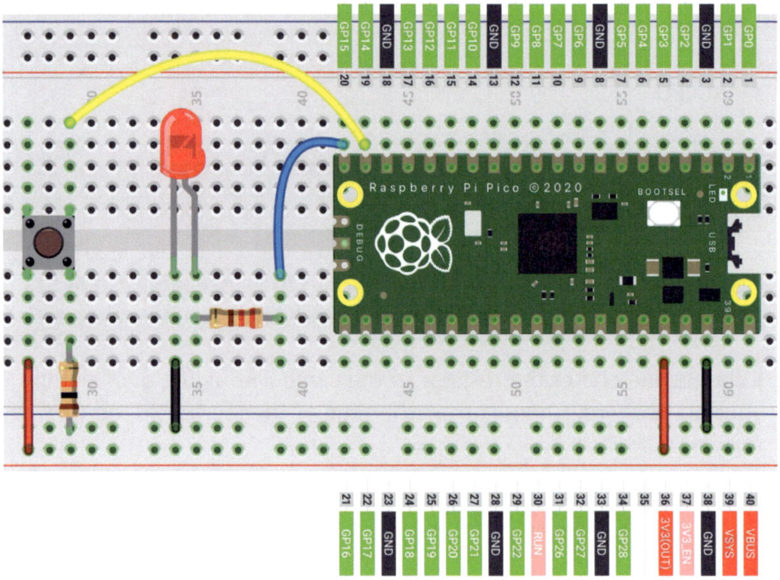

Figure 2.2.4.15: The wiring diagram for the momentary button and LED circuit.

Raspberry Pi Pico MicroPython Code

We aim to control the LED's state connected to the Pico with a Node-RED dashboard button. On the Pico, we can hold an LED connected to GP15 with this code:

```
from machine import Pin
from time import sleep

led = Pin(15, Pin.OUT)

while True:
    led.value(1) # LED ON
    sleep(1)
    led.value(0) # LED OFF
    sleep(1)
```

To hand over the LED control to the Node-RED flow, we'll add the MicroPython code related to the Wi-Fi connectivity and the MQTT topic subscription. You learn about both in the chapter "Frequency used patterns". Below, you can see the three blocks of code (LED, Wi-Fi, and MQTT) integrated into one functioning MicroPython script:

```
from machine import Pin
from time import sleep
from umqtt.simple import MQTTClient
import network

led = Pin(15, Pin.OUT)

# Configure your Wi-Fi credentials
   SSID     = '<your_wifi_network_SSID>'
   PASSWORD = '<your_wifi_network_password>'

wlan = network.WLAN(network.STA_IF)
wlan.active(True)
wlan.connect(SSID,PASSWORD)

while not wlan.isconnected():
    pass

print('Connected to Wi-Fi')
print(wlan.ifconfig())

# Setup MQTT
   SERVER      = '<your_MQTT_broker_IP_address>'
   CLIENT_ID   = b'<your_device_client_ID>'
   TOPIC       = b'led_state'
   USER        = b'<your_MQTT_username>'
   PASSWORD    = b'<your_MQTT_password>'

def mqtt_connect():
   client = MQTTClient( CLIENT_ID,
```

```
                            SERVER,
                            user=USER,
                            password=PASSWORD,
                            keepalive=3600
                           )
    client.connect()
    print('Connected to %s MQTT Broker'%(SERVER))
    return client

def reconnect():
    print('Failed to connect to the MQTT Broker. Reconnecting...')
    sleep(5)
    machine.reset()

def on_message(topic, msg):
     print("Received message:", msg.decode())
     if msg == b'on':
         led.value(1)
     elif msg == b'off':
         led.value(0)

def main(server, topic, username, passw):
    client = MQTTClient( CLIENT_ID,
                            SERVER,
                            user=USER,
                            password=PASSWORD,
                            keepalive=3600
                           )
     client.set_callback(on_message)
     client.connect()
     client.subscribe(topic)
     print("Connected to %s, subscribed to %s topic" % (server, topic))

     try:
         while True:
             client.check_msg()
             sleep(1)

     finally:
         client.disconnect()

main(SERVER, TOPIC, USER, PASSWORD)
```

In this script, the following operations are noteworthy:

1. **MQTT Connection**: It attempts to connect to an MQTT broker using the provided server (IP address), client ID, topic, username, and password. The "mqtt_connect" function is used for the initial connection and returns an MQTT client object. The reconnect function attempts a reconnection to the MQTT broker if the initial connection fails.

2. **Message Handling**: It defines a callback function on_message triggered every time a message is received on the subscribed topic. This function checks the content of the message. If the message is "on", it sets the LED to HIGH; if the message is "off", it sets the LED to LOW.

3. **Main Function**: In the "main" function, it creates an MQTT client object with the provided parameters, sets the message callback function, connects to the MQTT broker, and subscribes to the desired topic. It then enters an infinite loop, continuously checking for new messages. It also has error handling to disconnect from the broker gracefully if an error or interruption occurs.

This script is intended to run indefinitely, listening for MQTT messages and toggling the LED based on those messages until it is stopped or encounters an error.

Node-RED flow

Let's create the matching Node-RED flow. Start Node-RED, navigate your workspace and open a new flow editor. Construct the flow like this:

1. Drag and drop an 'mqtt out' node from the node palette onto your workspace.

2. Double-click on the 'mqtt out' node to configure it. Set the 'Server' to your MQTT broker and the 'Topic' to "led_state" to match the topic's name in the MicroPython script. Click 'Done' to save your settings.

3. Drag and drop two 'ui_button' nodes (found under the dashboard section) onto your workspace.

4. Double-click one of the two 'ui_button' nodes to configure it. We'll set this one up so the LED turns on when you click it.
   - Set 'Group' to your desired UI group. I have set it to "[LED control] LED control".
   - Set 'Label' to 'ON'.
   - Set 'Payload' to 'on'.
   - Set 'Name' to "LED control ON".
   - Click 'Done' to save your settings.

5. Double-click the other 'ui_button' node to configure it. We'll set this one up so the LED turns off when you click it.
   - Set 'Group' to your desired UI group. I have set it to "[LED control] LED control".

- Set 'Label' to 'OFF'.
- Set 'Payload' to 'off'.
- Set 'Name' to "LED control OFF".
- Click 'Done' to save your settings.

6. Connect both button nodes to the MQTT Out node.

Your flow should look like this:

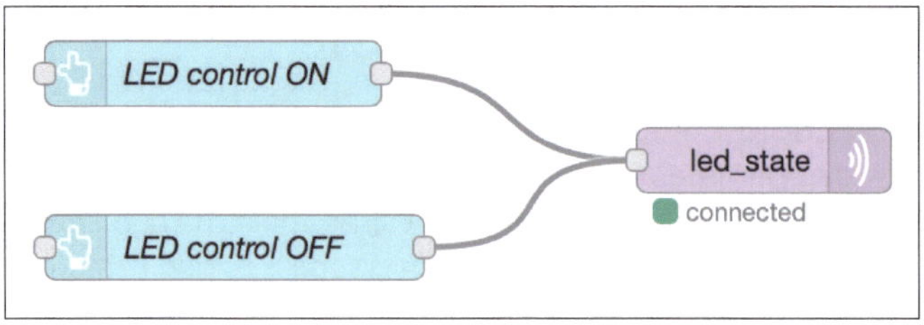

Figure 2.2.4.16: This flow creates a dashboard with two buttons.

Go ahead and deploy this flow. In the Dashboard, you will see the tab with the two LED buttons:

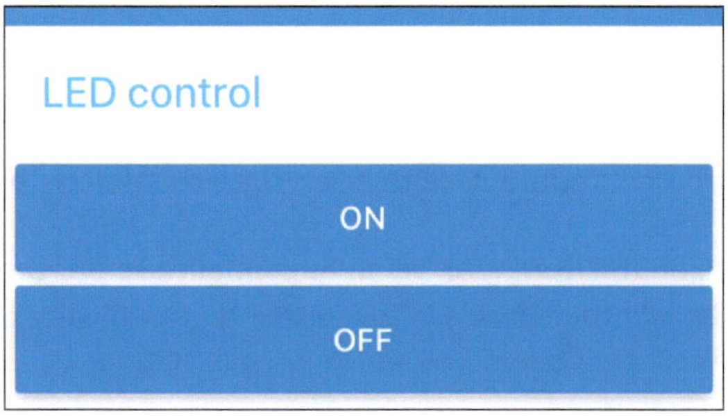

Figure 2.2.4.17: This LED control dashboard

When you click the buttons, the LED will turn on and off accordingly. Your Pico will check for new messages with the MQTT broker once per second. Therefore, there will be delays between the time you click a button and the LED's status change. This method of the microcontroller actively looking for updates from a remote system (in this case, the MQTT broker) is called "polling". For applications that require faster updates, you can reduce the check-in frequency.

However, MQTT, as implemented in both MicroPython and Node-RED, provides a way to get updates automatically without polling. Let's look at how this works next.

2.5. LED Control without Polling

It is possible to subscribe to a topic in Node-RED and have your MicroPython script be notified of topic updates without polling. The MQTT protocol, supported by Node-RED and MicroPython, provides this functionality.

To make this work, you must make a small change to the script from the previous section. The change required is in the "main" function, and you can see it below:

```
de main(server, topic, username, passw):
   client = MQTTClient( CLIENT_ID,
                        SERVER,
                        user=USER,
                        password=PASSWORD,
                        keepalive=3600
                       )
    client.set_callback(on_message)
    client.connect()
    client.subscribe(topic)
    print("Connected to %s, subscribed to %s topic" % (server, topic))

    try:
        while True:
                                client.wait_msg()
            #client.check_msg() # OK TO REMOVE
            #sleep(1)           # OK TO REMOVE

    finally:
        client.disconnect()
```

You will find the change in the "while True:" block of the section above, marked in bold. I have commented on the original content of the code. The new line of code calls the "wait_ msg" function on the client object. If a new message is available in the MQTT topic, then the callback function "on_message" will be called to get the message and do something with it.

Upload this version of the script to your Pico and press the on/off button a few times. You will notice that the LED is now much more responsive. It turns on and off almost simultaneously when you click the control buttons.

2.6. Combined

With the button and LED projects, you learned how to establish two-way communication between your Raspberry Pi Pico and your Node-RED instance. The software layer that makes this communication possible and easy is the MQTT broker. In most practical applications, you should have a two-way communication setup to work simultaneously.

In this chapter, you will learn how to combine the techniques you learned in the button and LED projects into a single project. In the end, you will have a Node-RED dashboard that shows you the state of the button and allows you to control the LED.

Wiring

You can reuse the circuit from the LED and button projects. I remind you that:

1. The LED anode is connected to Pico GP15.
2. The button sampling pin is connected to Pico GP14.

Here is my circuit on the breadboard:

Figure 2.2.6.18: This LED and button are connected to a Raspberry Pi Pico W.

Raspberry Pi Pico MicroPython code

The code for this project is a combination of the code you saw in the "solo" button and LED projects. I copy the code here and provide some highlights below the code. Please take the necessary time to review the code and ensure you understand it before continuing with the project's Node-RED part.

```
from machine import Pin
from time import sleep
from umqtt.simple import MQTTClient
import network

led = Pin(15, Pin.OUT)

# Configure your Wi-Fi credentials
   SSID     = '<your_wifi_network_SSID>'
   PASSWORD = '<your_wifi_network_password>'

wlan = network.WLAN(network.STA_IF)
wlan.active(True)
wlan.connect(SSID,PASSWORD)
```

```
while not wlan.isconnected():
    pass

print('Connected to Wi-Fi')
print(wlan.ifconfig())

# Setup MQTT
   SERVER        = '<your_MQTT_broker_IP_address>'
   CLIENT_ID     = b'<your_device_client_ID>'
TOPIC_LED     = b'led_state'
TOPIC_BUTTON  = b'button_state'
   USER          = b'<your_MQTT_username>'
   PASSWORD      = b'<your_MQTT_password>'

client        = MQTTClient( CLIENT_ID,
                            SERVER,
                            user=USER,
                            password=PASSWORD,
                            keepalive=3600
                            )
client.connect()

def button_callback(pin):
    if button.value():
        print('Button Pressed')
        client.publish(TOPIC_BUTTON, "1")
    else:
        print('Button Unpressed')
        client.publish(TOPIC_BUTTON, "0")

button = Pin(14, Pin.IN, Pin.PULL_UP)
button.irq(trigger=Pin.IRQ_FALLING | Pin.IRQ_RISING, handler=button_callback)

def led_callback(topic, msg):
    if msg == b'on':
        print('LED ON')
        led.value(1)
    elif msg == b'off':
        print('LED OFF')
        led.value(0)

client.set_callback(led_callback)
client.subscribe(TOPIC_LED)

while True:
    client.wait_msg()
```

This script will both "publish" and "subscribe" to the MQTT broker. This requires that the publisher and subscriber both have access to the MQTT "client" object. To make sharing possible in the simplest way, I have adjusted the script so that the client object is created as soon as the Pico connects to the Wi-Fi network. The client object is available through the "client" global variable, and as such, it is accessible from within the two functions: "button_callback" and "led_callback".

I also decided to use a dedicated topic for the LED and the button. The LED topic is "led_state", and the button topic is "button_state". These are the same names I used in the "solo" experiments.

Upload the new script to your Raspberry Pi Pico and continue with the Node-RED flow.

Node-RED flow(s)

In the heading for this section, I hinted that we may use one or more flows in this experiment. The two flows you created in the button and LED experiments should still work correctly with the combined MicroPython script. This is true because the new script and the existing flows use the same broker and topics for the button and the LED. Go ahead and give it a try.

First, the LED. Go to the Dashboard and look for the LED Control group and tab. Click the "ON" button and confirm that the LED on the Pico turns on.

Figure 2.2.6.19: Testing the LED control with the dashboard.

Continue with the button. Look for the Button group, then press and hold the button on the Pico and confirm that the text widget shows "Pressed" as the current state.

Figure 2.2.6.20: Testing the Button status with the dashboard.

It is probably more practical for this experiment to have all widgets in the same group so that you don't need to switch between tabs in the dashboard. Let's combine them.

Because you already have the working flows, you can easily combine them into a single flow by copying. Start by creating a new empty flow. I titled mine "LED + button".

Open the LED flow from the previous section, and using your mouse, click and select all of the nodes in the flow. You can also choose all nodes from the keyboard shortcut "Ctr-A" for Windows/Linux or "Cmd-A" for the Mac. With all nodes selected and highlighted, right-click on one to open the context menu and select "Copy selected nodes". You can also use the keyboard shortcut "Ctrl/Cmd-C" to copy.

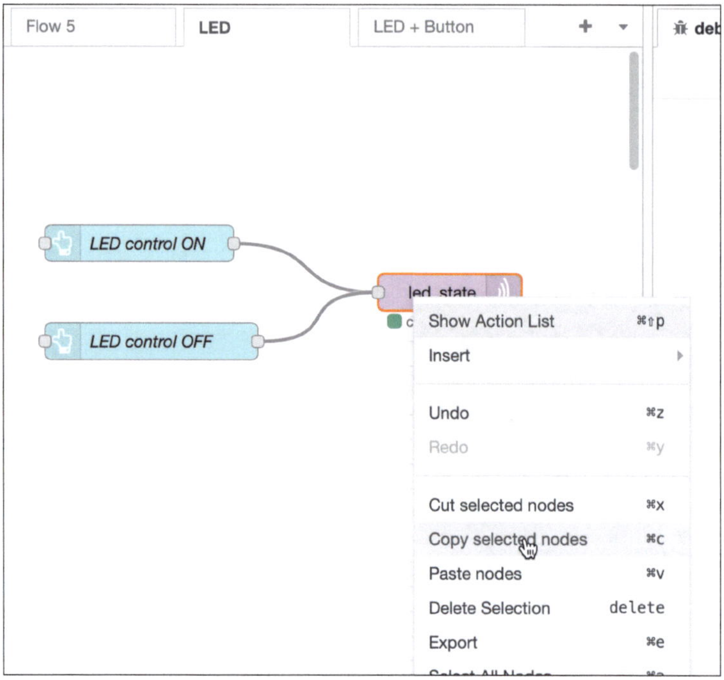

Figure 2.2.6.21: Copying the LED flow.

The LED flow is now in the clipboard, ready to be pasted to the new flow. Return to the new flow tab ("LED + button") and paste the clipboard contents. You can use the right-click context menu options "Paste nodes" or the keyboard shortcut "Ctrl/Cmd-V" (see below).

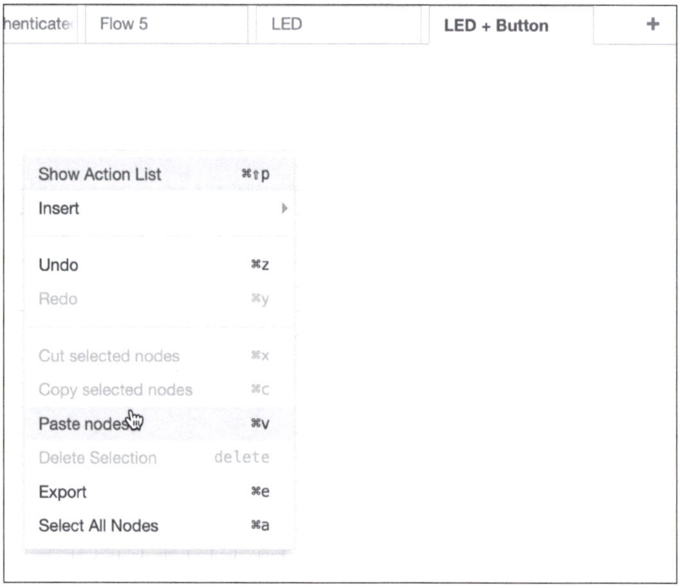

Figure 2.2.6.22: Pasting the LED flow.

The flow in the "LED + button" tab should look like this:

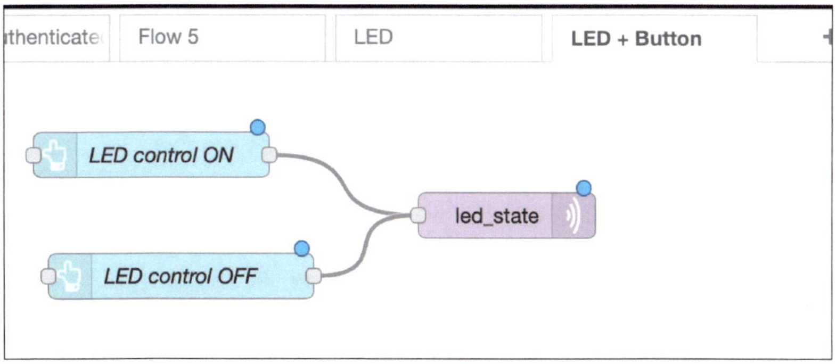

Figure 2.2.6.23: The LED + button flow with the LED button widgets.

Repeat the copy-and-paste process for the button flow and add the button functionality to the "LED + button" flow. The updated combined flow should look like this:

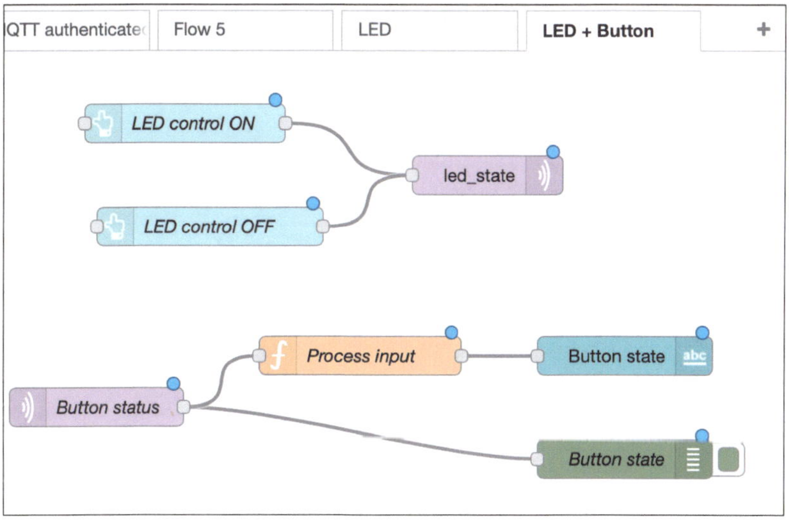

Figure 2.2.6.24: The LED + button flow, completed.

The combined flow contains all the nodes that are needed. However, you must still set up the dashboard widgets to appear in the correct group and tab. To do this:

1. Double-click on one of the widgets to open its properties.
2. Select "Add new dashboard group" from the Group dropdown.
3. Click the pencil button to open the Dashboard group edit window.
4. In the Name field, type "LED and button".

5. Select "Add new dashboard tab" from the Tab menu.

6. In the Tab properties window, type "LED + Button" in the Name field.

7. Click Add to return to the Dashboard group edit window.

8. Click Add to return to the widget node properties window.

9. Click Done to return to the flow.

The properties window for one of the button should look like this:

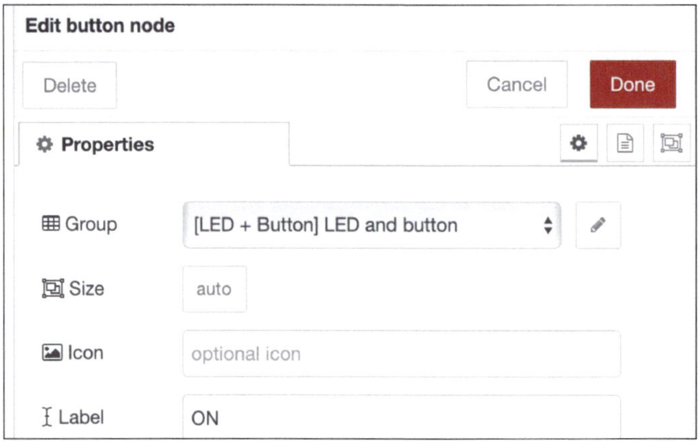

Figure 2.2.6.25: The new group and tab for the dashboard widget.

Repeat the process for the other two widgets. Since you have already done that, you will not need to create new tabs and groups for the remaining widgets. All three widgets of this dashboard should belong to the same group and tab, with the name "[LED + Button] LED and button". Go ahead and deploy the new combined flow. Browse to the Dashboard and select the "LED + Button" tab.

Your dashboard should look like this:

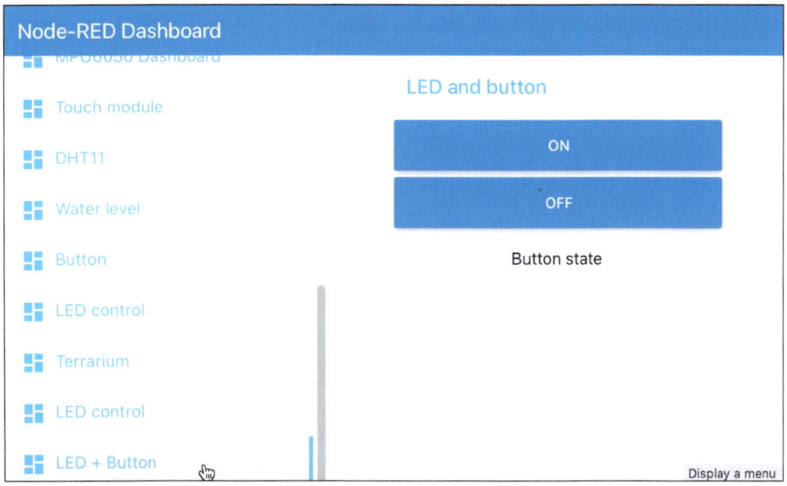

Figure 2.2.6.26: The new combined dashboard.

Click the dashboard buttons and press the physical button on the breadboard to confirm that everything works as expected (it should!).

Wrapping up

The projects in this "warming up" chapter have been relatively simple. However, they contain code and patterns that you will reuse in the following tasks. In particular, you now know how to set up bi-directional communications between a Node-RED flow and the Raspberry Pi Pico, using the MQTT broker as the middleware. You also learned how to reuse flows or part of flows into new flows using simple copy and paste.

Let's continue with more projects to learn how to use other hardware in your Node-RED projects, including Joysticks, RFID tags, and displays.

Chapter 2.3 • Inputs and Outputs

This chapter will teach you the use of various input and output devices with the help of five engaging projects.

You'll start by exploring the basics of a slide switch and a joystick. Both are simple and common components that you can use in various projects.

You'll also work with relays, enabling you to control high-voltage devices safely.

From there, the difficulty level increases with an RFID project that you can use to prototype basic "security" projects. Finally, we'll conclude this chapter with a project involving an IR receiver and transmitter, giving you the tools to set up your remote-controlled devices.

Each project in this chapter is a threefold learning experience. You'll build a Raspberry Pi Pico W circuit, develop a Node-RED flow, and write a MicroPython script. By the end of this chapter, you'll have a well-rounded skill set and the confidence to tackle more complex projects. You will use a similar pattern in the project chapters that follow.

3.1. Slide Switch

This section introduces you to slide switches and their application with the Raspberry Pi Pico. It includes a practical guide on how to wire a slide switch to the Raspberry Pi Pico and write a MicroPython script to monitor its state. Once you understand how to use a slide switch with the Pico, you will use the slide switch to control a slide switch widget on a Node-RED dashboard.

A slide switch is activated by sliding a knob or lever from one position to another. It is a digital input device that typically has two or three states, depending on the design. Slide switches are often used where a device requires a simple on-off or selector functionality, like turning a circuit on and off or selecting between two or more circuit paths. Unlike a momentary button, the selected state of a slide switch persists until the user changes the position of the knob.

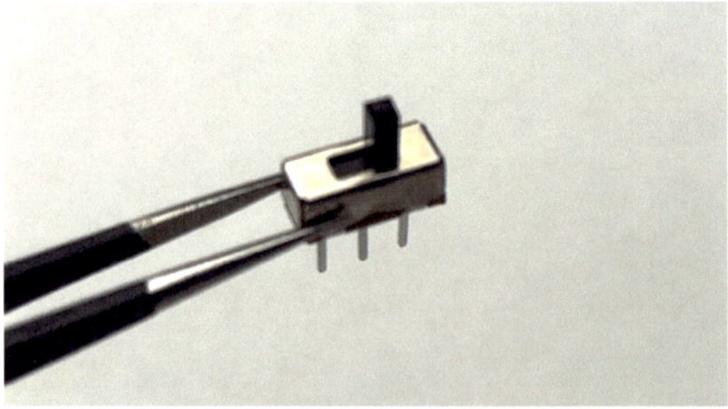

Figure 2.3.1.1: A slide switch with two states (like the one pictured above) has three pins.

Wiring

To connect a slide switch to the Raspberry Pi Pico, you will need:

- A slide switch. I am using the one in the Sunfounder kit that has three pins.
- Raspberry Pi Pico.
- A 10-kΩ resistor.
- A 104 ceramic capacitor.
- Breadboard.
- Jumper wires.

Here is the wiring diagram:

Figure 2.3.1.2: The slide switch wiring diagram

Start by inserting the slide switch into the breadboard. We will use the middle pin to read the slide switch's state. The simplest way to do the wiring (but not the best) is to connect the middle pin of the switch to a sampling pin on the Pico, then connect one of the remaining two pins to 3.3V and the other to GND. You will also need to connect the 10 KΩ pull down resistor between the middle pin of the switch and GND.

However, because the slide switch is an electro-mechanical device, it produces electrical noise when moving between the two positions. Electrical noise in switches is a phenomenon characterised by unwanted fluctuations in the electrical signal that occur during the actuation of the switch. These fluctuations manifest as transient voltage spikes, oscillations, or random electrical current or voltage variations. In electronic systems, such noise is often considered interference that can degrade the circuit's performance and lead to erroneous readings or malfunction.

The origin of electrical noise in switches can be attributed to various factors. One common source is mechanical bouncing, which occurs when the contacts of the switch make or break a connection. During this brief period, the contacts may oscillate between open and closed states multiple times before settling, generating high-frequency noise. Another noise source is the inductive and capacitive coupling from adjacent components or external electromagnetic fields, which can induce voltage spikes in the switch circuitry.

Electrical noise in switches is particularly problematic in sensitive electronic applications such as analog-to-digital converters, precision measurement systems, and communication devices. Even minor fluctuations can result in significant errors or data loss in these systems. Therefore, various techniques are employed to mitigate the impact of electrical noise. These include using debouncing circuits to filter out the noise generated by mechanical bouncing and shielding methods to protect against external electromagnetic interference.

Even though our simple Pico circuit is very tolerant to the noise that our slide switch produces, we can still be aware of the issue and make a small adjustment to the circuit to reduce this electrical noise. To deal with this noise, we will use a small ceramic capacitor (look for the marking "104" on its casing) to suppress the electrical noise and reduce the jitter.

Connect this capacitor in parallel to the resistor, with one pin connected to the middle pin of the slide switch and the other pin connected to GND. Because this capacitor is ceramic, it does not have a polarity so that you can connect it either way.

Here is my assembled circuit:

Figure 2.3.1.3: A slide switch connected to the Pico.

MicroPython script

Here's an example MicroPython script to monitor the state of a slide switch connected to GPIO pin GP14:

```
from machine import Pin

slide = Pin(14, Pin.IN, Pin.PULL_DOWN)

def check_slide():
   if slide.value():
      print('Right')
   else:
      print('Left')

def button_callback(pin):
    check_slide()

slide.irq(trigger=Pin.IRQ_FALLING | Pin.IRQ_RISING, handler=button_callback)

check_slide()

while True:
    pass
```

This script continuously reads the value of the switch and prints its state to the console. The logic is inverted because of the pull-down resistor: when the switch is open (off), the GPIO pin is pulled up to a high voltage (1), and when the switch is closed (on), it's connected to the ground and reads a low voltage (0).

To make the output of the script more intuitive, I use the words "Right" and "Left" instead of "On" and "Off". This way, when I slide the switch to the left, I see "Left" in the console.

In the second line of the script, note that I have enabled the internal pull-down resistor even though I have connected an external pull-down resistor. The internal and external pull-down resistors are redundant, and the circuit would work the same without one or the other. For example, you can remove the external resistor and change the position of the switch. As you see, it still works as expected. Alternatively, you can set up GP14 as an input without a pull-down resistor, like this: "slide = Pin(14, Pin.IN)". The switch will operate normally as the external pull-down resistor is still connected.

As you can see, this script uses a call-back function to deal with interrupt requests resulting from a change in the position of the slide switch instead of sampling. I have created the function "check_slide", which will check the state of the slide, decide if it is "left" or "right", and print it to the console. I did this so that the script can call it once (when the script starts) and every time an IRQ occurs to avoid code repetition and ensure that the switch's position is read and printed when the script starts. Without this, the script would start and not print anything to the console until the switch position changed for the first time.

If you compare this script to the one for the button in the previous chapter, you will notice it is essentially the same. A button and a slide switch are electrically identical.

Node-RED project

Now, let's add Node-RED to the mix. We'll create a Node-RED flow that reads the state of a slide switch from a Raspberry Pi Pico using MQTT and then updates a slide switch widget on the dashboard. Usually, the switch widget is used as an output to control something else. In this project, we'll use it as an input to depict the state of the hardware slider on the breadboard.

Before doing any work in Node-RED, you will need to edit the MicroPython script so that it publishes the state of the slide switch to the MQTT broker. This script is almost trivial based on your knowledge and experience from previous projects (especially the one with the button). Here is it:

```
from machine import Pin
from umqtt.simple import MQTTClient
from time import sleep
import network

slide = Pin(14, Pin.IN, Pin.PULL_DOWN)

def check_slide():
   if slide.value():
      print('Right')
      client.publish(TOPIC, "true")
   else:
      print('Left')
      client.publish(TOPIC, "false")

def button_callback(pin):
    check_slide()

slide.irq(trigger=Pin.IRQ_FALLING | Pin.IRQ_RISING, handler=button_callback)

# Configure your Wi-Fi credentials
   SSID     = '<your_wifi_network_SSID>'
   PASSWORD = '<your_wifi_network_password>'

wlan = network.WLAN(network.STA_IF)
wlan.active(True)
wlan.connect(SSID,PASSWORD)

while not wlan.isconnected():
    pass
```

```
print('Connected to Wi-Fi')
print(wlan.ifconfig())

# Configure MQTT
   SERVER        = '<your_MQTT_broker_IP_address>'
   CLIENT_ID     = b'<your_device_client_ID>'
   TOPIC         = b'slide_switch'
   USER          = b'<your_MQTT_username>'
   PASSWORD      = b'<your_MQTT_password>'

def mqtt_connect():
   client = MQTTClient(CLIENT_ID, SERVER, user=USER, password=PASSWORD,
keepalive=3600)
   client.connect()
   print('Connected to %s MQTT Broker'%(SERVER))
   return client

def reconnect():
   print('Failed to connect to the MQTT Broker. Reconnecting...')
   sleep(5)
   machine.reset()

try:
   client = mqtt_connect()
except OSError as e:
   reconnect()

check_slide()

while True:
    pass
```

In the "check_slide()" function, the script publishes values "true" and "false" to the MQTT topic "slide_switch". These values are essential because the switch widget in Node-RED can understand them. When "true" is passed to the switch widget as input, the widget's virtual handle will move to the left, and the switch will appear off. The switch will appear when "false" is passed, and its handle will move to the right.

Let's continue with the Node-RED flow. Create a new flow and drop two nodes: an MQTT-in node and a switch node. The flow will look like this:

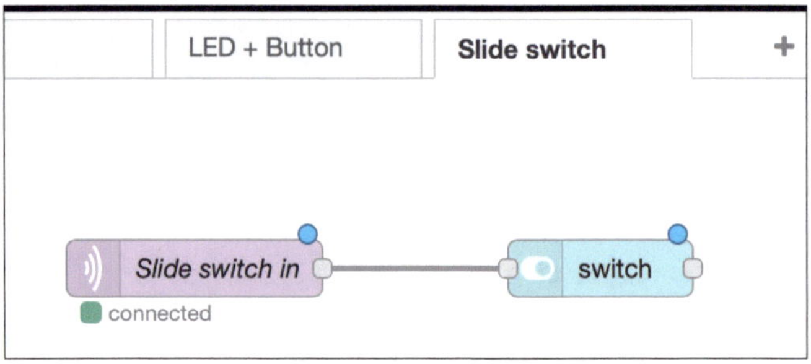

Figure 2.3.1.4: The slide switch flow.

Configure the MQTT-in node like this:

- Server: your MQTT server's IP address and port number.
- Action: Subscribe to a single topic.
- Topic: slide_switch.
- Name: Slide switch in

Configure the switch node like this:

- Group: Create a new group and tab. I used "[Slide switch Pico] Slide switch Pico"
- Label: switch
- Leave everything else at default.

Deploy the new flow and go to the dashboard. The project dashboard should look like this:

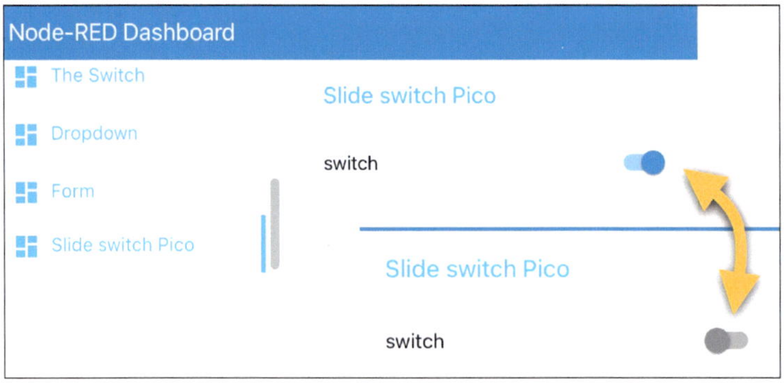

Figure 2.3.1.5: The switch dashboard.

As you change the position of the slide switch on the breadboard, you will notice that the switch widget in the dashboard will reflect its actual position.

3.2. Joystick

A joystick module is a versatile input device commonly used in interactive projects. It can provide two analog readings representing the X and Y axes, and it typically includes a button activated when you press down on the joystick itself.

In this chapter, you will learn to connect a joystick module to the Raspberry Pi Pico and read the X and Y coordinates and the state of the button using MicroPython.

The joystick module outputs three values: the X-axis, Y-axis and the button state. Because of this, the joystick module allows us to learn how to send more than one value to Node-RED using a single MQTT topic.

Figure 2.3.2.6: The joystick module.

Let's begin by setting up the circuit on the Raspberry Pi Pico side.

Wiring

The joystick module combines two analog potentiometers with a digital button. When you press the joystick knob down, you will hear a click, indicating that the button is pressed. Two analog pins output the x-axis and y-axis position information and one pin for the button. There are also VCC and GND pins.

Be careful! Many joystick modules are on the market, and their pin configurations may differ from what you will see in the wiring diagram below. In this diagram, the Vcc and GND pins are on opposite sides of the joystick header. The x-axis and y-axis pins are marked "VRx" and "VRy", and the button pin is marked as "SW". Depending on the manufacturer of the module, you might see other labelling. My actual hardware button has a different pin configuration (see photo below). As long as you identify the pin roles for your joystick, the MicroPython program below will work normally.

Follow these instructions to wire the joystick:

1. Connect the 3.3V pin of the Raspberry Pi Pico to the VCC pin of the joystick module using a jumper wire. I used the 3.3V red power rail of the breadboard for this.

2. Connect the GND pin of the Raspberry Pi Pico to the GND pin of the joystick module using another jumper wire. I used the GND black power rail of the breadboard for this.

3. Connect the GP26 pin of Raspberry Pi Pico to the joystick module's x-axis ("VRx") pin.

4. Connect the GP27 pin of Raspberry Pi Pico to the joystick module's y-axis ("VRy") pin.

5. Connect the GP17 pin of the Raspberry Pi Pico to the SW pin of the joystick module.

We will set up GP17 to use an internal pull-down resistor, so there is no need for an external pull-down resistor. Here is the wiring diagram:

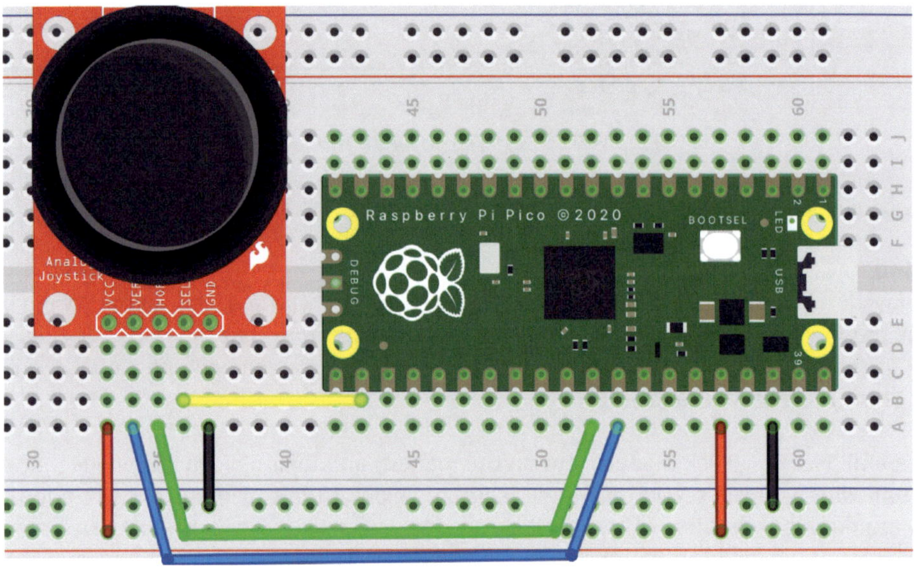

Figure 2.3.2.7: The joystick wiring diagram.

This is my assembled circuit on the breadboard. Notice how the pin layout of the joystick happens to be different from the reference wiring diagram. I have placed the joystick module disconnected to make it easier to see the pins.

Figure 2.3.2.8: The joystick on my breadboard. I placed it isconnected to make it easier to see its pin layout.

MicroPython script

Let's write a simple MicroPython script to read the values from the joystick module and print them to the serial monitor. Here is the script:

```
from machine import ADC, Pin
import utime

# Initialize ADC for the joystick's VRx and VRy pins
adc_x = ADC(26)
adc_y = ADC(27)

# Initialize Pin for the joystick's SW pin (button)
button = Pin(17, Pin.IN, Pin.PULL_UP)

while True:
    # Read X and Y coordinates
    x = adc_x.read_u16()
    y = adc_y.read_u16()

    # Read button state
    btn_state = button.value()

    print('X:', x, 'Y:', y, 'Button:', 'Pressed' if btn_state == 0 else 'Not
```

```
pressed')

    # Delay to avoid flooding console
    utime.sleep(0.1)
```

Because the joystick requires two analog input pins to read the x-axis and y-axis position, I have connected the VRx and VRy pins to GP26 and GP27. These pins are multifunctional and operate as ADC0 and ADC1, respectively.

In the script, we first create ADC objects for the joystick X and Y pins and a Pin object for the button. We set the button to be an input with an internal pull-down resistor. In the loop, we read the values of the X and Y axes using the "read_u16 method", which returns a 16-bit integer. The button value is read with the value method, which returns 0 when the button is pressed (because it's connected to the ground) and 1 when it's not. Finally, we print the values to the serial monitor.

Notice that in the script, variables "adc_x" and "adc_y" are initialised using the "ACD()" function. You can learn more about this function in the MicroPython documentation (see https://docs.micropython.org/en/latest/esp8266/tutorial/adc.html).

In the while loop, the script reads the values from the joystick module and prints them to the serial monitor, one at a time.

Copy the script to a new Thonny tab and upload it to your Pico. Move the joystick in different directions and observe the changes in the x-axis- and y-axis changes. Press the button and notice the change in the button value. The output will look like this:

```
X: 33448 Y: 33112 Button: Not pressed
X: 33304 Y: 32455 Button: Not pressed
X: 33480 Y: 32615 Button: Not pressed
X: 34152 Y: 33240 Button: Not pressed
X: 33112 Y: 32279 Button: Not pressed
X: 34008 Y: 33240 Button: Not pressed
```

The MicroPython script can read the joystick values and print them to the monitor, one at a time. We can use the knowledge from the previous projects and send these three values to a Node-RED flow, one at a time, using three MQTT topics. However, there is a better way, and this joystick allows us to learn it: we'll bundle the three values into a single message so we can send it to Node-RED using a single topic.

MicroPython script with JSON

There are many ways to go about bundling data, but I have chosen to work with JSON for this experiment. JSON (JavaScript Object Notation) is a lightweight data-interchange format that is easy for humans to read and write and for machines to parse and generate.

JSON is a text format that is entirely language-independent but uses conventions familiar to programmers of the C family of languages, including C, C++, C#, Java, JavaScript, Perl, Python, and others. These properties make JSON an ideal data-interchange language.

A basic JSON structure can either be a collection of key-value pairs, known as an 'object' (not to be confused with a class object from object-oriented programming), or an ordered list of values, known as an 'array'.

Here is an example of JSON data:

```
{
  "name": "John Doe",
  "age": 30,
  "isStudent": false,
  "courses": ["math", "english", "programming"],
  "address": {
    "street": "123 Main St",
    "city": "Anytown",
    "country": "USA"
  }
}
```

In this example:

- "name": "John Doe" is a key-value pair.
- "courses": ["math", "english", "programming"] is a key-array pair.
- "address": { ... } is a key-object pair, and the object contains further key-value pairs.

This simple and intuitive format makes JSON helpful in storing and exchanging data in various applications. We can use JSON in our joystick script with the help of the MicroPython "json" module.

Here is the updated script (ignoring all communication requirements for the moment):

```
from machine import ADC, Pin
import utime
import json

# Initialize ADC for the joystick's VRx and VRy pins
adc_x = ADC(26)
adc_y = ADC(27)

# Initialize Pin for the joystick's SW pin (button)
button = Pin(17, Pin.IN, Pin.PULL_UP)

while True:
```

```
    # Read X and Y coordinates
    x = adc_x.read_u16()
    y = adc_y.read_u16()

    # Read button state
    btn_state = button.value()

    # Create a dictionary to hold the data
    data = {
        'x_value': x,
        'y_value': y,
        'button_value': 'Pressed' if btn_state == 0 else 'Not Pressed'
    }

    print('X:', x, 'Y:', y, 'Button:', 'Pressed' if btn_state == 0 else 'Not
pressed')

    # Convert the dictionary to a JSON string
    data_json = json.dumps(data)

    print(data_json)

    # Delay to avoid flooding console
    utime.sleep(0.1)
```

Upload this script to your Pico and examine the output in the serial monitor. You will see something like this:

```
X: 33208 Y: 32359 Button: Not pressed
{"button_value": "Not Pressed", "y_value": 32359, "x_value": 33208}
X: 34024 Y: 33240 Button: Not pressed
{"button_value": "Not Pressed", "y_value": 33240, "x_value": 34024}
X: 34024 Y: 33208 Button: Not pressed
{"button_value": "Not Pressed", "y_value": 33208, "x_value": 34024}
```

The text within the curly brackets is the JSON message that the script will send to the MQTT topic. The Node-RED flow can extract the x, y, and button information from this single message.

Let's finalise the MicroPython script to combine the JSON code you see in the script above with the communications code from the previous projects.

MicroPython script with JSON and communications

Here is the completed script:

```
from machine import ADC, Pin
import utime
import json
from umqtt.simple import MQTTClient
import network

# Initialize ADC for the joystick's VRx and VRy pins
adc_x = ADC(26)
adc_y = ADC(27)

# Initialize Pin for the joystick's SW pin (button)
button = Pin(17, Pin.IN, Pin.PULL_UP)

# Configure your Wi-Fi credentials
   SSID     = '<your_wifi_network_SSID>'
   PASSWORD = '<your_wifi_network_password>'

wlan = network.WLAN(network.STA_IF)
wlan.active(True)
wlan.connect(SSID,PASSWORD)

while not wlan.isconnected():
    pass

print('Connected to Wi-Fi')
print(wlan.ifconfig())

# Configure MQTT
   SERVER          = '<your_MQTT_broker_IP_address>'
   CLIENT_ID       = b'<your_device_client_ID>'
   TOPIC           = b'joystick'
   USER            = b'<your_MQTT_username>'
   PASSWORD        = b'<your_MQTT_password>'

def mqtt_connect():
   client = MQTTClient(CLIENT_ID, SERVER, user=USER, password=PASSWORD,
keepalive=3600)
   client.connect()
   print('Connected to %s MQTT Broker'%(SERVER))
   return client

def reconnect():
   print('Failed to connect to the MQTT Broker. Reconnecting...')
```

```
    sleep(5)
    machine.reset()

try:
    client = mqtt_connect()
except OSError as e:
    reconnect()

while True:
    # Read X and Y coordinates
    x = adc_x.read_u16()
    y = adc_y.read_u16()

    # Read button state
    btn_state = button.value()

    # Create a dictionary to hold the data
    data = {
        'x_value': x,
        'y_value': y,
        'button_value': 'Pressed' if btn_state == 0 else 'Not Pressed'
    }

    print('X:', x, 'Y:', y, 'Button:', 'Pressed' if btn_state == 0 else 'Not
pressed')

    # Convert the dictionary to a JSON string
    data_json = json.dumps(data)

    print(data_json)
    client.publish(TOPIC, data_json)

    # Delay to avoid flooding console
    utime.sleep(1)
```

This script operates based on your existing knowledge. I should point out a couple of changes I have made compared to previous versions of the joystick script:

1. I have increased the sleep time to one second to prevent flooding the MQTT broker with messages.

2. This script is using the polling method. It could use the interrupt method instead. However, you must be mindful of creating three different interrupts (one per analog input and one for the button). If you are up to a challenge, give it a try!

Upload this script to your Pico, and let's continue with the Node-RED flow.

Node-RED flow

On the Node-RED side, we'll create a flow that can receive the JSON message, extract the x-axis, y-axis and button values from it, and display those values in the dashboard using two slider widgets (for the x-axis and y-axis values) and a text widget for the button.

Starting backward from the dashboard, I'd like to create this:

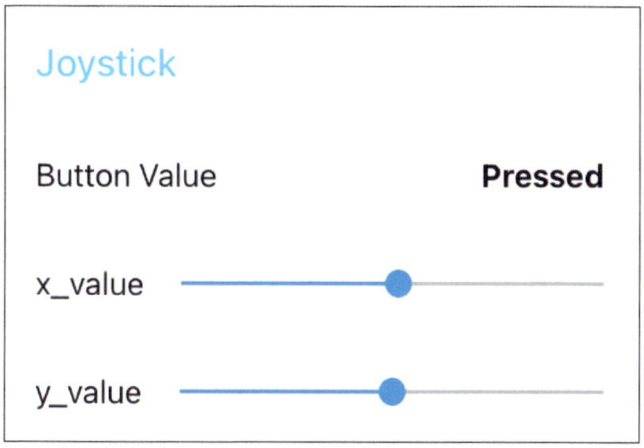

Figure 2.3.2.9: The dashboard for the Joystick. As I move the joystick, the dot on the sliders moves accordingly.

This flow will consist of these nodes:

- An "mqtt in" node to subscribe to the "pico/joystick" MQTT topic.
- A "json" node to convert the JSON string from the MQTT message payload into a JavaScript object.
- A "function" node to break down the values in the JSON object into individual components and pass them on to the dashboard widget nodes.
- A "ui_text" node to display the button_value on the Node-RED Dashboard.
- Two "ui_slider" nodes to display the x_value and y_value on the Node-RED Dashboard.
- Debug nodes to show values in the debug tab.

Here is the joystick flow:

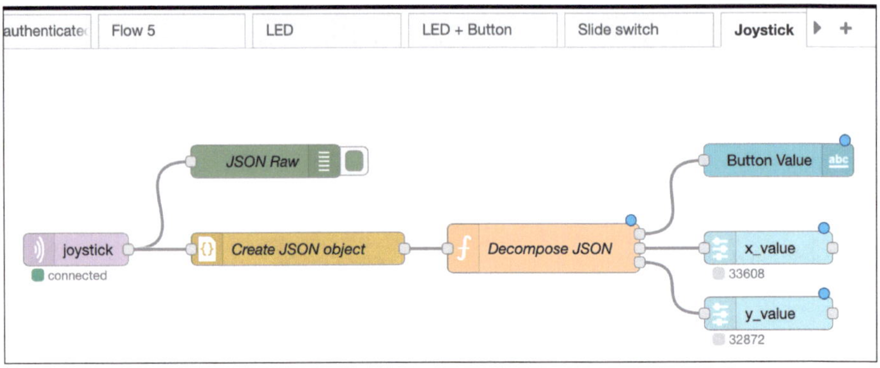

Figure 2.3.2.10: The joystick flow.

This is the first time we encounter the JSON node in this book, so let's take a moment to learn more about it. The JSON node in Node-RED converts between JSON and JavaScript objects. JSON, or JavaScript Object Notation, is a widely used, lightweight data-interchange format that is easy for humans to read and write and for machines to parse and generate.

The JSON node in Node-RED has two primary purposes:

- Stringify JavaScript objects: The JSON node can convert a JavaScript object into a JSON string. This is useful when sending data over a network or storing it in a file, as JSON is a widely recognised format that many programming languages can interpret. See below for a detailed explanation of the term "stringify".

- Parse JSON strings: The JSON node can also convert a JSON string into a JavaScript object. This is useful when receiving data in JSON format, for example, from an API or MQTT message, and you want to work with that data within your Node-RED flow.

These two actions can be selected in the JSON node properties under the "Action" field: "Convert between JSON String & Object" will do either depending on the input, "Always convert to JSON String" will always stringify, and "Always convert to JavaScript Object" will always parse. In this project, we'll use the JSON node to parse a JSON string (as it is published to the MQTT topic by the MicroPython script) so that the function node can process the JSON object.

Here is the configuration of the JSON node in this flow:

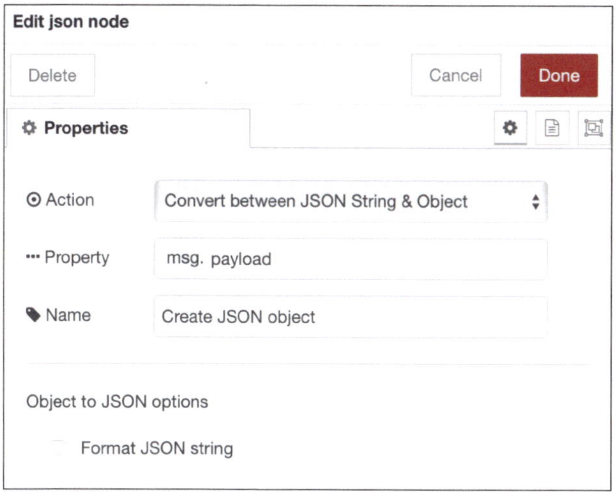

Figure 2.3.2.11: The configuration of the JSON node.

This node will output a JSON object that the Javascript in the function node can understand and work with. In the function node (named "Decompose JSON"), copy this Javascript:

```
var button  = { payload: msg.payload.button_value };
var x_value = { payload: parseInt(msg.payload.x_value) };
var y_value = { payload: parseInt(msg.payload.y_value) };
return [button, x_value, y_value];
```

This Javascript is simple. It looks in the payload (a JSON object received from the JSON node) and takes each of the three values within it. The x_value and y_value are strings that represent integers, so this Javascript uses the "parseInt" function to convert the strings into integers. This is important because the following slider nodes require integers, not strings.

Also, notice that the JavaScript has a "return" instruction in the last line that returns an array, not a single value. This array contains three values. Therefore, this function node requires three outputs. You can see the three output boxes in the flow screenshot. To configure the function node with three outputs, click on the Setup tab and change the output counter to "3", as in the screenshot below:

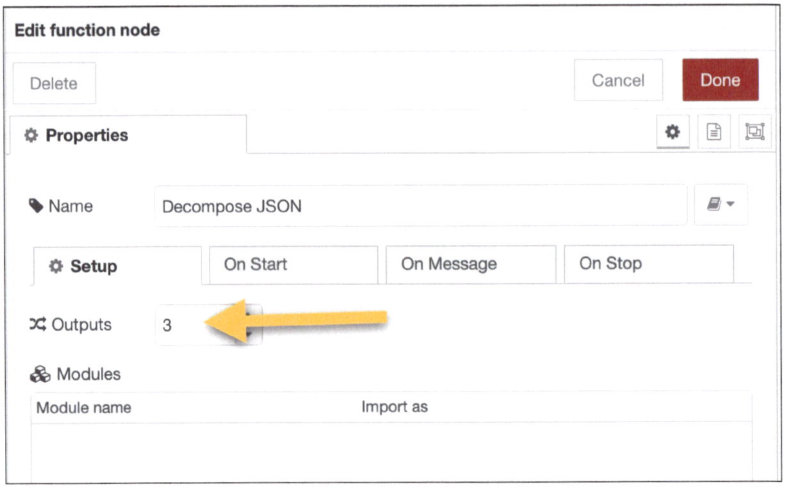

Figure 2.3.2.12: This function node has three outputs.

From top to bottom, the three outputs will export the values in the return statement array from left to right. Therefore, the "button" value is exported from the top output, the "x_value" from the middle output, and the "y_value" from the bottom output.

Deploy this flow and play with the joystick and the button. Remember that values are updated once per second because of the delay in the MicroPython script. See how the slider dots move as per the position of the joystick.

This experiment showed you how to bundle and unbundle values in JSON messages. In the future, you will use this technique extensively in your MicroPython and Node-RED projects.

What is "stringify"?

The term "stringify" in JavaScript refers to converting a JavaScript object into a JSON-formatted string. This is commonly done using the JSON.stringify() method. The primary purpose of stringifying an object is to make it easily transferable over a network or stored in a text-based format like a file or a database. Once the data reaches its destination or is retrieved from storage, it can be parsed into a JavaScript object using JSON.parse().

Let's look at an example. Consider a JavaScript object that represents a person:

```
const person = {
  name: "John",
  age: 30,
  email: "john@example.com"
};
```

To stringify this object, you would use the JSON.stringify() method as follows:

```
const stringifiedPerson = JSON.stringify(person);
```

After this operation, stringifiedPerson will hold the following JSON-formatted string:

```
"{\"name\":\"John\",\"age\":30,\"email\":\"john@example.com\"}"
```

This stringified version can now be sent over a network or saved to a file. To revert it back to a JavaScript object, you would use JSON.parse():

```
const parsedPerson = JSON.parse(stringifiedPerson);
```

Now, parsedPerson will be a JavaScript object identical to the original person object.

3.3. Relay

In the warm-up chapter, you learned how to control an LED using two buttons in a Node-RED dashboard. In this project, you will use this knowledge to control a 5 V relay. With this relay, you can turn on and off larger loads, such as desk lights and fans. Programmatically, a relay works the same as an LED.

First, let's take a moment to understand what a relay is and how it works. After this, you can use your knowledge from the chapter on LEDs to learn how to control an LED programmatically. Finally, you will work on a simple Node-RED flow that can control an LED connected to your Raspberry Pi Pico.

Figure 2.3.3.13: A typical 5V relay, top and bottom.

Understanding relays

A relay is an electromechanical switch that allows control over a high-power device using a lower-power signal. It comprises an electromagnet, a set of output terminals, and a flexible mechanical part that moves when the electromagnet is energised. When a voltage is applied to the electromagnet, it attracts the flexible mechanical part, causing it to make or break a circuit connected to the output terminals.

Operating Principles of a Relay

Relays are primarily used to isolate circuits. The relay's input (coil) and output (contact) sides are electrically isolated. This allows a low-voltage circuit (e.g., a 5 V circuit from a Raspberry Pi Pico W) to control a higher voltage or current circuit.

When voltage is applied to the coil, a magnetic field is created, which pulls the relay's switch closed, allowing current to flow through the output side of the relay. When the voltage is removed, the magnetic field collapses, the switch opens, and the current flow on the output side stops.

Comparison of a 5 V relay and an LED

You have already learned to control an LED with your Raspberry Pi Pico and turn it on and off. You can use this knowledge to control a relay so that it activates and deactivates the high-voltage circuit.

Both 5V relays and LEDs require a current to function and can be controlled using a Raspberry Pi Pico W. They can act as output devices in a circuit, reacting to signals from the Pico. In both cases, you must ensure the current and voltage requirements are met to operate them safely and effectively.

When programming for control, relays and LEDs can both be switched on and off using digital write commands. However, LEDs can also use PWM (Pulse Width Modulation) signals to control the brightness, which isn't applicable in the case of relays.

Of course, relays and LEDs differ significantly in their purpose and operation. An LED is a semiconductor device that emits light when an electric current passes through it. A relay is a switch that controls the flow of electricity in a circuit using an electromagnet that controls a mechanical switch.

Wiring

In this experiment, we will use the 5 V relay that comes with the Sunfounder Kepler kit. We will use an LED with its 220 Ω current limiting resistor for the load. We will use a lithium-ion rechargeable battery with its circuit charger to provide power for the relay high-voltage circuit. Both of those components are also part of the Kepler kit.

We must also use a transistor, a resistor, and a diode to control the relay. Before looking at the actual connections of these components, let's understand why we must use these three components to safely "drive" (or "control") a relay.

Transistor

A transistor is a fundamental component in electronics. It is a three-terminal device usually made from a semiconductor material. The three terminals are known as the emitter, base, and collector for a bipolar junction transistor, which includes NPN and PNP types.

The transistor's primary role is to regulate current or voltage flow and act as a switch or an amplifier in a circuit. The base current controls the larger current flowing from the collector to the emitter.

Transistors are frequently used as electronic switches in microcontroller circuits. A small base current, controlled by the microcontroller, can turn the transistor on or off, similar to how a switch operates. This property is beneficial for controlling loads requiring higher current or voltage than a microcontroller can provide, such as motors, relays, or high-power LEDs. This is how we will use the transistor in this project.

As an amplifier, a transistor can apply a small signal to the base and produce a larger replica of this signal at the collector-emitter terminals. This is useful in various applications, including signal processing and audio amplification.

Transistors can also be used for level shifting. This is necessary when interfacing components that operate at different voltage levels. For instance, a 3.3V microcontroller output can be level-shifted to a 5 V signal using a transistor, suitable for driving a 5 V device.

You probably already know that transistors are the building blocks of digital logic gates, which form the basis of digital circuits and microcontrollers. By combining transistors in various configurations, it is possible to create AND, OR, NOT, NAND, NOR, and XOR gates, which can be used to implement complex digital logic functions.

In this project, we will use an NPN transistor (S8050) to operate as a switch and an amplifier. A small current or voltage applied at the base terminal controls a larger current that flows from the collector to the emitter. When you connect a relay directly to a microcontroller, you risk drawing too much current from its digital output pin, potentially damaging it.

The S8050 transistor mitigates this risk, allowing the relay's coil current to be sourced from a separate power supply, not directly from the microcontroller. The microcontroller only needs to supply a small base current to the transistor within its output capacity.

Resistor

The 1 kΩ (1 kilo-ohm; 1000 ohms) resistor is placed between the microcontroller's digital output pin and the transistor's base terminal. Its purpose is to limit the current entering the base terminal of the transistor. Without the resistor, excessive current from the microcontroller could flow into the base terminal when the output pin is high, potentially damaging the transistor.

Diode

The diode is connected across the relay's coil terminals in reverse bias. This diode is known as a "flyback" or "freewheeling" diode. When the relay de-energises, the diode protects the circuit from a reverse voltage spike (or back EMF) generated by the collapsing magnetic field around the coil. The reverse voltage could damage the transistor or the microcontroller without this diode.

Wiring instructions

Let's wire the components of this project on the breadboard. Follow these instructions and refer to the wiring diagram and the photograph below.

1. Connect the emitter of the NPN transistor (S8050) to the ground (GND) of the separate power supply for the relay. You can locate the emitter by holding the transistor with the flat side towards you. The emitter is the left-most pin.

2. Connect the collector of the transistor to one end of the relay coil.

3. Connect the other end of the relay coil to the power supply's positive terminal (VCC). Because we use a 5 V relay, we must provide 5 V power to energise its coil. On the Raspberry Pi Pico, 5V is available from the VBUS pin, pin 40 (the first pin to the right side of the USB connector).

4. Connect the diode's anode to the transistor's collector. You can determine the anode and cathode of a diode by looking for the white stripe. The end with the white stripe is the cathode, and the other side is the anode.

5. Connect the cathode of the diode to the positive terminal of the power supply (VCC from VBUS).

6. Connect one end of the 1KΩ resistor to the base of the transistor.

7. Connect the other end of the 1 kΩ resistor to the microcontroller›s digital output pin that you will use to control the relay. We will use GP15 in this project.

8. Connect one pin of the 220 Ω resistor to one of the two output pins of the relay and the other to the anode of the LED (the anode is the long pin).

9. Connect the cathode of the LED to the negative pin of the battery or the battery module).

10. Connect the positive pin of the battery or the battery module to the switch's common ("COM") terminal. Two pins are common on the right side of the relay as you hold the device with the writing facing you. A circle marking on the top side of the relay package indicates the side of the relay with the common pins.

Note that the 5 V relay included in the Sunfounder Kepler kit has a coil that can be energised with the 3.3 V supply available on pin 36 of the Raspberry Pi Pico. In my testing, I operated the relay using either the 5 V supply from pin 40 or the 3.3 V supply from pin 36.

I realise that the wiring is more complex when compared to the previous projects, so take the time you need to complete this task correctly. Again, refer to the wiring diagram and photograph below to help you with this task.

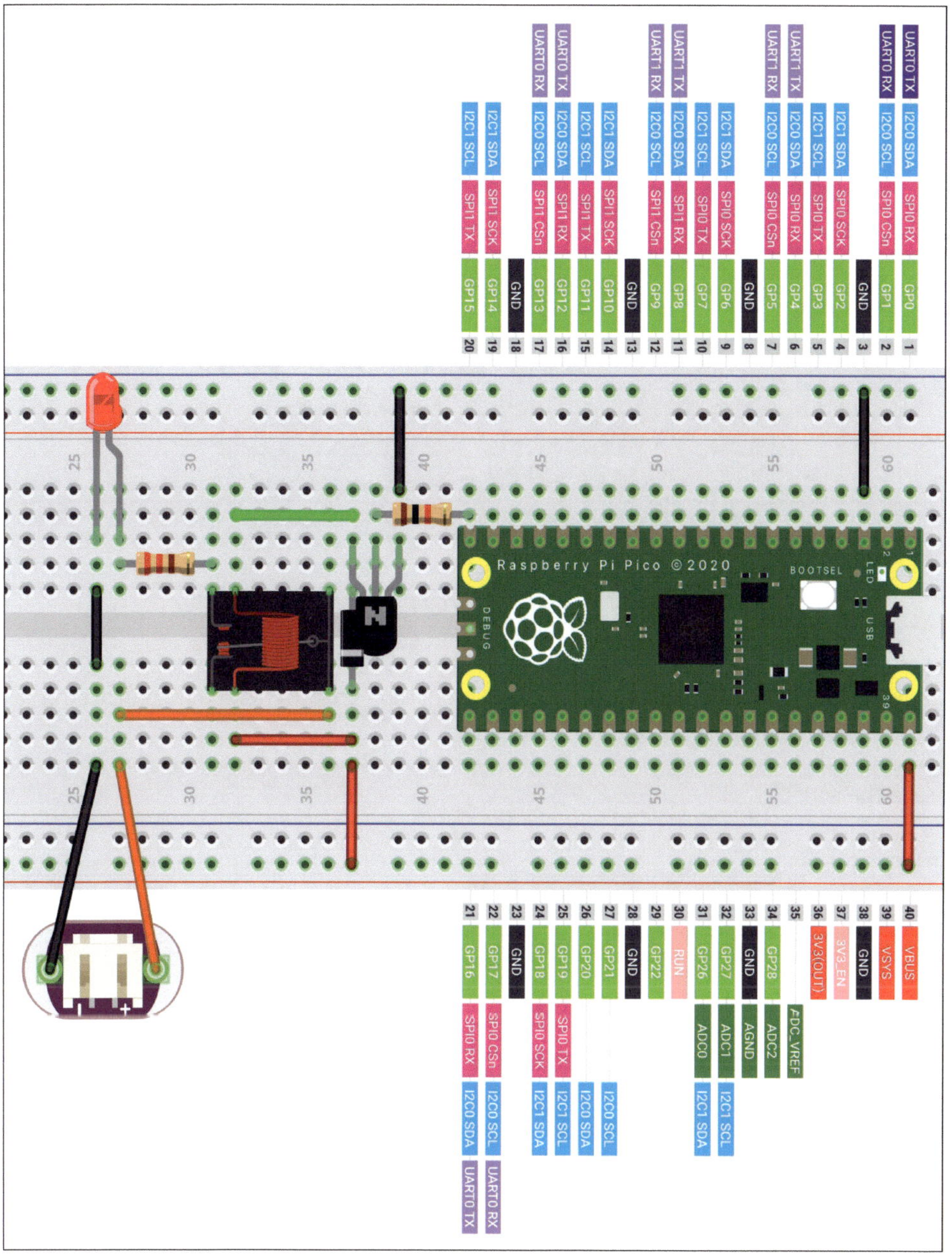

Figure 2.3.3.14: The wiring diagram for the relay project.

Here is the circuit on my breadboard.

Figure 2.3.3.15: My relay circuit on the breadboard.

MicroPython script

With the circuit completed, let's look at the MicroPython script. This script controls the relay state by toggling GP15 to HIGH and LOW.

Here is the script:

```
import machine
import utime

relay = machine.Pin(15, machine.Pin.OUT)

while True:
    relay.value(1)
    utime.sleep(2)
    relay.value(0)
    utime.sleep(2)
```

As you can see, this script is as simple as MicroPython scripts go. It defines the "relay" variable as a proxy for GP15 and defines its type as an output. In the infinite "while" loop, GP15 is turned on for two seconds and off for two seconds.

Upload this script to the Pico and see how the relay turns the LED on and off every two seconds. Each time the relay changes its state, it will make a "click" sound. Because the relay is an electromechanical device, it is very slow compared to a transistor. Do not cycle between on/off states too quickly. A rate of one or two changes per second is comfortable for a relay like the one you use in this project.

Node-RED project

Let's set up a simple Node-RED flow that will allow you to control the load on the relay by using a slide switch in the Node-RED dashboard. The user interface will look like this:

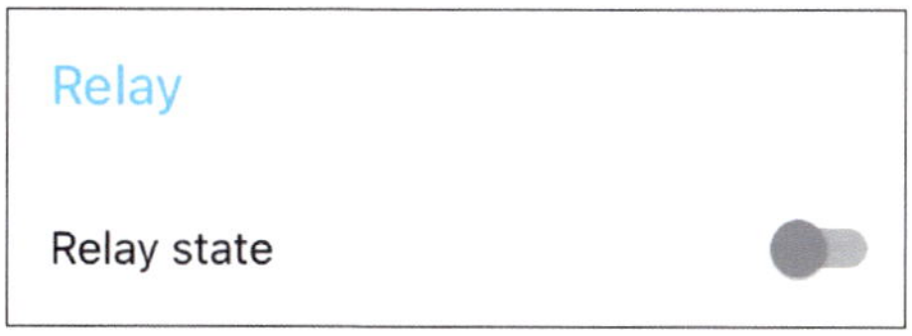

Figure 2.3.3.16: We'll control the relay with this slide switch.

When you click the slide switch to turn it on, the relay will turn its load on, and vice versa.

Here is the Node-RED flow in the editor:

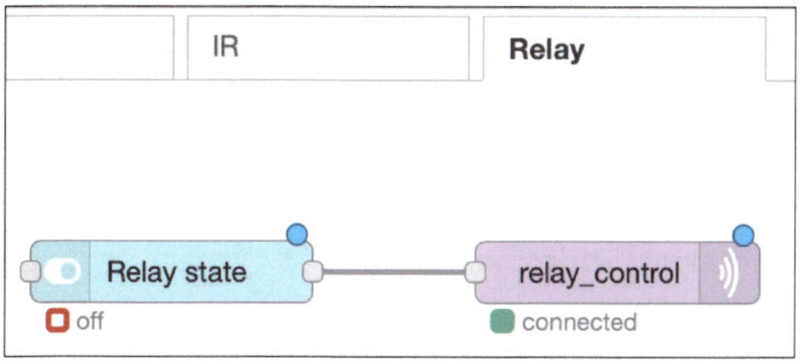

Figure 2.3.3.17: The Node-RED flow that controls the relay.

For the slide switch node, set its off and on payload properties like this:

- On Payload: "on" (string).
- Off Payload: "off" (string).

For the MQTT-out node, set the MQTT topic to "relay_control". Go ahead and deploy the flow. It will not work (yet) because we must first add the MQTT and control code to the MicroPython script.

Node-RED MicroPython script

Let's look at the MicroPython script that will subscribe to the MQTT topic "relay_control" and set the relay control pin of the Raspberry Pico W to the appropriate state, depending on the value of the message. This script borrows heavily from the script "led_mqtt_mo_polling.py" you saw in the LED control project in the "warm up" chapter. The only difference is that the MQTT topic in the script version below is "relay_control". You can see the entire script below:

```
from machine import Pin
from time import sleep
from umqtt.simple import MQTTClient
import network
```

```
#from random import *

led = Pin(15, Pin.OUT)

# Configure your Wi-Fi credentials
   SSID     = '<your_wifi_network_SSID>'
   PASSWORD = '<your_wifi_network_password>'

wlan = network.WLAN(network.STA_IF)
wlan.active(True)
wlan.connect(SSID,PASSWORD)

while not wlan.isconnected():
    pass

print('Connected to Wi-Fi')
print(wlan.ifconfig())

# Setup MQTT
   SERVER        = '<your_MQTT_broker_IP_address>'
   CLIENT_ID     = b'<your_device_client_ID>'
TOPIC         = b'relay_control'
   USER          = b'<your_MQTT_username>'
   PASSWORD      = b'<your_MQTT_password>'

def mqtt_connect():
   client = MQTTClient(CLIENT_ID, SERVER, user=USER, password=PASSWORD,
keepalive=3600)
   client.connect()
   print('Connected to %s MQTT Broker'%(SERVER))
   return client

def reconnect():
   print('Failed to connect to the MQTT Broker. Reconnecting...')
   sleep(5)
   machine.reset()

def on_message(topic, msg):
    print("Received message:", msg.decode())
    if msg == b'on':
        led.value(1)
    elif msg == b'off':
        led.value(0)


def main(server, topic, username, passw):
```

```
    client = MQTTClient(CLIENT_ID, server, user=username, password=passw,
keepalive=3600)
    client.set_callback(on_message)
    client.connect()
    client.subscribe(topic)
    print("Connected to %s, subscribed to %s topic" % (server, topic))

    try:
        while True:
            client.wait_msg()
    finally:
        client.disconnect()

main(SERVER, TOPIC, USER, PASSWORD)
```

Set the Wi-Fi and MQTT credentials and IP address to match your setup and upload the script to your Pico. Go to the Relay dashboard and toggle the slide switch on and off. The LED (which simulates the relay's load) will turn on and off accordingly.

You can use the relay in this circuit to control much larger loads than the LED, including an LED desk light or strip a fan (or other motors). However, be careful! As you start working with larger loads, the power requirement increases, and so do the risks for injury, including deadly electrocution! Do not work with any loads over 5V unless you clearly understand what you are doing!

3.4. RFID

The RFID-RC522 is a popular module for RFID (Radio Frequency Identification) applications. This section will guide you using the RFID-RC522 with a Raspberry Pi Pico W using MicroPython. The aim is to read an RFID card's code and print it on the console.

Radio Frequency Identification (RFID) technology uses electromagnetic fields to identify and track tags attached to objects automatically. An RFID tag comprises a small radio transponder, a radio receiver, and a transmitter. When triggered by an electromagnetic interrogation pulse from a nearby RFID reader device, the tag transmits digital data back to the reader, usually an identifying inventory number.

There are two main types of RFID systems: active and passive.

Active RFID systems: These have a battery within the RFID tag that can be used as a partial or complete power source for the tag's circuitry and antenna. Some active RFID tags contain a battery used only if the tag is in the presence of a reader, while others use the battery to broadcast a signal continuously.

Passive RFID systems: These rely on the reader as the power source. The reader sends out a Radio Frequency (RF) signal that gets picked up by the tag. The RF signal energises the tag, powering its circuits and enabling it to send information.

The process of operation for a passive RFID system, which is commonly used for applications like access control, works as follows:

1. **Activation:** The RFID reader generates an electromagnetic field that "wakes up" the passive RFID tag.
2. **Data Transfer:** Once activated, the RFID tag uses the energy harvested from the reader's electromagnetic field to power its internal circuits and return its stored information.
3. **Data Reception:** The RFID reader receives the transmitted signal, which carries the tag's data (e.g., a unique identification number).
4. **Data Processing:** The reader then passes this received data to a computer system for processing and action.

This whole process happens in real-time, allowing for immediate identification or action. Unlike barcodes, the RFID system can work in various environmental conditions and does not require a line of sight. RFID technology has various applications, such as inventory management, access control, contactless payment, and pet identification.

The RFID-RC522 module

The RFID-RC522 is an inexpensive yet versatile radio frequency identification (RFID) module. It operates at a frequency of 13.56MHz and is commonly used for contactless cards, smart cards, and NFC applications. This module is based on the MFRC522 integrated circuit, developed by NXP Semiconductors, known for its impressive features and capabilities. It is used widely in Arduino and other microcontroller projects because of its relatively simple usage and low cost.

Figure 2.3.4.18: An RFID-RC522 module with a key fob and a card (both passive).

The RFID-RC522 can both read and write data to RFID tags. It communicates with a host system, such as a microcontroller or single-board computer, using the Serial Peripheral Interface (SPI), simplifying connecting the module and transferring data.

One of the critical features of the RFID-RC522 module is its support for MIFARE cards, a proprietary technology based on ISO/IEC 14443 Type A with a typical read/write distance of 10 cm (4 inches). MIFARE cards are widely used worldwide, making the RFID-RC522 module suitable for various applications such as access control, payment systems, and identification.

The RFID-RC522 module is also commonly bundled with two types of RFID tags - a card and a key fob - both operating at the same 13.56MHz frequency. These tags are passive, meaning they don't require their own power supply. When brought into the proximity of the RFID-RC522 reader, the tags draw power from the reader's electromagnetic field, power up their internal chip, and send back their unique identification number.

Despite the complexity behind its operation, using the RFID-RC522 is surprisingly straightforward, thanks to the availability of libraries and code examples for popular platforms like Arduino and Raspberry Pi. Whether creating a secure door lock, a time tracking system, or dipping your toes into the world of RFID for the first time, the RFID-RC522 is a powerful and accessible module that can bring your project to life.

Wiring the RFID-RC522 to Raspberry Pi Pico

Let's start by wiring up RFID-RC522 to the Raspberry Pi Pico. Refer to the table below for wiring instructions:

RFID-RC522 Pins	Raspberry Pi Pico Pins
SDA/CS	GP0
SCK	GP2
MOSI	GP3
MISO	GP4
IRQ	Not Connected
GND	GND
RST	GP5
3.3 V	3V3

Ensure all connections are firm and secure. Below is the wiring diagram:

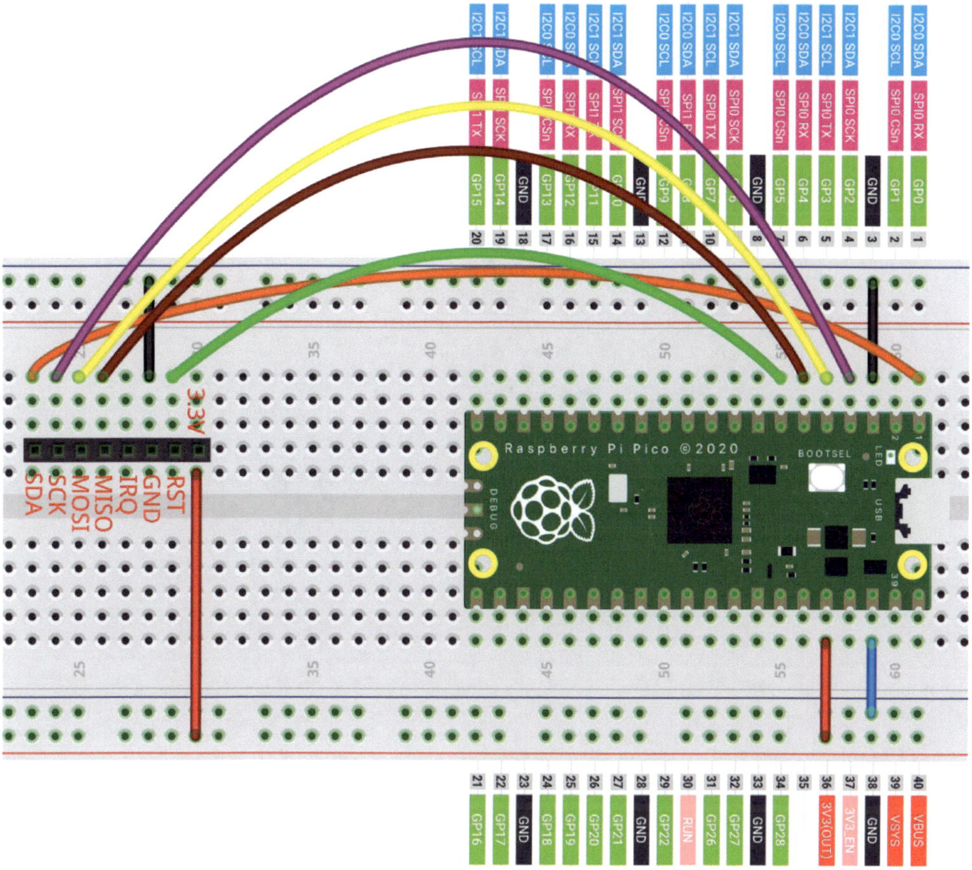

Figure 2.3.4.19: The wiring diagram for the RFID project.

Here is my RFIO-RC-RC522 connected to the Raspberry Pi Pico:

Figure 2.3.4.20: An RFID-RC522 module connected to a Raspberry Pi Pico.

MicroPython Code to Read an RFID Card's Code

To work with the RFID-RC522 in MicroPython, we'll use the mfrc522.py module. Download the script from the micropython-mfrc522 GitHub repository and upload it to your Raspberry Pi Pico.

Manually install a third-party module

If you are not familiar with the process of manual installation of a third-party MicroPython module to Thonny so that you can use it in your MicroPython scripts, then follow these instructions:

1. First, you need to download the Python file (.py) for the module you wish to use. Usually, these are found on the module's GitHub page or official website. Note that you need the MicroPython version of the module, as regular Python modules may not work with MicroPython due to its reduced standard library. For this project, you must install the file with filename "mfrc522.py" from the author's GitHub repository at https://github.com/wendlers/micropython-mfrc522.

2. Start Thonny and connect your MicroPython device (Raspberry Pi Pico, in your case). Make sure your device is connected to the appropriate port.

3. In Thonny, select "File" → "Open..." and navigate to the location where you saved the .py file for the module. Once the file is open in Thonny, select "File" → "Save copy..." → "MicroPython device". This will upload the .py file to your MicroPython device.

4. Now, you can use the module in your code! Just import it as you would any other Python module.

You may have wondered if you can use the Thonny package manager to install the MFRC522 module. After all, searching for "mfrc522" in the package manager will see one hit. This result, however, is not compatible with the Raspberry Pi Pico; it only works on the "regular" Raspberry Pi single-board computer.

It's important to note that while Thonny's Plugin Manager can install Python packages from PyPi, not all will be compatible with MicroPython. MicroPython runs in a more restricted environment than regular Python, and many Python packages rely on features or libraries that aren't available in MicroPython. So, even if you can install a package with the Plugin Manager, it might not work on your MicroPython device. Always ensure you're using Micro-Python-specific modules or modules known to be compatible with MicroPython.

After installing the mfrc522.py, you should confirm that it is available to use by looking for it in the file system of the Raspberry Pi Pico, as in the screenshot below:

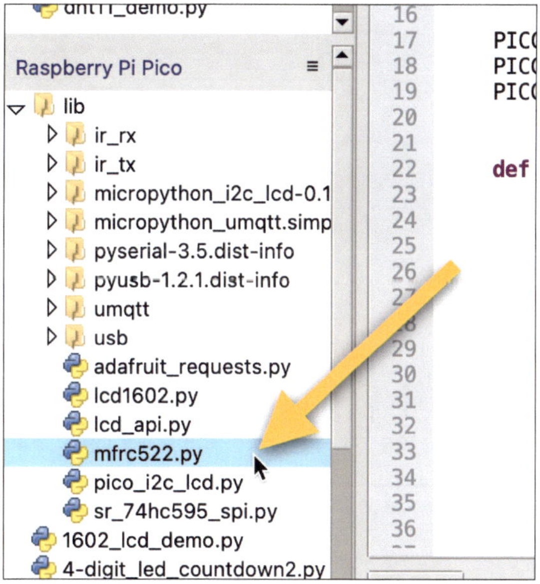

Figure 2.3.4.21: The mfrc522.py is stored on my Raspberry Pi Pico file system.

Read data from the RFID tag.

Once the module is on your Raspberry Pi Pico, the following script can be used to read data from an RFID tag. For this purpose, we'll use this MicroPython script:

```
from mfrc522 import MFRC522
import utime

reader = MFRC522(spi_id=0,sck=2,miso=4,mosi=3,cs=0,rst=5)

print("Bring TAG closer...")
print("")

while True:
    reader.init()
    (stat, tag_type) = reader.request(reader.REQIDL)
    if stat == reader.OK:
        (stat, uid) = reader.SelectTagSN()
        if stat == reader.OK:
            card = int.from_bytes(bytes(uid),"little",False)
            print("CARD ID: "+str(card))

utime.sleep_ms(500)
```

This script continually polls for new RFID cards. When one is found, it reads the UID and prints it. The data is printed in decimal format. After you upload the script to your Pico, bring one of the tags near the reader. Below, you can see the output of my reader in the serial monitor:

```
Bring TAG closer...

CARD ID: 496165579
CARD ID: 496165579
CARD ID: 496165579
CARD ID: 496165579
CARD ID: 496165579
```

Node-RED flow

Let's use this script to set up a Raspberry Pi Pico / Node-RED system so that when the RFID reader reads a known RFID card, the Node-RED flow reacts by opening a lock. We will simulate the lock with dashboard "template" node widgets. We'll set up this widget to show a big green circle when the RFID tag is recognised and a red circle when it is not.

Here is the Dashboard user interface for this totally insecure RFID lock:

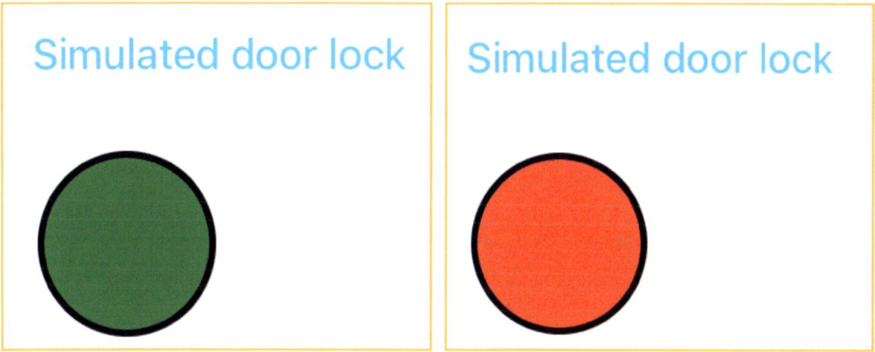

Open Locked

Figure 2.3.4.22: The Dashboard of this prototype RFID lock system.

MicroPython script

Let's modify our MicroPython RFID script so that it can read the ID of a tag, compare it against a known ID, and notify the Node-RED flow with the result of the comparison. If the result is "open", the dashboard will show a green circle (oval in the above image), and if it is "close", it will show a red circle.

Here is the script that implements this functionality:

```
from mfrc522 import MFRC522
from umqtt.simple import MQTTClient
from time import sleep
import network

# The numbers below refer to Raspberry Pi Pico GPIOs.
# The spi_id number refers to the SPI channel in use, which is 0.
reader = MFRC522(spi_id=0,sck=2,miso=4,mosi=3,cs=0,rst=5)

# Configure your Wi-Fi credentials
SSID     = '<your_wifi_network_SSID>'
PASSWORD = '<your_wifi_network_password>'

wlan = network.WLAN(network.STA_IF)
wlan.active(True)
wlan.connect(SSID,PASSWORD)

while not wlan.isconnected():
    pass

print('Connected to Wi-Fi')
print(wlan.ifconfig())

# Configure MQTT
```

```
SERVER        = '<your_MQTT_broker_IP_address>'
CLIENT_ID     = b'<your_device_client_ID>'
TOPIC         = b'rfid_status'
USER          = b'<your_MQTT_username>'
PASSWORD      = b'<your_MQTT_password>'

def mqtt_connect():
   client = MQTTClient(CLIENT_ID, SERVER, user=USER, password=PASSWORD,
keepalive=3600)
   client.connect()
   print('Connected to %s MQTT Broker'%(SERVER))
   return client

def reconnect():
   print('Failed to connect to the MQTT Broker. Reconnecting...')
   sleep(5)
   machine.reset()

try:
   client = mqtt_connect()
except OSError as e:
   reconnect()

while True:
    reader.init()
    (stat, tag_type) = reader.request(reader.REQIDL)
    if stat == reader.OK:
        (stat, uid) = reader.SelectTagSN()
        if stat == reader.OK:
            card = int.from_bytes(bytes(uid),"little",False)
            print("CARD ID: "+str(card))
            if card == 560681490:
                print("Card ID: "+ str(card)+" PASS: Open door")
                client.publish(TOPIC, "open")
            else:
                print("Card ID: "+ str(card)+" UNKNOWN CARD! Lock door")
                client.publish(TOPIC, "close")
    sleep(1)
```

My kit contains two RFID cards. The one with ID "560681490" is the one that I use to open the door, while the other card (or any other card) will lock the virtual door. Of course, you must change this code to match the one that belongs to your RFID card.

Node-RED flow

Let's turn our attention to the Node-RED flow. This flow consists of three nodes:

- "MQTT-in" to get the lock status from the "rfid_status" topic.
- "Template" to draw the red or green circle graphically conveys the lock's status.
- A "debug" node to print the MQTT-in output to the debug tab.

You can see the flow below:

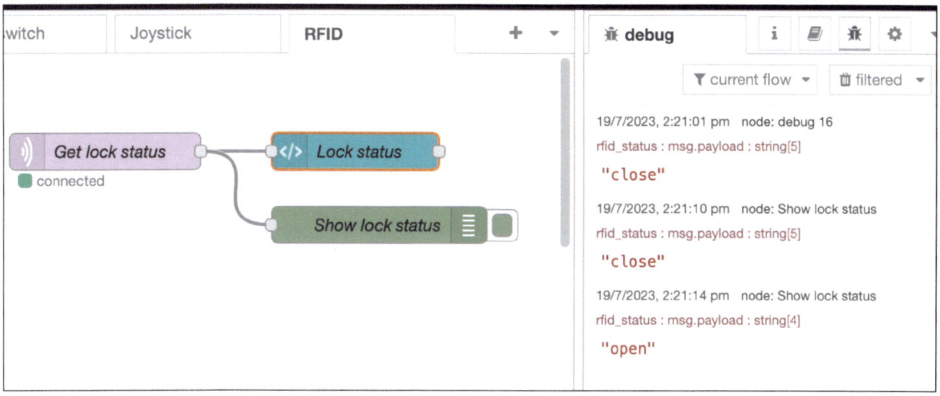

Figure 2.3.4.23: The door lock flow.

The most interesting node in this flow is the template. You learned about the template node in the chapter on the Node-RED dashboard, but here is a quick refresher. You can use Node-RED's dashboard template node to create custom HTML, including dynamic elements. In this project, we'll use the template node to display a red circle when the input message payload is "locked" and a green circle when it's "open". We will use SVG to create the circles. SVG, or Scalable Vector Graphics, is an XML-based vector image format for two-dimensional graphics with support for interactivity and animation, and it is a very efficient way to create simple graphics and user interfaces.

This is the SVG code that I wrote for the template node:

```
<svg height="100" width="100">
    <circle cx="50"
        cy="50"
        r="40"
        stroke="black"
        stroke-width="3"
        fill="{{msg.payload==='close' ? 'red' : 'green' }}" >
</svg>
```

As you can see, this SVG code contains executable Javascript in the "fill" attribute of the "circle" tag. When the payload that the template node receives from the MQTT-in node is "close", the circle is filled with red and green.

Simple, right?

Copy this code in the template field of the template node. It looks like this:

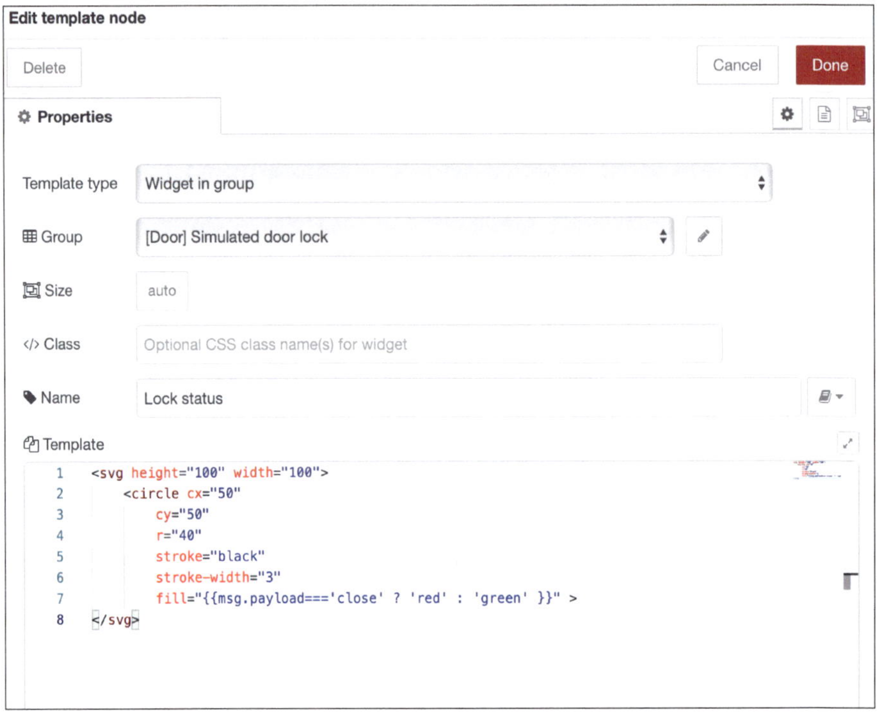

Figure 2.3.4.24: The SVG code in the Template node.

I have assigned this widget node to a new group and tab.

Go ahead and deploy this flow. With your Raspberry Pi Pico already running the RFID code, try reading your "open" and "close" cards with the RFID reader and confirm that the dashboard circle changes between green and red.

Congratulations, you have created a rudimentary electronic locking system!

3.5. IR Receiver and Transmitter

Infra-red (IR) communication is a widely used wireless communication method for short-range applications. It utilizes light from the infrared spectrum to transmit information. Many devices, such as televisions, air conditioners, and other appliances, use IR remote controls for user interaction.

In this chapter, you will learn how to interface an IR remote control receiver with the Raspberry Pi Pico using MicroPython. This will allow your Raspberry Pi Pico to receive commands from an IR remote control and display the received commands on the console. You will also learn to send commands from the IR remote control device to a Node-RED flow.

Figure 2.3.5.25: The Infrared receiver ("1") and transmitter ("2") that we'll use in this project.

Infrared communications, a primer

Infrared (IR) communication uses infrared signals to transmit data. Infrared waves are a type of electromagnetic radiation with wavelengths longer than those of visible light, which makes them invisible to the human eye. This form of communication is commonly used in remote control systems for televisions, air conditioners, and other appliances.

An IR communication system typically has two main components: the transmitter and the receiver. The transmitter emits an infrared signal with a specific frequency and pattern, representing the encoded data. On the other hand, the receiver captures these signals, decodes them, and converts them into electrical signals for further processing.

- IR Transmitter: An IR transmitter consists of an LED that emits light in the infrared spectrum. The LED is rapidly turned on and off to transmit data, creating pulses of infrared light. These pulses are modulated (switched on and off) at a specific frequency to represent binary data, with a pulse representing a binary one and a lack of a pulse representing a binary 0.

- IR Receiver: The receiver module contains a photodiode (a photodetector capable of converting light into current or voltage) sensitive to infrared light. This photodiode detects the modulated IR signals, demodulates them (converts them back to binary data), and sends this binary data to the microcontroller for further processing.

Infrared communication with microcontrollers

Microcontrollers like the RP2040 and the Arduino can use IR communication for various applications. To enable this, they interact with IR transmitter and receiver communications:

- Sending IR Signals (Transmitter): When the microcontroller needs to send an IR signal, it generates a sequence of binary data (often in the form of a specific protocol like NEC, RC5, etc.) and outputs this to the IR transmitter module. The transmitter module then modulates this data onto an IR LED, effectively 'flashing' the LED on and off in a pattern representing the data.

- Receiving IR Signals (Receiver): When the IR receiver module detects a modulated IR signal from a remote control or other IR transmitter, it demodulates this signal and presents it as a series of binary data. The microcontroller then reads this binary data. The microcontroller uses this data to perform specific actions, such as changing a setting or activating a device.

In the context of microcontrollers like the RP2040 and Arduino, these binary sequences often represent commands from a remote control. For example, each button on a remote control will have a unique binary code. When a user presses a button, the corresponding code is sent to the microcontroller, which performs the action associated with that button.

Through this combination of hardware (transmitters and receivers) and software (libraries and scripts), microcontrollers can use IR communication for various applications, from simple remote controls to more complex communication systems.

Wiring the IR Receiver to Raspberry Pi Pico

The table below connects the IR receiver to the Raspberry Pi Pico. To identify the pins on the IR receiver, hold the device with the lens (i.e. the bulging side) towards your face.

IR Receiver Module Pin	Raspberry Pi Pico Pin
VCC (right pin)	3V3
GND (middle pin)	GND
OUT (left pin)	GP17

Here is the wiring diagram:

Figure 2.3.5.26: The IR receiver connected to the Raspberry Pi Pico W.

This is what the assembled circuit looks like:

Figure 2.3.5.27: The IR receiver connected to the Raspberry Pi Pico W.

To send commands to the receiver, you will need the IR transmitter. In the recommended kit for this book, the Kepler Kit for the Raspberry Pi Pico W by Sunfounder, the IR receiver and transmitter are bundled together, ensuring compatibility. It is possible to use a third-party transmitter, however, it might not work because of the variety of communications protocols that are used in infrared devices. Point the front side of the transmitter (where the transparent diode is placed) towards the receiver. Remember, infrared communications require line-of-sight between the receiver and transmitter.

Also, remember to check that the transmitter's batteries are correctly inserted. I am astonished by how often I forget about this detail.

MicroPython Code to Read Commands from the IR Remote

Let's look at the programming side of this project. We will use a MicroPython module (library) written by Peter Hinch called "micropython_ir". You can find this module on GitHub at https://github.com/peterhinch/micropython_ir. In my searches, I was not able to find this module in PiPy, so you will need to install it manually, as you learned in the previous experiment with the RFID. This module is well documented, and you can learn everything you need to know on the project home page on GitHub. As you will see, the module is straightforward to use in a MicroPython script.

Once you install the micropython_ir module, you can use the following script to read data from an IR Remote. This is a simple script that only shows the received data but does not decode it:

```
import utime
from machine import Pin, freq
from ir_rx.print_error import print_error
from ir_rx.nec import NEC_8

pin_ir = Pin(17, Pin.IN)

# Define the callback function
def callback(data, addr, ctrl):
    if data < 0:  # NEC protocol sends repeat codes.
        pass
    else:
        print('Data received: ', data)
        print('Address: ', addr)
        print('Control bit: ', ctrl)

# Initialize the IR_RX object
ir = NEC_8(pin_ir, callback)
ir.error_function(print_error)  # Show debug information

try:
    while True:
        pass
except KeyboardInterrupt:
    ir.close()
```

This script uses the NEC protocol on the specified GPIO pin to set up an infrared receiver module. It defines a callback function to handle received data, address, and control bits. The program then enters an infinite loop, continuously receiving and decoding infrared signals and printing the received data to the console. The loop can be terminated by pressing Ctrl+C or clicking the Stop button in the Thony IDE, which closes the infrared receiver module.

Upload the script to the Pico and press a few buttons on the transmitter. Look at the output of the script in the serial monitor. Here is some output from my session:

```
>>> %Run -c $EDITOR_CONTENT
Data received:  24
Address:  0
Control bit:  0
Data received:  94
Address:  0
Control bit:  0
Data received:  90
Address:  0
Control bit:  0
Data received:  24
Address:  0
Control bit:  0
Data received:  12
Address:  0
Control bit:  0
```

This output shows the numeral of the commands the receiver captured but does not decode the command to the button I pressed. Command "24" is not meaningful without knowing which button I pressed to issue that command (it was number "2" on the transmitter keypad).

MicroPython script with decoding

To identify the transmitted command, you can add a new function with a decode function. When a code is recognised as a valid remote control button press, you can select an appropriate function, such as to turn on/off lights, motors, etc. Below is a simple script that can receive a command from the transmitter, decode it, and print the button pressed on the keypad (not only the command code).

```
import time
from machine import Pin, freq
from ir_rx.print_error import print_error
from ir_rx.nec import NEC_8

pin_ir = Pin(17, Pin.IN)
```

```
def decodeKeyValue(data):
    if data == 0x16:
        return "0"
    if data == 0x0C:
        return "1"
    if data == 0x18:
        return "2"
    if data == 0x5E:
        return "3"
    if data == 0x08:
        return "4"
    if data == 0x1C:
        return "5"
    if data == 0x5A:
        return "6"
    if data == 0x42:
        return "7"
    if data == 0x52:
        return "8"
    if data == 0x4A:
        return "9"
    if data == 0x09:
        return "+"
    if data == 0x15:
        return "-"
    if data == 0x7:
        return "EQ"
    if data == 0x0D:
        return "U/SD"
    if data == 0x19:
        return "CYCLE"
    if data == 0x44:
        return "PLAY/PAUSE"
    if data == 0x43:
        return "FORWARD"
    if data == 0x40:
        return "BACKWARD"
    if data == 0x45:
        return "POWER"
    if data == 0x47:
        return "MUTE"
    if data == 0x46:
        return "MODE"
    return "ERROR"

# User callback
```

```
def callback(data, addr, ctrl):
    if data < 0:  # NEC protocol sends repeat codes.
        pass
    else:
        print(decodeKeyValue(data))

ir = NEC_8(pin_ir, callback)  # Instantiate receiver
ir.error_function(print_error)  # Show debug information

try:
    while True:
        pass
except KeyboardInterrupt:
    ir.close()
```

This script depends on the function decodeKeyValue to compare the received command against a list of possible matches and return the matched button string. This string identifies the button pressed to produce the command.

Here is some output from my session:

```
>>> %Run -c $EDITOR_CONTENT
2
3
4
5
```

In this example, you can see that buttons 2, 3, 4, and 5 were pressed.

Let's add Wi-Fi and MQTT connectivity to this script to prepare it for interaction with the Node-RED flow we'll create next. Drawing from the previous projects, the new version of the IR receiver MicroPython code looks like this:

```
import time
from machine import Pin, freq
from ir_rx.print_error import print_error
from ir_rx.nec import NEC_8
from umqtt.simple import MQTTClient
import network

pin_ir = Pin(17, Pin.IN)

client = None  # Create the client object here so that it is
               # available to the callback function, but initialise it
               # later
```

```
def decodeKeyValue(data):
    if data == 0x16:
        return "0"
    if data == 0x0C:
        return "1"
    if data == 0x18:
        return "2"
    if data == 0x5E:
        return "3"
    if data == 0x08:
        return "4"
    if data == 0x1C:
        return "5"
    if data == 0x5A:
        return "6"
    if data == 0x42:
        return "7"
    if data == 0x52:
        return "8"
    if data == 0x4A:
        return "9"
    if data == 0x09:
        return "+"
    if data == 0x15:
        return "-"
    if data == 0x7:
        return "EQ"
    if data == 0x0D:
        return "U/SD"
    if data == 0x19:
        return "CYCLE"
    if data == 0x44:
        return "PLAY/PAUSE"
    if data == 0x43:
        return "FORWARD"
    if data == 0x40:
        return "BACKWARD"
    if data == 0x45:
        return "POWER"
    if data == 0x47:
        return "MUTE"
    if data == 0x46:
        return "MODE"
    return "ERROR"

# User callback
```

```
def callback(data, addr, ctrl):
    if data < 0:  # NEC protocol sends repeat codes.
        pass
    else:
        decoded_value = decodeKeyValue(data)
        print(decoded_value)
        client.publish(TOPIC, decoded_value)

ir = NEC_8(pin_ir, callback)  # Instantiate receiver
ir.error_function(print_error)  # Show debug information

# Configure your Wi-Fi credentials
SSID     = '<your_wifi_network_SSID>'
PASSWORD = '<your_wifi_network_password>'

wlan = network.WLAN(network.STA_IF)
wlan.active(True)
wlan.connect(SSID,PASSWORD)

while not wlan.isconnected():
    pass

print('Connected to Wi-Fi')
print(wlan.ifconfig())

# Configure MQTT
   SERVER         = '<your_MQTT_broker_IP_address>'
   CLIENT_ID      = b'<your_device_client_ID>'
   TOPIC          = b'IR_received_command'
   USER           = b'<your_MQTT_username>'
   PASSWORD       = b'<your_MQTT_password>'

def mqtt_connect():
   client = MQTTClient(CLIENT_ID, SERVER, user=USER, password=PASSWORD,
keepalive=3600)
   client.connect()
   print('Connected to %s MQTT Broker'%(SERVER))
   return client

def reconnect():
   print('Failed to connect to the MQTT Broker. Reconnecting...')
   sleep(5)
   machine.reset()

try:
   client = mqtt_connect()
```

```
except OSError as e:
   reconnect()

try:
    while True:
        pass
except KeyboardInterrupt:
    ir.close()
```

Upload the script to your Pico and continue with the Node-RED flow.

Node-RED flow

Now that we can decode the commands from the IR transmitter, we can continue with a simple Node-RED flow that does something interesting with this information. This Node-RED flow features a numerical keypad generated by a template node widget in the dashboard. It looks like this:

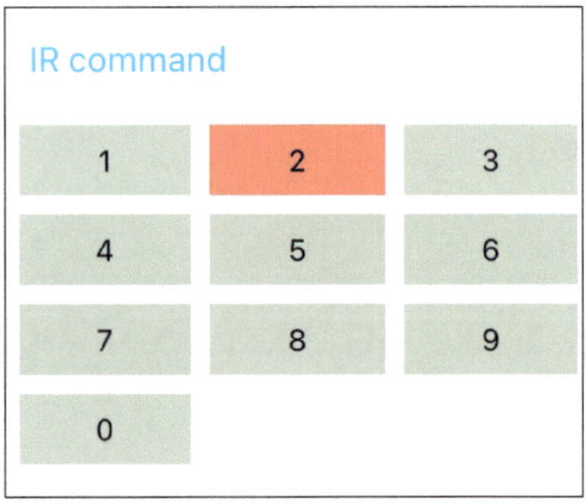

Figure 2.3.5.28: The template widget creates a numerical keypad in the dashboard.

When the flow receives a message from the Pico (which contains the decoded commands received from the IR transmitter), the template widget button in the keypad is highlighted differently than the rest. Here is the flow:

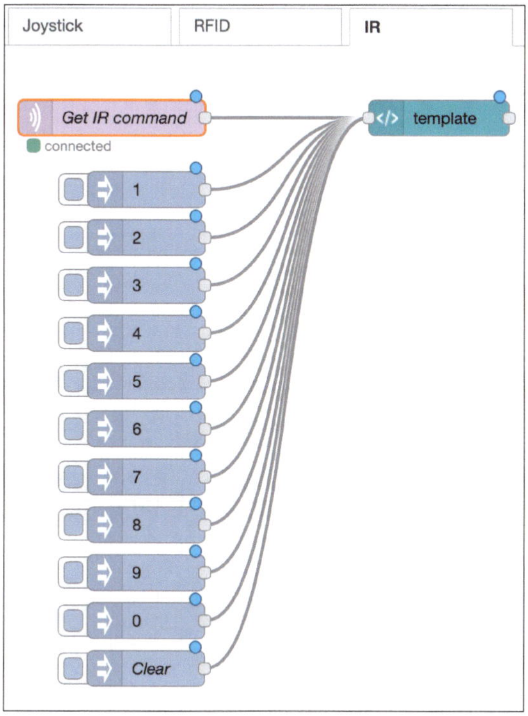

Figure 2.3.5.29: The Node-RED IR receiver flow.

This flow contains only two essential nodes: the MQTT-in node, which listens for new values in the MQTT topic with the name "IR_received_command", and the template node, which generates the keypad in the dashboard.

I have also included 11 inject nodes to help me test the template node. Each inject node sends a string that contains numbers from 0 to 9 to the template. The last inject node sends an empty string, which clears the keypad (so that no number is highlighted). It is important to remember that the template node expects string input (such as the string "1"), not numbers (such as the integer "1"). Therefore, if you want also to implement the testing inject nodes, configure them like this:

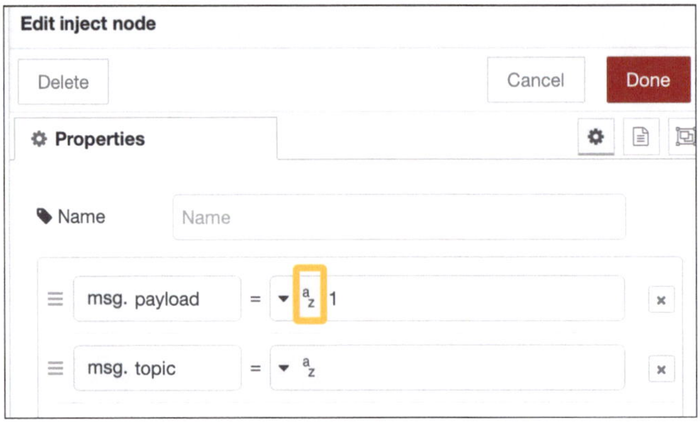

Figure 2.3.5.30: The payload type is a string ("az"), not a "number".

Let's focus on the template node. The code that creates the keypad in the dashboard and highlights the appropriate number depending on the numerical string that is received is this:

```
<div>
    <style>
        .keypad {
            display: grid;
            grid-template-columns: repeat(3, 1fr);
            gap: 10px;
            margin-bottom: 10px;
        }

        .key {
            background-color: #CCD7C5;
            padding: 10px;
            text-align: center;
            cursor: pointer;
        }

        .highlight {
            background-color: #EE9480;
        }
    </style>

    <div class="keypad">
        <div class="key" ng-class="{'highlight': number === '1'}"
ng-click="send(1)">1</div>
        <div class="key" ng-class="{'highlight': number === '2'}"
ng-click="send(2)">2</div>
        <div class="key" ng-class="{'highlight': number === '3'}"
ng-click="send(3)">3</div>
```

```
        <div class="key" ng-class="{'highlight': number === '4'}"
ng-click="send(4)">4</div>
        <div class="key" ng-class="{'highlight': number === '5'}"
ng-click="send(5)">5</div>
        <div class="key" ng-class="{'highlight': number === '6'}"
ng-click="send(6)">6</div>
        <div class="key" ng-class="{'highlight': number === '7'}"
ng-click="send(7)">7</div>
        <div class="key" ng-class="{'highlight': number === '8'}"
ng-click="send(8)">8</div>
        <div class="key" ng-class="{'highlight': number === '9'}"
ng-click="send(9)">9</div>
        <div class="key" ng-class="{'highlight': number === '0'}"
ng-click="send(0)">0</div>
    </div>
</div>

<script>
    (function(scope) {
        scope.$watch('msg.payload', function(value) {
            scope.number = value;
        });
    })(scope);
</script>
```

Let's break down how the template code provided works:

HTML Structure:

The HTML code inside the Template node represents the structure and layout of the numerical keypad widget. It consists of a <div> container with a " keypad " class that holds the individual keys.

CSS Styling:

The CSS styles defined within the <style> tags control the appearance of the keypad and its keys. The "keypad" class sets the grid layout with three columns and a gap between the keys. The "key" class defines the common styles for each key, including background colour, padding, text alignment, and cursor.

Highlighting Keys:

The ng-class directive in each key's <div> element dynamically adds the "highlight" class to the key based on the condition defined within the curly braces. This class applies the red background colour to the key. For example, if number === '1', the "highlight" class will be added to the key with the value "1".

Click Event and Data Sending:

The ng-click directive is used to bind a click event to each key. When a key is clicked, the send() function is called with the corresponding number as the argument. For instance, when the key "1" is clicked, send(1) is executed.

JavaScript Logic:

The JavaScript code enclosed in the <script> tags provides the necessary functionality for updating the highlighted key based on the incoming msg.payload value. The code uses a $watch function to monitor changes to "msg.payload". When a new value is received, the anonymous function inside the $watch function is executed. It sets the number variable in the scope to the value of "msg.payload".

This template code defines the HTML structure and styling for the numerical keypad widget. It handles the highlighting of keys by dynamically applying the "highlight" class based on the "msg.payload" value received. When a key is clicked, the "ng-click" directive triggers the "send()" function, allowing you to send the corresponding number to the next node in your Node-RED flow.

Go ahead and deploy the flow. If you have not already uploaded the MQTT-enabled version of the MicroPython code, do so now. Press the numerical keys on the transmitter and confirm that the correct key in the dashboard keypad is highlighted. When you press any non-numerical keys, the highlight of the dashboard keys is cleared for all keys (non-numerical buttons work like the empty string inject node).

Well done! Completing this project taught you how to use infrared wireless communications between your Raspberry Pi Pico. You also saw a more advanced application of the template node in Node-RED that combines HTML, CSS and Javascript. Such templates will allow you to create more elaborate user interfaces in the Node-RED dashboard.

Chapter 2.4 • Displays and LEDs

This chapter will teach you the use of displays and LED devices, as always, with the help of projects.

First up, you'll learn the use of the I²C LCD display. With this display, your Raspberry Pi Pico can provide information in the form of alphanumeric text. With this simple display, your projects can show real-time data, messages, and simple graphics. You'll learn to set up an I²C LCD display, connect it to your Raspberry Pi Pico W, and control it using Node-RED and MicroPython.

Next, you'll explore how to control multiple LEDs with just a few pins on your Raspberry Pi Pico W using the 74HC595N shift register. This project will teach you the multiplexing technique to control many components with minimal hardware. You'll build a Raspberry Pi Pico W circuit, develop a Node-RED flow, and write a MicroPython script to bring your LED array to life.

Finally, you'll get hands-on with the WS2812 Strip RGB LED strip, a favourite among makers for creating vibrant, colourful displays. Whether you want to add flair to your home, create a mood light, or even design a flashy wearable, this project will show you how.

Each project in this chapter comes complete with a Node-RED flow, a Raspberry Pi Pico W circuit, and a MicroPython script. By the end of this chapter, you'll have a deeper understanding of these technologies and be well-equipped to tackle even more complex projects. So, let's roll up our sleeves and get started!

4.1. I²C LCD

This project will teach you to use a 16×2 LCD interfaced via an I²C backpack with the Raspberry Pi Pico. Our objective is to demonstrate how the 16×2 LCD can be used for output display purposes in the MicroPython environment and how it can interact with a Node-RED flow. The project is split into two parts to break down its complexity.

The first part involves setting up the hardware connection between the 16×2 LCD with the I²C backpack and the Raspberry Pi Pico. This is followed by introducing the programming aspects using MicroPython to control the LCD and display a simple text message.

In the second part, we step up the complexity by integrating a Node-RED flow. This exercise involves creating a Node-RED flow that consistently fetches and dispatches the current time and date information to the Raspberry Pi Pico, which displays the received data on the 16×2 LCD.

By the end of this project, you will be proficient in using the 16x2 LCD with an I²C backpack, programming the Raspberry Pi Pico in MicroPython for display purposes, and utilising Node-RED for IoT applications involving time and date display.

Figure 2.4.1.1: A 16x2 LCD. Top: front. Bottom: back, showing the I^2C backpack.

The 16×2 LCD

The 16×2 LCD with a 1602 controller is a popular display module in electronics, specifically for microcontroller-based projects. It provides an efficient, compact, cost-effective solution for displaying alphanumeric characters and special symbols.

The term "16×2" in the module name indicates the LCD's ability to display 16 characters per line across two lines, accommodating 32 characters. The heart of this display module is the 1602 controller, a standard chipset that manages the operation of the LCD, including handling data and commands sent from a microcontroller and driving the appropriate pixels on the screen.

One notable characteristic of this module is its versatility — it can interface with various microcontrollers, such as the Arduino and Raspberry Pi Pico, through parallel or serial communication protocols. The module's pin configuration allows it to operate in either 4-bit or

8-bit parallel mode, thus offering flexibility regarding the number of GPIO pins required for interfacing. Moreover, using an I²C backpack can reduce the pin requirement to just two, making it even more microcontroller-friendly.

If you want to display sensor data, provide a simple user interface, or output debugging information, the 16×2 LCD with a 1602 controller is a powerful tool for microcontroller-based projects. Over the subsequent sections, we will delve deeper into its characteristics, operational principles, and integration with popular microcontrollers like Arduino and Raspberry Pi Pico.

16×2 LCD using parallel connection

When interfacing a 16×2 LCD with a microcontroller using a parallel connection, between 6 to 11 GPIO (General Purpose Input Output) pins are required. The six essential pins are Register Select (RS), Enable (E), and four data lines (D4, D5, D6, D7). If the LCD is used in 8-bit mode, two more data lines (D0, D1, D2, D3) are needed. An additional Read/Write (R/W) pin is utilised to read LCD data.

The parallel interface offers high speed due to simultaneous data transfer on the data lines. However, the key drawback is the large number of GPIO pins required, which may not be available on a compact microcontroller or if multiple peripheral devices are connected.

To learn more about connecting an LCD to an Arduino Uno using the parallel connection method, head to my guide page on the Tech Explorations website at https://txplo.re/2x16_lcd_parallel.

16×2 LCD with an I²C backpack

An I²C backpack or I²C adapter is an interface board that can be attached to the back of an LCD module. It integrates a small microcontroller that communicates with the main microcontroller via the I²C protocol, requiring only two data lines (SDA for data and SCL for clock). This considerably reduces the number of GPIO pins from 6-11 down to just 2.

The I²C backpack also has a built-in potentiometer for adjusting the LCD's contrast and can optionally control the LCD backlight, often through Pulse Width Modulation (PWM).

While an I²C interface may not transfer data as fast as a parallel interface, its speed is more than adequate for an LCD text display. Its main advantage is the substantially reduced required GPIO pins, making it easier to interface with compact microcontrollers and freeing up pins for other peripherals.

While parallel connection for a 16x2 LCD allows for faster data transfer, an I²C backpack is a more GPIO-efficient option that retains adequate speed for typical LCD text display applications. In this project, we'll use the LCD module with the I²C backpack to take advantage of the more straightforward wiring available.

Hardware Connection

Connect the LCD with the I²C backpack to the Raspberry Pi Pico. Follow the process below, and please reference the wiring diagram and photograph below.

1. Connect the VCC pin on the I²C backpack to the VBUS (5V) pin on the Raspberry Pi Pico. This is pin 40 on the Pico, as the LCD requires 5V for the backlight. If you connect the LCD VCC to 3.3V, it will still operate. However, the text will not be visible (or it will be very faint and hard to see).

2. Connect the GND pin on the I²C backpack to a Ground (GND) pin on the Raspberry Pi Pico.

3. Connect the SDA pin on the I²C backpack to the GP14 (SDA) pin on the Raspberry Pi Pico. This is I²C channel 1.

4. Connect the SCL pin on the I²C backpack to the GP15 (SCL) pin on the Raspberry Pi Pico. This is I²C channel 1.

Here is the wiring diagram:

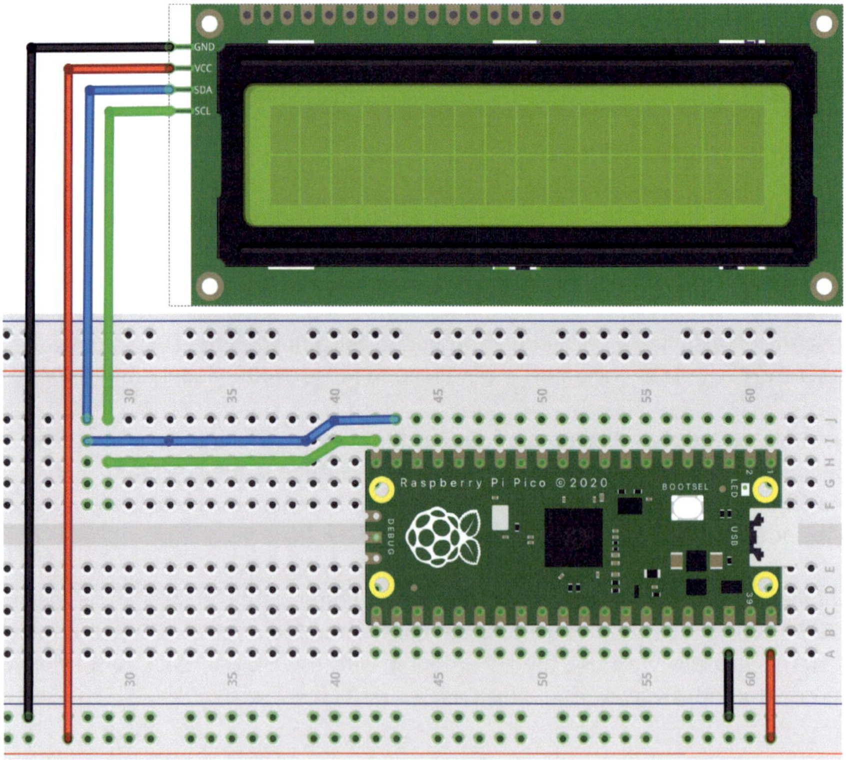

Figure 2.4.1.2: The wiring diagram for the LCD project.

And here is my circuit on the breadboard:

Figure 2.4.1.3: My LCD project circuit on the breadboard.

Let's look at the MicroPython script that displays text in the LCD.

The MicroPython script

Let's look at a MicroPython script that will allow us to test the screen. This script uses two library files from https://github.com/T-622/RPI-PICO-I2C-LCD:

- lcd_api.py from https://github.com/T-622/RPI-PICO-I2C-LCD/blob/main/lcd_api.py
- pico_i2c_lcd.py from https://github.com/T-622/RPI-PICO-I2C-LCD/blob/main/pico_i2c_lcd.py

Copy these files into the "lib" folder on the Raspberry Pi Pico. Then, continue with the demo script:

```
from machine import I2C, Pin
from time import sleep
from pico_i2c_lcd import I2cLcd
i2c = I2C(1, sda=Pin(14), scl=Pin(15), freq=400000)

I2C_ADDR = i2c.scan()[0]
lcd_device = I2cLcd(i2c, I2C_ADDR, 2, 16)
```

```
lcd_device.putstr("Hello world")
counter = 0

while True:
    lcd_device.move_to(0, 1)
    lcd_device.putstr(str(counter))

    counter += 1
    sleep(1)
```

This MicroPython code uses an I²C (Inter-Integrated Circuit) interface to communicate with an LCD (Liquid Crystal Display).

I²C primer for the Raspberry Pi Pico

I²C (Inter-Integrated Circuit) is a two-wire communication protocol that allows multiple devices to share the same bus for communication. Devices on the bus can either be controllers, which initiate communication, or peripherals, which respond to controller commands. The Raspberry Pi Pico supports the I²C protocol and can function as a controller or a peripheral device.

The Raspberry Pi Pico supports two I²C channels:

- I2C Channel 0 (I2C0)
- I2C Channel 1 (I2C1)

Each channel can be configured independently and communicate with different I²C devices or share the same devices on the bus.

The Raspberry Pi Pico has designated GPIO pins for I²C communication associated with each I²C channel:

- For I²C Channel 0, GP0 and GP1 pins are used. GP0 acts as the SDA (Serial Data) line, while GP1 is the SCL (Serial Clock) line.

- For I²C Channel 1, GP2 and GP3 pins are the SDA and SCL lines, respectively.

It is noteworthy that while these are the default configurations, the flexible I/O system in the Raspberry Pi Pico allows for mapping of the I²C interfaces to other GPIO pins if needed.

The I²C protocol, with its efficient and flexible communication characteristics, combined with the Raspberry Pi Pico's support, provides a robust platform for interfacing with various peripherals such as sensors, display modules, and more. In this project, we'll use I²C channel 1 (I2C1) with SDA and SCL pins 14 and 15 to use shorter wires.

Code review

Let's look at the example MicroPython code, line-by-line.

- from machine import I2C, Pin

This line imports the I²C and Pin classes from the machine module in MicroPython.

- from time import sleep

This line imports the sleep function from the time module in MicroPython. "sleep()" creates a delay for a specific number of seconds.

- from pico_i2c_lcd import I2cLcd

This line is importing the I2cLcd class from the pico_i2c_lcd module. This module is a MicroPython driver for controlling LCDs via I²C.

- i2c = I2C(1, sda=Pin(14), scl=Pin(15), freq=400000)

This line is initialising an I²C object on bus number 1 with a clock frequency of 400 kHz. It specifies that the SDA (Serial Data) pin is GPIO 14 and the SCL (Serial Clock) pin is GPIO 15 on the Raspberry Pi Pico.

- I2C_ADDR = i2c.scan()[0]

This line scans the I²C bus for any connected devices and stores the address of the first found device in the I2C_ADDR variable. The "scan()" function is handy for two reasons: 1. It will find the address of the first I²C device attached to the Pico, so you don't need to find it in the documentation or with the help of a scanner script.

- lcd_device = I2cLcd(i2c, I2C_ADDR, 2, 16)

This line creates an I2cLcd object, which represents the LCD. It specifies that the LCD is connected to the previously initialised I²C interface, has the address stored in I2C_ADDR, and has two rows of 16 characters.

- lcd_device.putstr("Hello world")

This line writes the string "Hello world" to the LCD.

- counter = 0

This line initialises a variable called counter and sets its initial value to 0.

- while True:

This line starts an infinite loop.

- lcd_device.move_to(0, 1)

Inside the loop, this line moves the cursor on the LCD to the start (column 0) of the second row (row 1).

- lcd_device.putstr(str(counter))

This line writes the current value of the counter to the LCD at the cursor position.

- counter += 1

This line increments the value of the counter by one.

- sleep(1)

This line creates a delay of 1 second before the next loop iteration. This effectively means that the counter value is updated on the LCD every second.

Experiment with Node-RED and Raspberry Pi Pico W

In this project, we'll create a Node-Red flow that takes the current date and time and sends the text to the Raspberry Pi Pico W. The Raspberry Pi Pico will print the date in row one of the LCD and the time in row two. The time and date is updated every second. You can see the result of this project in the photograph below:

Figure 2.4.1.4: The LCD showing information received from the Node-RED flow.

Here is the Node-RED flow that sends the date and time information to the Raspberry Pi Pico:

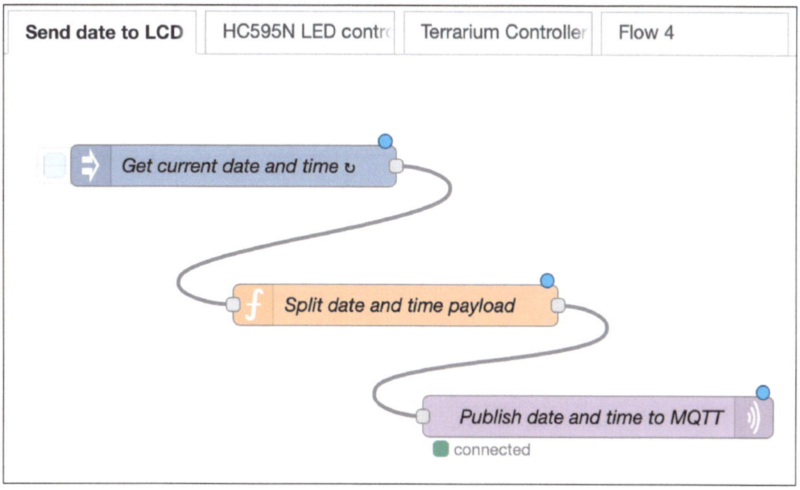

Figure 2.4.1.5: The Node-RED flow that sends date and time to the Pico.

This Node-RED flow starts with an "inject" node triggered every second. The payload of this node is the current date and time in the form of a Node-RED timestamp:

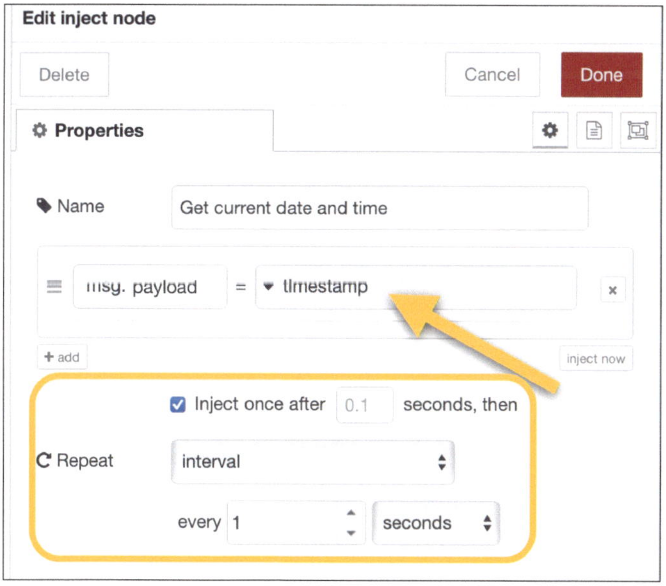

Figure 2.4.1.6: The Inject node captures the current date and time every second.

This date and time are then formatted into a more readable form with a "function" node. This node separates the date and time, puts them in an object, and then publishes to the MQTT topic with the name "datetime". Here is the Javascript in the function node:

```
var date = new Date(msg.payload);
msg.payload = {
    date: date.toLocaleDateString(),
    time: date.toLocaleTimeString()
}
return msg;
```

Go ahead and deploy this flow, then continue with the MicroPython script.

MicroPython script for Raspberry Pi Pico

You can see the MicroPython script for this project below. A short discussion follows.

```
from machine import I2C, Pin
from pico_i2c_lcd import I2cLcd
from umqtt.simple import MQTTClient
import ujson
import network
import time

# I2C parameters
i2c = I2C(1, sda=Pin(14), scl=Pin(15), freq=400000)

# Scan for available I2C devices and get the first one
I2C_ADDR = i2c.scan()[0]

# Create LCD instance
lcd = I2cLcd(i2c, I2C_ADDR, 2, 16)

# Wi-Fi credentials
SSID     = '<your_wifi_network_SSID>'
PASSWORD = '<your_wifi_network_password>'

print('Starting...')

# Activate the network interface
wlan = network.WLAN(network.STA_IF)
wlan.active(True)
wlan.connect(SSID,PASSWORD)

print('Trying to connect to Wifi...')

# Wait until connection is established
while not wlan.isconnected():
    print('.', end='')
    time.sleep(1)
    pass
```

```
print('\nConnected to Wi-Fi')
print(wlan.ifconfig())

# MQTT parameters
SERVER          = '<your_MQTT_broker_IP_address>'
CLIENT_ID       = b'<your_device_client_ID>'
TOPIC      = b'datetime'
USER            = b'<your_MQTT_username>'
PASSWORD        = b'<your_MQTT_password>'

# Callback function for MQTT subscriptions
def sub_cb(topic, msg):
    # Decode and load JSON message
    datetime = ujson.loads(msg)

    # Extract date and time from JSON
    date = datetime["date"]
    time = datetime["time"]

    # Print date and time on LCD
    lcd.move_to(0, 0)     # Move to first row
    lcd.putstr(date)      # Display date

    lcd.move_to(0, 1)     # Move to second row
    lcd.putstr(time)      # Display time

# Function to connect and subscribe to MQTT broker
def connect_and_subscribe():
    client = MQTTClient(CLIENT_ID, SERVER, user=USER, password=PASSWORD,
    keepalive=3600)
    client.set_callback(sub_cb)
    client.connect()
    client.subscribe(TOPIC)
    return client

# Main function
def main():
    client = connect_and_subscribe()

    # Keep waiting and processing MQTT messages
    while True:
        client.wait_msg()

main()
```

This code should be familiar from previous projects. The code that handles the text printed on the LCD is bundled in the "sub_cb" call-back function. Within this function, we use "ujson.loads(msg)" to decode the received message from JSON format and store the resulting array in the "datetime" variable.

In the following two lines, we save the date and time strings in individual variables "date" and "time". Finally, we print the strings to the LCD using the "move_to" (for positioning the cursor) and "putstr" (for printing).

What else can you do with an LCD?
This list will inspire you if you are looking for projects that combine an LCD, a Raspberry Pi Pico, and Node-RED. Here are ten projects that feature the components you learned about in this chapter:

1. **Weather Station**: You can use an LCD connected to a Raspberry Pi Pico to display real-time weather information. Node-RED can fetch data from an API like OpenWeatherMap, process it and then send the information, such as temperature, humidity, and weather conditions, to the Pico, which will display it on the LCD.

2. **Home Automation Display**: The LCD can display the status of various home automation devices—for example, the status of lights, thermostat settings, or even door locks. Node-RED can gather the status information from various sensors and then relay that data to the LCD.

3. **RSS Feed Reader**: You can create an RSS feed reader where the news headlines are displayed on the LCD. Node-RED can fetch and process the RSS feed from various sources, and the headlines can be sent to the LCD for display.

4. **Social Media Notifications**: You can display real-time notifications from social media like Twitter or Facebook. Node-RED can use these platforms' APIs to fetch new posts or messages and display them on the LCD.

5. **Email Notification**: The LCD can display notifications of new emails. Node-RED can check for new emails using protocols like IMAP or POP and show the subject and sender on the LCD.

6. **Internet Radio Player Information**: If you have an internet radio player, you can use Node-RED and Raspberry Pi Pico to display the current song's title and artist. Node-RED can interface with the radio player to retrieve this information and send it to the LCD for display.

7. **Stock Market Information**: You can display real-time stock market information. Node-RED can fetch data from a stock market API, and then the current stock prices and other related information can be displayed on the LCD.

8. **Network Monitor**: You can display network information such as IP address, download/upload speed, or ping times on the LCD. Node-RED can fetch this information from the system or a network monitoring tool and display it on the LCD.

9. **Calendar and Event Reminder**: You can display today's date, day of the week and reminders for upcoming events. Node-RED can fetch this information from a calendar API or a local calendar file and send it to the Raspberry Pi Pico to display on the LCD.

10. **Real-Time Sensor Data**: If you have different sensors (like temperature, humidity, pressure, etc.) connected to your Raspberry Pi, you can display the real-time data from these sensors on the LCD. Node-RED can gather the data from the sensors and relay it to the LCD for display.

Each project involves fetching data from a source (either local sensors or online services), processing it using Node-RED and then sending it to the Raspberry Pi Pico for display on the LCD. These examples show the versatility of Node-RED, Raspberry Pi Pico, and an LCD in creating interactive and informative projects.

4.2. Control 8 LEDs with the 74HC595N

In this section, you will create a Node-RED flow with a dashboard containing eight switch widgets to control the state of the LEDs connected to the HC595N shift register IC. I assume you have Node-RED installed on your Raspberry Pi Pico and running correctly. If not, please refer to the Node-RED installation guide for Raspberry Pi.

Before we begin, let's look at the HC595N shift register IC.

Figure 2.4.2.7: The HC595N shift register in a DIP-16 package.

HC595N Shift Register IC Overview

The HC595N shift register is a widely used integrated circuit that increases the number of digital outputs available from a microcontroller. It can control multiple outputs while only utilising a few pins from the microcontroller. You will find this shift register integrated circuit in many Arduino, ESP32 and Raspberry Pi Pico projects as it is well understood and documented, low cost, and robust. While it only ties up three digital output pins on the microcontroller side, it gives us eight additional pins to connect to digital output devices like LEDs.

It is also possible to daisy-chain HC595N chips to get multiples of eight digital outputs while still utilising the original three digital pins on the microcontroller side. In other words, we can use two HC595N chips to give 16 digital outputs to our project while using only three pins to drive them. We can use three HC595N chips for 24 digital outputs (still using three microcontroller pins to drive the shift registers), and so on.

With this, you can understand why this shift register chip is so popular with makers.

Let's start with a simple example of using a single HC595N chip to drive eight LEDs with the Raspberry Pi Pico.

Pins

The (74)HC595N chip is available in a DIP package with 16 pins. You can find the datasheet with all relevant details at this URL: https://txplo.re/hc595n

Below, you can see the pin diagram for this chip that I have copied from page 3 of the datasheet:

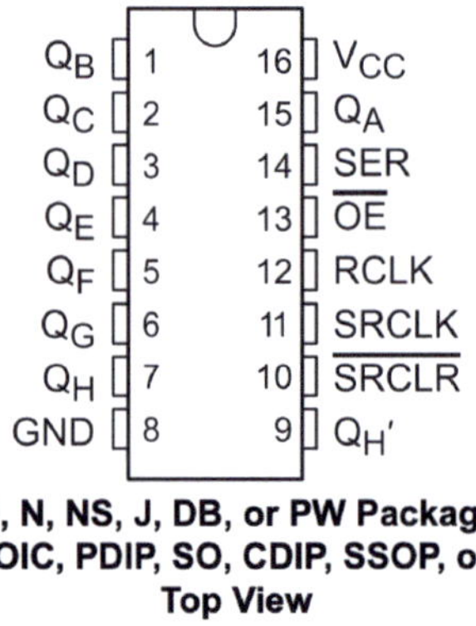

Figure 2.4.2.8: The pin diagram for the HC595N chip.

Pins with names starting with "Q" are outputs. You can connect those pins to a load, such as an LED, that can be driven with a digital HIGH/LOW signal. As you can see, there are eight total "Q" pins: QA, QB, QC, QD, QE, QF, QG, QH. A ninth "Q" pin, marked as QH', is used for daisy-chaining another HC595N chip. You connect the QH' pin (pronounced "QH bar") of the first HC595N chip to the data pin (pin "SER" in the diagram above) of the second HC595N chip to allow the two chips to work together. In this project, we will only use a single HC595N chip. Still, as you can see, it is relatively easy to double, triple and quadruple your microcontroller's digital output pins using the daisy chain method.

You can see a list with descriptions of all the pins below (I am using the PINs from the pin diagram above):

- **1 to 7, 15 and 9.** The output pins.
- **SER, or "Serial input":** Connects to the data output pin of the microcontroller. We can use any digital output pin on the microcontroller to send one bit at a time to the shift register.
- **11. SRCLK, or "Shift Register Clock"**: Connects to the clock pin of the microcontroller. When this pin rises to "high", the bits are pushed into the shift register. Again, we can use any digital output pin on the microcontroller as the clock pin.
- **10. SRCLR, or "Shift Register Clear"**: when this pin is pulled to "low", the contents of the shift register are reset to zero. In normal operation, we tie this pin to HIGH.
- **12. RCLK, or "Register Clock / Latch":** Controls when the data is transferred to the outputs. When this pin goes "high", the bit contents of the register are copied to the storage/latch register. Eventually, the contents of the storage/ latch register will appear at the "Q" outputs.
- **13. Output Enable Pin (OE):** Controls the enabling/disabling of the outputs (optional). When this pin is "high", the output pins are disabled by setting their impedance state too high. When this pin is "low", the output pins function normally.
- **16. Vcc:** This pin can be connected to 5V or 3.3V, depending on the microcontroller. We will connect it to the 3.3V pin on the Raspberry Pi Pico.
- **8. GND:** Connects to the ground of the circuit.

Example Projects

In this chapter, we 're uing a single HC595N chip to control eight LEDs, but what can you do with the HC595N shift register? Here is a list of simple projects you can try.

LED Matrix Control

The HC595N can be used to control many LEDs arranged in a matrix. Connecting multiple HC595N ICs in series allows you to drive a substantial LED array with just a few pins from the microcontroller.

Seven-Segment Display: For displaying numbers on a seven-segment display, the HC595N is an excellent choice. Multiple digits can be controlled by chaining the shift registers together.

Extended GPIO: In a project requiring additional digital outputs, such as controlling motors or reading from numerous sensors, the HC595N can extend the microcontroller's GPIO capabilities.

Simple project

Let's use an HC595N chip to create a simple project in which we control the state of eight LEDs. This is an excellent opportunity to become familiar with this chip before using it in a more elaborate project where you will add a Node-RED flow. Below, you can see the wiring diagram for this project:

Figure 2.4.2.9: Using one HC595N chip to control eight LEDs.

Do the wiring according to these instructions:

1. Place the chip on the breadboard with the circular cutoff side towards the Raspberry Pi Pico (as shown in the wiring diagram above).

2. Use short jumper wires to connect power.
   - Pin 16 is Vcc. Connect it to the 3.3V rail, which draws power from the Raspberry Pi Pico 3.3V(Out) pin (pin 36).
   - Pin 8 is GND. Connect it to the GND power rail. Connect any Raspberry Pi Pico GND pins to the same rail on the breadboard.

3. Connect HC595N pin 10 (SRCLR) to the 3.3 V rail. We will not need to clear the register programmatically in this project, so we will keep this pin tied to "high".

4. Connect HC595N pin 13 to the GND rail. With this, we will always enable the output "Q" pins.

5. Connect HC595N pin 14 (SER) to Raspberry Pi Pico pin GPIO 20. We will use this pin to output bits from the Raspberry Pi Pico to the shift register.

6. Connect HC595N pin 12 (RCLK) to Raspberry Pi Pico pin GPIO 18. We'll use this pin as the latch.

7. Connect HC595N pin 11 (SRCLK) to Raspberry Pi Pico pin GPIO 16. We'll use this pin as the clock.

8. Leave HC595N pin nine unconnected, as we will use a single shift register chip.

9. Connect the LEDs, starting with HC595N pin 15 ("QA"). Connect pin 15 to the LED anode (long pin). Then, connect a 220 Ω resistor to the LED's cathode and the GND rail. Repeat the same process for the rest of the LEDs.

By the end of this process, you will have a circuit like this:

Figure 2.4.2.10: My HC595N circuit on the breadboard.

Double-check the circuit and correct any errors in the wiring. When confident that the circuit is correct, continue with the MicroPython script.

MicroPython script

This MicroPython script will turn the eight LEDs on and off in a sequence containing two passes. In the first pass, the LEDs will turn on, one at a time, until all eight are on. In the second pass, the LEDs will turn off, one at a time.

Here is the script and a line-by-line description of how it works.

```
import machine
import time

ser   = machine.Pin(20,machine.Pin.OUT)
rclk  = machine.Pin(18,machine.Pin.OUT)
srclk = machine.Pin(16,machine.Pin.OUT)

def hc595_shift(dat):
    rclk.low()
    time.sleep_ms(5)
    for bit in range(7, -1, -1):
```

```
        srclk.low()
        time.sleep_ms(5)
        value = 1 & (dat >> bit)
        ser.value(value)
        time.sleep_ms(5)
        srclk.high()
        time.sleep_ms(5)
    time.sleep_ms(5)
    rclk.high()
    time.sleep_ms(5)

num = 0

while True:
    for i in range(16):
        if i < 8:
            num = (num<<1) + 1
        elif i>=8:
            num = (num & 0b01111111)<<1  #0b01111111
        hc595_shift(num)
        print("{:0>8b}".format(num))
        time.sleep_ms(200)
```

Let's examine this code.

Importing libraries

- import machine: This line imports the machine module, which contains classes to control different hardware components on the Raspberry Pi Pico.

- import time: This line imports the time module to add delays within the code.

Initialising pins

- ser = machine.Pin(20, machine.Pin.OUT): Configures pin 20 as an output, connected to the serial data pin (SER) of the 74HC595.

- rclk = machine.Pin(18, machine.Pin.OUT): Configures pin 18 as an output, connected to the register clock pin (RCLK) of the 74HC595.

- srclk = machine.Pin(16, machine.Pin.OUT): Configures pin 16 as an output, connected to the shift register clock pin (SRCLK) of the 74HC595.

Defining shift function

- def hc595_shift(dat): Defines a function hc595_shift that takes one argument, dat, and shifts it through the 74HC595. The "dat" argument contains an eight-bit word to be shifted into the shift register.

Inside the "shift" function

- rclk.low(): Clears the register clock pin.
- time.sleep_ms(5): Waits for five milliseconds.
- for bit in range(7, -1, -1): Loops through the bits of the "dat" variable from the most significant to least significant.
- srclk.low(): Clears the shift register clock pin.
- time.sleep_ms(5): Waits for five milliseconds.
- value = 1 & (dat >> bit): Extracts the current bit from the "dat" variable.
- ser.value(value): Sends the extracted bit to the 74HC595's data pin.
- time.sleep_ms(5): Waits for five milliseconds.
- srclk.high(): Sets the shift register clock pin, shifting in the current bit.
- rclk.high(): Sets the register clock pin, latching the shifted data.

Main loop

- num = 0: Initializes a variable num with 0.
- while True:: Creates an infinite loop.
- for i in range(16): Loops 16 times.
- if i < 8:: If the current loop index is less than 8:
- num = (num << 1) + 1: Shifts num left by one bit and adds 1.
- elif i >= 8:: If the current loop index is greater than or equal to 8:
- num = (num & 0b01111111) << 1: Masks num with 0b01111111 and then shifts it left by one bit.
- hc595_shift(num): Calls the "hc595_shift" function with num, shifting it through the 74HC595.
- print("{:0>8b}".format(num)): Prints num as a zero-padded 8-bit binary string.
- time.sleep_ms(200): Waits for 200 milliseconds.

The main loop creates a pattern where the first eight iterations turn on individual bits one by one, and the subsequent eight iterations turn off individual bits one by one, using the 74HC595 shift register.

Upload the script to your Raspberry Pi Pico and observe how the LEDs turn on/off during the 2×8 cycles. As I mentioned, this script uses a "for loop" to generate the 8-bit pattern written to the shift register and eventually controls the state of the eight LEDs.

Using an array to store bit patterns

Before bringing Node-RED into the mix, I will show you a technique that uses an array to store bit patterns for the LEDs. Using an array, we can create arbitrary bit patterns that can set the state of the LEDs in ways that may not be possible with simple loops.

Below, you can see a line of code from the next MicroPython code example you are about to work with:

```
led_patterns = [0b10000001, 0b01000010, 0b00100100, 0b00011000, 0b00011000,
0b00100100, 0b01000010, 0b10000001]
```

The variable "led_patterns" stores an array of eight bytes. Each cell in this array stores eight bits, and each bit controls the state of one of the LEDs in our circuit. For example, in the first cell, the bit pattern is "10000001". When this pattern is pushed (or written) to the shift register, the first and last connected LED will be turned on, and the rest will be off.

Of course, this array can store as many cells as you want. This example contains eight cells, but you could hold nine, fifty, or more.

Here is the entire MicroPython script, followed by a line-by-line description (I will not repeat the explanation of the code that is shared between the two scripts):

```
import machine
import time

# Define pins
ser = machine.Pin(20, machine.Pin.OUT)
rclk = machine.Pin(18, machine.Pin.OUT)
srclk = machine.Pin(16, machine.Pin.OUT)

# Define LED patterns in an array
led_patterns = [0b10000001, 0b01000010, 0b00100100, 0b00011000, 0b00011000,
0b00100100, 0b01000010, 0b10000001]

# Function to shift data into 74HC595
def hc595_shift(dat):
    rclk.low()
    time.sleep_ms(5)
    for bit in range(7, -1, -1):
        srclk.low()
        time.sleep_ms(5)
        value = 1 & (dat >> bit)
        ser.value(value)
        time.sleep_ms(5)
        srclk.high()
        time.sleep_ms(5)
    time.sleep_ms(5)
    rclk.high()
    time.sleep_ms(5)

# Main loop to iterate through the LED patterns
while True:
    for pattern in led_patterns:
        hc595_shift(pattern)
        print("{:0>8b}".format(pattern)) # Print the pattern in binary
        time.sleep_ms(200)
```

Let's examine this code.

- The "led_patterns" array stores eight binary patterns representing the LED configurations.
- The "for pattern" in "led_patterns:" loop iterates through each pattern in the led_patterns array.
- The "hc595_shift(pattern)" function call sends each pattern to the 74HC595 shift register, displaying the corresponding LED configuration.
- The "time.sleep_ms(200)" line adds a 200-millisecond delay between each pattern so that you can see the change.

This code will cycle through the eight different LED patterns defined in the led_patterns array, displaying them one by one using the 74HC595 shift register. Two of eight LEDs will be turned on in each cycle, moving outwards and inwards.

Now that we have a circuit with a working HC595N shift register let's add Node-RED.

Node-Red and the HC595N shift register project

Let's implement a Node-RED flow that can control the circuit's LEDs that you have assembled on the breadboard from the first part of this chapter. The Node-RED flow uses a slider widget in the dashboard. You will be able to control the LEDs by moving the slider.

The dashboard with the slider looks like this:

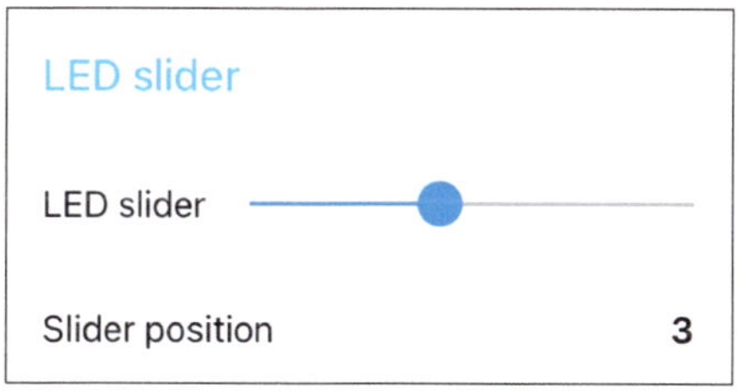

Figure 2.4.2.11: We'll use this slider to control the LEDs.

The slider has eight positions. As you move it through each position, the corresponding position number (from "0" to "7") is sent to the Raspberry Pi Pico W via the MQTT broker. The Pico will use this number to fetch the corresponding bit pattern from the array, send it to the shift register, and set the state of the LEDs.

Let's implement the Node-RED flow and then continue with the MicroPython script. Here is the Node-RED flow:

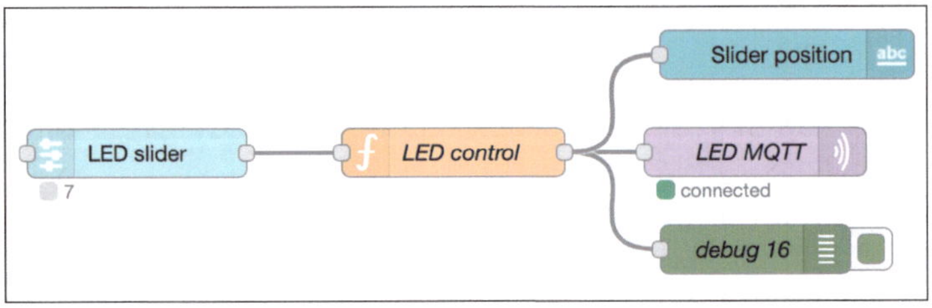

Figure 2.4.2.12: The Node-RED flow for this experiment.

Here are the details for each of the nodes:

- "LED slider" is a "slider" node.
  - Add it to a new Group and Tab titled "LED slider".
  - Label: LED slider
- Range:
  - min: 0
  - max: 7
  - step: 1
- Output: continuously while sliding
- When changed, send:
  - Payload: current value
  - Topic: msg.payload
- "Slider position" is a "text" node that shows the output of the function node in the dashboard.
  - Group: same group and tab that contains the "LED slider" node.
  - Label: "Slider position"
  - Value format: {{msg.payload}}
- "LED control" is a "function" node.
  - Name: LED control
  - On Message Javascript code:

```
var intToSend = 0;
if (msg.payload == 0) intToSend = 0;
if (msg.payload == 1) intToSend = 1;
if (msg.payload == 2) intToSend = 2;
if (msg.payload == 3) intToSend = 3;
if (msg.payload == 4) intToSend = 4;
if (msg.payload == 5) intToSend = 5;
if (msg.payload == 6) intToSend = 6;
if (msg.payload == 7) intToSend = 7;
return { payload: intToSend};
```

- "LED MQTT" is an "mqtt out" node.
  - Server: set to your working MQTT server
  - Topic: LED_shift
  - Name: LED MQTT

- Add a regular debug node connected to the output of the "LED control" node so that you can see the function node output.

When you have completed setting up the flow, deploy it. Load the Dashboard in your browser and slide the slider widget left and right. Check that the slider position text widget shows the position number as you move the slider. Whenever the slider changes position, its position number will be sent to the Pico via the MQTT broker.

Once you get the dashboard working, let's continue with the MicroPython script.

MicroPython code

The MicroPython script you see below is based on the second script we experimented with earlier in this chapter (the one with the array). The script connects the Pico to Wi-Fi, subscribes to the MQTT topic "LED_shift", and pushes the appropriate bit pattern to the shift register when a new message is received from the broker. Here is the entire script:

```
import machine
import utime
from umqtt.simple import MQTTClient
import ujson
import network
import time

# Define pins
ser   = machine.Pin(20, machine.Pin.OUT)
rclk  = machine.Pin(18, machine.Pin.OUT)
srclk = machine.Pin(16, machine.Pin.OUT)

# Define LED patterns in an array
led_patterns = [0b10000001, 0b01000010, 0b00100100, 0b00011000,
                0b00011000, 0b00100100, 0b01000010, 0b10000001, 0b00000000]

# Wi-Fi credentials
SSID      = '<your_wifi_network_SSID>'
PASSWORD  = '<your_wifi_network_password>'

# MQTT parameters
SERVER        = '<your_MQTT_broker_IP_address>'
CLIENT_ID     = b'<your_device_client_ID>'
TOPIC         = b'LED_shift'
USER          = b'<your_MQTT_username>'
PASSWORD      = b'<your_MQTT_password>'

# Function to shift data into 74HC595
def hc595_shift(dat):
    rclk.low()
```

```
    time.sleep_ms(5)
    for bit in range(7, -1, -1):
        srclk.low()
        time.sleep_ms(5)
        value = 1 & (dat >> bit)
        ser.value(value)
        time.sleep_ms(5)
        srclk.high()
        time.sleep_ms(5)
    time.sleep_ms(5)
    rclk.high()
    time.sleep_ms(5)

def mqtt_connect():
   client = MQTTClient(CLIENT_ID, SERVER, user=USER, password=PASSWORD, 
keepalive=3600)
   client.connect()
   print('Connected to %s MQTT Broker'%(SERVER))
   return client

def reconnect():
   print('Failed to connect to the MQTT Broker. Reconnecting...')
   sleep(5)
   machine.reset()

def on_message(topic, msg):
    print("Received message:", msg.decode())
    hc595_shift(led_patterns[int(msg.decode())])

def main(server, topic, username, passw):
    client = MQTTClient(CLIENT_ID, server, user=username, password=passw, 
keepalive=3600)
    client.set_callback(on_message)
    client.connect()
    client.subscribe(topic)
    print("Connected to %s, subscribed to %s topic" % (server, topic))

    try:
        while True:
            client.check_msg()
            time.sleep(0.5)

    finally:
        client.disconnect()

main(SERVER, TOPIC, USER, PASSWORD)
```

Let's break this code down and explain its operation, line by line. To summarise this discussion, I will not explain the parts of the code related to Wifi and the MQTT broker as you have seen them in previous projects.

This script creates a Raspberry Pi Pico to connect to an MQTT broker and subscribe to a specific topic (LED_shift). When a message is received on that topic, it selects one of the predefined LED patterns, which is then shifted into a 74HC595 shift register to control a set of LEDs.

The script uses the "umqtt.simple" library to handle MQTT communication, which allows the Pico to interact with a Node-RED flow that manages the LED patterns. It also includes functions to handle connections, reconnections, and received messages.

The LED patterns array led_patterns holds nine different binary patterns representing various LED configurations. The message received from the MQTT broker is expected to be an integer index that selects one of these patterns, which is then sent to the 74HC595 to display the corresponding LED configuration.

From this script, I'd like to highlight the following lines of code:

- **9-11**: Define pins for serial data ("ser"), register clock ("rclk"), and shift register clock ("srclk") of the 74HC595 shift register. This script uses the same pins for these roles as this chapter's first two MicroPython scripts.

- **14-16**: Define the "led_patterns" array with nine 8-bit LED patterns. Feel free to edit the patterns in this array or add new ones in additional cells and experiment.

- **30-41**: Define a function "hc595_shift(dat)" to shift data into the 74HC595 shift register. This is the same function you saw in the second MicroPython script in this chapter. The "dat" parameter is a bit pattern that contains eight bits. In this script, the source of the bit patterns is the "led_patterns" array. However, it is easy to modify this script so that the bit pattern is received from the Node-RED flow and further outsources functionality from the MicroPython script to Node-RED.

- **53-56**: Define a function "on_message(topic, msg)" to handle received messages from the MQTT broker and shift the corresponding LED pattern.

Once you are satisfied that you understand this script, go ahead and upload it to your Raspberry Pi Pico W. Wait for a few seconds for the Pico to connect to the network and the MQTT broker, and then move the dashboard slider. The LEDs on the breadboard should light up following the bit pattern in the array cell corresponding to the slider position integer.

4.3. WS2812 RGB LED Strip

The WS2812 is an intelligent control LED light source that integrates the control circuit and RGB chip in a package of 5050 components. Each LED acts as a separate addressable unit, allowing you to control the colour and brightness of individual LEDs on the strip.

This project teaches you how to use a WS2812 RGB 8 LED strip with the Raspberry Pi Pico. As always, we'll apply the new knowledge to a Node-RED project after learning to use the WS2812 RGB strip with the Raspberry Pi Pico and MicroPython.

Hardware overview

The WS2812 RGB strip is a chain of individually addressable LEDs. Each LED can be independently controlled and set to any colour by adjusting the intensity of the red, green, and blue channels. Each RGB module is managed by a controller circuit that deceives and decodes data to control the chip's individual red, green, and blue LEDs. The controller uses a single-wire serial communication protocol based on pulse width modulation (PWM) to define bits. According to this protocol:

- 0 bit: a short high pulse followed by a long low pulse.
- 1 bit: a long high pulse followed by a short low pulse.
- Reset Pulse: a low pulse for a specific period to mark the end of data transmission.

The data transmission order is Green-Red-Blue for each LED, and the data is sent in a cascade manner to control multiple LEDs.

Here are some additional details about the WS2812 RGB strip that are interesting to note:

- The LED strip operates at a voltage of 5 V. The Raspberry Pi Pico can provide a 5V supply voltage from the VBUS pin.

- Each LED draws up to 60 mA when set to full brightness and white (all RGB channels on). The USB power supply in this experiment draws power from the computer USB port and should provide sufficient power for all eight RGB LEDs to turn on. However, if you plan to use a strip with more LEDs, consider the power requirements carefully and perhaps use an external power supply.

- Each LED is individually addressable with 24-bit colour (8 bits for each Red, Green, and Blue), resulting in 16.7 million colour combinations.

- The strip has three connection points: 5 V, GND, and DATA. The data connection is a one-wire interface that sequentially transmits data to the strip.

Wiring

Let's connect the RGB LED strip to the Raspberry Pi Pico. Here are the connections (also refer to the wiring diagram below):

- Connect the 5 V pin of the LED strip's VBUS pin to the Raspberry Pi Pico.

- Connect the GND pin of the LED strip to a GND pin on the Raspberry Pi Pico.
- Connect the DATA pin of the LED strip to a GPIO pin on the Raspberry Pi Pico. Any GPIO pin can be used; I have used GPIO28 because it is conveniently located near the VBUS and GND pins.

You can see the wiring in the diagram below:

Figure 2.4.3.13: The RGB LED strip wiring.

You can see my circuit on the breadboard in the photograph below. My RGB LED strip came with flexible jumper wires, making connecting to the breadboard very easy. I used three solid-core jumper wires to secure my RGB LED strip against the breadboard.

Figure 2.4.3.14: The RGB LED strip on my breadboard.
The yellow jumper wires serve as tie-downs.

Once you have completed the wiring, continue with the programming.

Programming the WS2812 Strip with MicroPython

The NeoPixel library in MicroPython can be used to control the WS2812 strip. This library is built to the MicroPython firmware you have installed in your Raspberry Pi Pico, so you don't need to install anything else. You can find detailed documentation on the MicroPython website at https://txplo.re/upython_neopixel.

Let's start with something simple, such as using the neopixel library to make one of the RGB LED modules on the strip display red. Here is the MicroPython script that can do this:

```
from machine import Pin
import neopixel

# create NeoPixel driver on GPIO28 for 8 pixels
np = neopixel.NeoPixel(Pin(28), 8) # Using GP28 because it is convenient.

# set the first pixel to red
np[0] = (255, 0, 0)

# write data to all pixels
np.write()
```

In this code, the neopixel.NeoPixel class is used to create an object that controls the LED strip. The constructor of this class takes two parameters: the pin that the DATA line of the strip is connected to and the number of LEDs in the strip.

The colour of each LED is set using a tuple of three values between 0 and 255 that represent the intensity of the red, green, and blue channels, respectively. In this example, we want to instruct the RGB LED module in position "0" (the LED closest to the side of the strip where the three wires are located) to show full red. To do this, we write the tuple "(255, 0, 0)" to the "np" array cell "0". The first parameter of the tuple controls the red LED in the RGB LED device, and the value 255 will display red at its maximum intensity. The other two colours (green and blue) remain turned off.

The "np.write()" function writes the colour data to the LED strip, causing the LEDs to change colour.

Go ahead and upload this script to your Pico. Notice how the first RGB LED in the strip lights up red, and the rest are off.

More lights effects

Turning on one LED is a good start, but we won't stop there. Let's look at a new script that can "write" the red colour to all RGB LEDs on the strip, one at a time, to create a simple light effect.

The following script creates a simple animation that cycles through each LED in the strip:

```
from machine import Pin
import neopixel
import time
```

```
np = neopixel.NeoPixel(Pin(28), 8)

while True:
    for i in range(8):
        np[i] = (255, 0, 0)
        np.write()
        time.sleep(0.5)
        np[i] = (0, 0, 0)
```

This code uses a "for" loop to iterate over each LED in the strip. The "time.sleep" function creates a delay between each animation step. The colour of each LED is set to red and then turned off before moving to the next LED.

Remember always to call "np.write()" after setting the LED colours to update the actual LEDs.

Let's try one more experiment because we bring Node-RED into the mix. A MicroPython script will generate a random colour for each LED on the WS2812 strip every second. The script uses the "urandom" module to generate random numbers:

```
from machine import Pin
import neopixel
import urandom
import time

# create NeoPixel driver on GPIO0 for 8 pixels
np = neopixel.NeoPixel(Pin(28), 8)

while True:
    for i in range(8):  # 8 for the number of LEDs in the strip
        # generate random RGB values
        r = urandom.randint(0, 100)
        g = urandom.randint(0, 100)
        b = urandom.randint(0, 100)

        # set the LED to the random color
        np[i] = (r, g, b)

    # write data to all pixels
    np.write()

    # wait for a second
    time.sleep(0.1)
```

The "urandom.randint(0, 100)" function generates a random integer between 0 and 100 in this script. I decided to use "100" instead of the maximum "255" to reduce eye strain, as the maximum values produce extremely bright light!

This is done for each colour channel (red, green, blue) to create a random colour. This colour is set for each LED in the strip, and the "np.write()" function is used to update the LEDs. The script pauses briefly before starting the loop, creating a continuous random colour show.

Control the RGB LED strip with a Node-RED flow

Let's create a Node-RED flow that can control a WS2812 RGB LED strip with eight LEDs, like the one in the previous experiments in this chapter. We'll set up the Node-RED flow to use four slider widgets. We'll use three sliders to set the red, green and blue light intensities for the RGB LEDs, and the fourth slider is for the delay speed. You can see the RGB LED control widgets on the dashboard below:

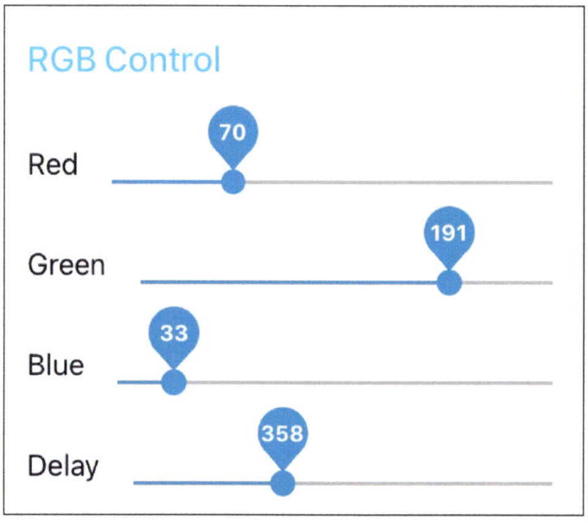

Figure 2.4.3.15: This Node-RED dashboard allows full colour and blink rate control of the RGB LED strip.

The red, green and blue sliders have an integer range from 0 to 255, and the delay slider from 0 (no blink) to 1000 (one-second blink).

When a change is made to the widgets, the information from all widgets is sent to the Raspberry Pi Pico via an MQTT broker as a single JSON message. On the Raspberry Pi Pico side, the MicroPython program will fetch the four values from the broker and use them to display the defined colour in the strip. If the delay value exceeds zero, the LED strip will blink at the rate determined by the "delay" value.

Here is the Node-RED flow in the editor:

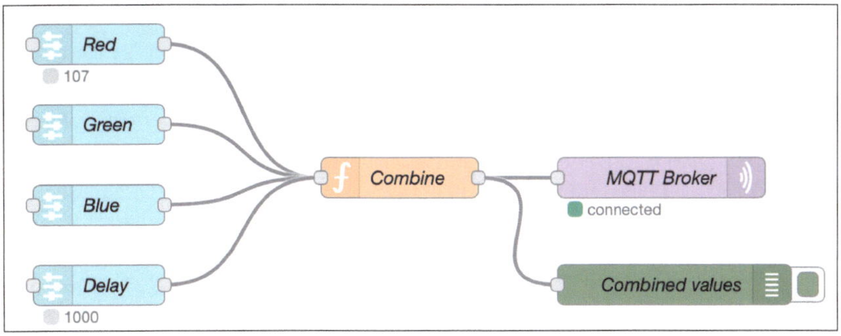

Figure 2.4.3.16: This Node-RED flow creates a dashboard t hat allows us to control the RGB LED strip.

Let's look at the details for each of the nodes. The settings are identical except for the labels and payload topics for the Red, Green and Blue slider nodes. See below:

- Dashboard Group: Create a new dashboard group and tab titled "RGB LED Control" and "RGB Control", respectively.
- Label: "Red" (then "Green" and "Blue" for the other two sliders).
- Range:
  - min: 0
  - max: 255
  - step: 1
- Output: Only on release.
- When changed:
  - Payload: current value
  - Topic: (string) red (then "green" and "blue" for the other two sliders).
- Name: "Red" (then "Green" and "Blue" for the other two sliders).

For the Delay slider nodes:

- Dashboard Group: Same as the other three slider widgets.
- Label: Delay
- Range:
  - min: 0
  - max: 1000
  - step: 1
- Output: only on release.
- Payload:
  - Current value
  - Topic: (string) delay
- Name: Delay

In the Combine function node, we'll copy the Javascript code responsible for receiving the values from the sliders and combining them into a single JSON payload that is then sent to the MQTT broker. This is the Javascript (followed by a line-by-line explanation of this code):

```
var colors = context.get('colors') || {red:0, green:0, blue:0, delay:0};

if(msg.topic === 'red') colors.red = msg.payload;
if(msg.topic === 'green') colors.green = msg.payload;
if(msg.topic === 'blue') colors.blue = msg.payload;
if(msg.topic === 'delay') colors.delay = msg.payload;

context.set('colors', colors);

msg.topic = 'RGB_LED_Control';
msg.payload = colors;
return msg;
```

Here's the line-by-line explanation of the provided code:

1. var colors = context.get('colors') || {red:0, green:0, blue:0, delay:0};
   - This line retrieves the 'colours' object from the Node-RED context if it exists. If not, it initialises a new object with red, green, blue, and delay properties, all set to 0.

2. if(msg.topic === 'red') colors.red = msg.payload;
   - If the incoming message's topic is 'red', the red property of the 'colours' object is updated with the payload value from the message.
   - The same operation repeats for the "green", "blue" and "delay" widgets.
3. context.set('colors', colors);
   - This line stores the updated 'colours' object in the Node-RED context, making it accessible to other nodes or subsequent messages in the same flow.

4. msg.topic = 'RGB_LED_Control';
   - This line sets the topic of the outgoing message to 'RGB_LED_Control'. This line is important because we'll use it in the "MQTT out" node to automatically set the MQTT topic where the payload will be sent. Therefore, the "MQTT out" node will use the "msg.topic" value to set the MQTT topic to publish its payload.

5. msg.payload = colors;
   - This line sets the payload of the outgoing message to the 'colors' object, containing the updated red, green, blue, and delay values.

6. return msg;
   - Finally, the modified message object is returned, sending it to the next node in the flow.

Set up the flow and deploy it. When you are ready, continue with the MicroPython script for the Pico.

MicroPython script

On the Raspberry Pi Pico W side, we will use the following script to receive the payload with the RGB and delay values from the MQTT broker and use those values to control the RGB LED strip. Let's look at the script below and follow up with a line-by-line description of its operation:

```
import machine
import time
from umqtt.simple import MQTTClient
import network
import neopixel

# Constants
NUM_LEDS = 8
PIN = 28 # GPIO pin connected to WS2812

# Initialize the NeoPixel
np = neopixel.NeoPixel(machine.Pin(PIN), NUM_LEDS)

# Wi-Fi credentials
   SSID      = '<your_wifi_network_SSID>'
   PASSWORD  = '<your_wifi_network_password>'

# MQTT parameters
SERVER     = '192.168.20.18'    # Replace with your MQTT broker IP
CLIENT_ID  = b'pico'
TOPIC      = b'RGB_LED_Control'
USER       = b'<your_MQTT_username>'  # Replace with your MQTT username
PASSWORD   = b'<your_MQTT_password>'  # Replace with your MQTT password

# MQTT Callback function
def on_message(topic, msg):
    print("Received message:", msg)
    data  = eval(msg) # Convert string message to dictionary
    red   = data['red']
    green = data['green']
    blue  = data['blue']
    delay = data['delay']

    # Set all LEDs to the received color
    for i in range(NUM_LEDS):
        np[i] = (red, green, blue)
    np.write()

    # Blink the LEDs with the received delay
    time.sleep_ms(delay)
```

```
    for i in range(NUM_LEDS):
        np[i] = (0, 0, 0)
    np.write()
    time.sleep_ms(delay)

# Function to connect and subscribe to MQTT broker
def connect_and_subscribe():
    client = MQTTClient(CLIENT_ID, SERVER, user=USER, password=PASSWORD,
keepalive=3600)
    client.set_callback(on_message)
    client.connect()
    client.subscribe(TOPIC)
    return client

# Main function
def main():
    # Activate the network interface
    wlan = network.WLAN(network.STA_IF)
    wlan.active(True)
    #wlan.connect(SSID,PASSWORD)
    wlan.connect(SSID, PASSWORD, channel=12)

    print('Starting... Trying to connect to Wifi...')
    while not wlan.isconnected():
        print('.', end='')
        time.sleep(1)

    print('\nConnected to Wi-Fi')
    print(wlan.ifconfig())

    client = connect_and_subscribe()
    print('Connected to broker')


    # Keep waiting and processing MQTT messages
    while True:
        client.wait_msg()

# Run the main function
if __name__ == "__main__":
    main()
```

The functions of this program should be familiar to you as much of its code is shared with scripts in earlier projects in this book. This script connects to a Wi-Fi network using the specified SSID and password and subscribes to the "RGB_LED_Control" topic of the MQTT broker. When messages are received on this topic, the script extracts the message's red, green, blue, and delay values, sets the LEDs to the corresponding colour, and blinks the LEDs with the specified delay. The script provides a complete solution to control an RGB LED strip over MQTT and Wi-Fi remotely.

The part of this program you have not seen in other projects in this book is inside the "on_message" function. The code there does the following:

1. Define a function that will be called when receiving an MQTT message.
2. Print the received message and convert it to a dictionary.
3. Extract the message's red, green, blue, and delay values.
4. Set the LEDs to the received color and write to the NeoPixel strip.
5. Blink the LEDs with the received delay value.

Upload this code to your Pico. Assuming that the matching Node-RED flow is already deployed, you can move the sliders in the dashboard and see the RGB LED strip on the breadboard responding with the matching colour and blink delay.

You can expand on this project by, for example, creating dashboard buttons with preset colours or light effects. For example, you can set up buttons to make the strip display a specific colour and another button to display a flash show or animation.

Chapter 2.5 • Motors

This chapter will teach you the use of the servo motor and the DC motor alongside the Raspberry Pi Pico W and Node-RED.

The first project is about Servo motors. These are the go-to choices for precise angular movement. Whether rotating a camera or moving a robotic arm, servo motors offer precise control. You'll learn to wire a Servo motor to your Raspberry Pi Pico W, control it through Node-RED, and write a MicroPython script to manage its movements.

Then, you'll shift gears and work with DC motors, the workhorses of the motor world. These motors provide continuous rotation and are commonly used in applications like fans, wheels, and conveyor belts. You'll set up a circuit with your Raspberry Pi Pico W, create a Node-RED flow to control the motor and write a MicroPython script to fine-tune its speed and direction.

Each project in this chapter is a complete learning package featuring a Node-RED flow, a Raspberry Pi Pico W circuit, and a MicroPython script. By the time you turn the last page of this chapter, you'll have a solid grasp of motor control, ready to bring dynamic movement to your future projects.

5.1. Servo Motor

In this project, you will learn how to control a 9*g* servo motor using the Raspberry Pi Pico and MicroPython. The servo motor is a type of motor that can be directed to rotate to specific angles. This feature makes it suitable for numerous applications like robotics, home automation, etc.

Let's start by understanding some of the details of servo motors and how they work.

An introduction to servo motors

A servo motor is a rotary or linear actuator that precisely controls angular or linear position, velocity, and acceleration. With a regular DC motor, it is possible to control its rotor's speed and direction of movement. However, it is impossible to control its position accurately without additional external hardware. A servo motor consists of a DC motor coupled to a sensor for position feedback, controlled by a closed-loop control system. This closed-loop control system makes it possible to achieve precise movements with a servo motor.

A servo motor's DC motor is driven by applying a voltage to its terminals, creating a magnetic field that rotates the motor shaft. A sensor, usually a potentiometer, is attached to the motor shaft to provide real-time feedback about the shaft's position. This information is returned to the control circuit, forming a closed-loop system. The control system processes the signal from the feedback sensor and compares it with the desired position. Any deviation from the desired position results in a corrective signal being sent to the motor, adjusting its position.

A microcontroller, like the Arduino UNO or the Raspberry Pi Pico, controls a servo motor by sending a PWM signal. This signal consists of a series of pulses at a fixed frequency, with the width of each pulse determining the position of the motor shaft. The frequency is typically 50 Hz for standard servos, like the one we will use in this chapter. The pulse width varies between 1 ms and 2 ms. For example, if we send a 1ms pulse, the shaft will rotate to one extreme position, while a 2ms pulse will cause the shaft to rotate to another extreme position. A 1.5 ms pulse, on the other hand, will cause the shaft to move to its centre position.

Servo motor experiment

In this experiment, you will learn how to control the position of a servo motor's shaft using MicroPython on the Raspberry Pi Pico. The small servo motor in the Sunfounder kit draws only a tiny amount of power within the Raspberry Pi Pico's capabilities. If you plan to use a larger servo motor, connect it to an external power supply capable of supplying the required power.

Let's start with the wiring and then continue with the MicroPython script.

A servo motor typically has three wires:

- Ground, usually black or brown.
- Power, usually red.
- Signal, usually orange, white or yellow.

Let's connect our small servo motor to the Pico. Here are the step-by-step instructions, followed by a wiring diagram:

1. Connect the Power wire from the servo to the VBUS pin of the Raspberry Pi Pico. This is physical pin 40 on the Pico and can provide 5V power to the servo motor.

2. Connect the Ground wire from the servo to a GND pin on the Raspberry Pi Pico.

3. Connect the Signal wire from the servo to a GPIO pin on the Raspberry Pi. I will be using GPIO18 because it makes it more convenient to make the connections.

Here is the wiring diagram:

Figure 2.5.1.1: Wiring of a small servo motor to the Pico.

And this is my servo motor connected to my Raspberry Pi Pico. I have used a small male header to facilitate the connection and avoid using additional jumper wires.

Figure 2.5.1.2: I use a yellow male header to wire the servo motor to the Pico.

The connection is complete, so let's turn to the MicroPython script.

Servo control with MicroPython

MicroPython comes with support that allows us to create PWM signals for servo motors and other devices easily. To make the PWM signals, MicroPython provides the PWM, which we'll use in the example script later in this chapter.

The Pulse Width Modulation (PWM) module in MicroPython generates PWM signals, allowing control over devices like servo motors, LEDs, and more. It provides a way to create a signal with a specific duty cycle and frequency.

Let's look into how to use the PWM module and its most essential functions. This will help you understand the code in the example script later in this chapter. If you want to learn more about this module, you can refer to the online documentation at https://txplo.re/upython_pwm.

PWM modulation quick start guide

As you probably know, PWM is a modulation technique that generates a waveform with a fixed frequency but a varying duty cycle. The duty cycle represents the percentage of time the signal is ON in one cycle. In MicroPython, we can use the PWM module to create a PWM signal on a microcontroller like the Raspberry Pi Pico.

Let's look at the steps in creating a PWM signal, followed by an example.

Step 1. Initialisation

To create a PWM object, initialise it with a pin and specify frequency and duty cycle optionally. Example code:

```
from machine import Pin, PWM
pwm = PWM(Pin(18)) # Initialize PWM on pin 18
```

Step 2. Frequency Control

You can set or retrieve the frequency of the PWM signal. The default frequency varies by hardware but can be easily adjusted. Example code:

```
pwm.freq(50)      # Set frequency to 50 Hz
current_freq = pwm.freq() # Get current frequency
```

Step 3. Duty Cycle Control

The duty cycle is defined as a 10-bit value ranging from 0 (0%) to 1023 (100%). Example code:

```
pwm.duty(512)       # Set duty cycle to 50%
current_duty = pwm.duty() # Get current duty cycle
```

Step 4. Deinitialisation

To stop the PWM signal, you can deinitialize it, which releases the pin for other uses. Example code:

```
pwm.deinit()
```

Fade In/Out PWM example

You can create effects like fading an LED in and out by varying the duty cycle over time. Here is a simple example code snippet:

```
import time

for i in range(1024):
    pwm.duty(i)
    time.sleep_ms(2)
```

This script will work on the Raspberry Pi Pico (and any other microcontroller that supports MicroPython). As I mentioned earlier, by default, it will work at a 10-bit resolution, allowing us to specify 1024 shaft positions (from 0 to 1023). On the Raspberry Pi Pico, the resolution of the PWM (Pulse Width Modulation) can be set to be up to 16 bits. This means that the duty cycle of the PWM signal can be controlled with a resolution of 65536 different levels. It allows for fine-grained output control, making it suitable for applications requiring precise signal adjustments.

Below, you can see a complete example that shows how to set the PWM resolution to 16-bit on the Pico:

```
from machine import Pin, PWM
import time

# Define the pin where PWM signal will be generated
pwm_pin = Pin(18)

# Initialize the PWM object on the specified pin with 16-bit resolution
pwm = PWM(pwm_pin)

# Set the frequency (for example, 50 Hz)
pwm.freq(50)

# Set the duty cycle using the full 16-bit resolution (0 to 65535)
pwm.duty_u16(32768) # 50% duty cycle

# Optional: Ramp the duty cycle from 0 to 100%
for i in range(65536):
    pwm.duty_u16(i)
    time.sleep(0.001)

# Deinitialize to stop the PWM
pwm.deinit()
```

The difference between the two examples is that in the 16-bit example, we use the "pwm.duty_u16" function to set the duty cycle, whereas in the 10-bit example, we use the "pwm.duty" function.

Full example script

Now that we have the basics let's put everything together in a script. This script uses the "pwm.duty_u16" function to produce a PWM signal that controls the attached servo motor precisely. The script will rotate the motor shaft to the two extremes and then cycle its position from one extreme to another in steps of 10°.

You can see the script below. After the script, you will find a line-by-line description of how it works.

```
from machine import Pin, PWM
import utime

# Define the GPIO pin and the frequency (50Hz is common for servo motors)
GPIO_PIN = 18
FREQ     = 50

# Create a PWM object
pwm = PWM(Pin(GPIO_PIN))

# Set the PWM frequency
pwm.freq(FREQ)

def set_angle(angle):
    # Duty cycle for servo is between 2.5% (0 degrees) and 12.5% (180 degrees)
    duty = 2.5 + (angle / 180) * 10
    # Update duty cycle
    pwm.duty_u16(int(duty * 65535 / 100))


# Set the servo to 90 degrees
set_angle(90)
utime.sleep(1)

# Set the servo to 180 degrees
set_angle(180)
utime.sleep(1)

# Set the servo to 0 degrees
set_angle(0)
utime.sleep(1)


# Make the servo motor continuously rotate from 0 to 180 degrees and back
while True:
    for angle in range(0, 180, 10): # Go from 0 to 180 degrees
        set_angle(angle)
        utime.sleep(0.5)
    for angle in range(180, -1, -10): # Go from 180 back to 0 degrees
        set_angle(angle)
        utime.sleep(0.5)


# Turn off PWM
pwm.deinit()
```

Below is a line-by-line description of the provided MicroPython script that controls a servo motor on the Raspberry Pi Pico.

1. Importing Libraries: The required modules, Pin and PWM from "machine" and" utime", are imported.

2. Pin and Frequency Definition: GPIO pin 16 and a frequency of 50 Hz (common for servo motors) are defined.

3. Creating PWM Object: A PWM object is created on the defined GPIO pin.

4. Setting PWM Frequency: The frequency of the PWM signal is set to 50 Hz.

5. Defining the set_angle Function: This function takes an angle as an argument and calculates the corresponding duty cycle to control the servo motor.
   - Calculating Duty Cycle: The duty cycle for a standard servo motor is calculated based on the angle. It ranges from 2.5% for 0 degrees to 12.5% for 180 degrees.
   - Setting Duty Cycle: The duty cycle is then set using the full 16-bit resolution.

6. Setting Servo to 90°: The servo is set to 90° and held there for 1 second.

7. Setting Servo to 180 °: The servo is set to 180° and held there for 1 second.

8. Setting Servo to 0°: The servo is set to 0° and held there for 1 second.

9. Continuous Rotation Loop: The script then enters a continuous loop where:
   - Forward Loop: The servo moves from 0 to 180 degrees in increments of 10, pausing for 0.5 seconds at each step.
   - Reverse Loop: The servo moves from 180 to 0 degrees in decrements of 10, pausing for 0.5 seconds at each step.

10. Deinitializing PWM: (This line is never reached due to the infinite loop.) The PWM is turned off by deinitialising the object.

The script will continuously rotate the servo motor between 0° and 180°, demonstrating control of a servo using PWM on the Raspberry Pi Pico.

Servo motor control with Node-RED

Let's create a simple Node-RED flow that can control a servo motor connected to the Raspberry Pi Pico. The flow will implement a slider that we can move left and right. The position of the servo motor shaft will be copying the slider's position on the dashboard. To make the user interface easy for the human user, we'll set the slider to use the right degrees from 0° to 180° instead of raw PWM values, as we did with the last example of the MicroPython script.

The dashboard will look like this:

Figure 2.5.1.3: This dashboard includes a slider and gauge widgets.

Below, you can see the Node-RED flow. It consists of a slider widget that we use to set the servo motor angle, a gauge widget to show the set angle in the dashboard, a debug node to show the selected angle in the debug tab, and an MQTT-out node to send the set angle value to the MQTT broker.

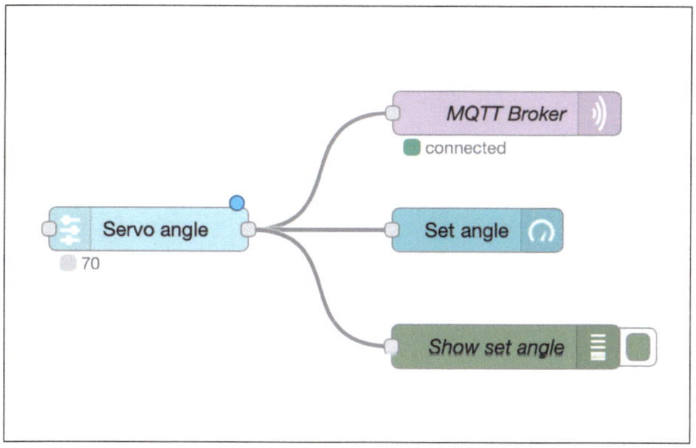

Figure 2.5.1.4: This flow for our Node-RED servo motor controller.

The MicroPython script fetches angle values from the MQTT broker and moves the servo motor shaft to match that angle value. As you can see, this script combines the previous MicroPython example in this project and the Wi-Fi and MQTT part of earlier experiments in this book.

```
from machine import Pin, PWM
from time import sleep
from umqtt.simple import MQTTClient
import network
```

```
# Define the GPIO pin and the frequency
# (50Hz is common for servo motors)
GPIO_PIN = 18
FREQ     = 50

# Create a PWM object
pwm = PWM(Pin(GPIO_PIN))

# Set the PWM frequency
pwm.freq(FREQ)

def set_angle(angle):
    # Duty cycle for servo is between 2.5% (0 degrees)
    # and 12.5% (180 degrees)
    duty = 2.5 + (angle / 180) * 10
    # Update duty cycle
    pwm.duty_u16(int(duty * 65535 / 100))


# Configure your Wi-Fi credentials
   SSID      = '<your_wifi_network_SSID>'
   PASSWORD  = '<your_wifi_network_password>'

wlan = network.WLAN(network.STA_IF)
wlan.active(True)
wlan.connect(SSID,PASSWORD)

while not wlan.isconnected():
    pass

print('Connected to Wi-Fi')
print(wlan.ifconfig())

# Setup MQTT
SERVER          = '192.168.20.18'  # Replace with your MQTT broker IP
CLIENT_ID       = b'PICO'
TOPIC_TIMESTAMP = b'servo_angle'

USER      = b'<your_MQTT_username>'  # Replace with your MQTT username
PASSWORD  = b'<your_MQTT_password>'  # Replace with your MQTT password

def mqtt_connect():
   client = MQTTClient(CLIENT_ID, SERVER, user=USER, password=PASSWORD,
keepalive=3600)
   client.connect()
   print('Connected to %s MQTT Broker'%(SERVER))
```

```
    return client

def reconnect():
    print('Failed to connect to the MQTT Broker. Reconnecting...')
    sleep(5)
    machine.reset()

def on_message(topic, msg):
     angle = int(msg.decode())
     print("Received message:", angle)
     set_angle(angle)
     sleep(0.5)

def main(server, topic, username, passw):
     client = MQTTClient(CLIENT_ID, server, user=username, password=passw,
keepalive=3600)
     client.set_callback(on_message)
     client.connect()
     client.subscribe(topic)
     print("Connected to %s, subscribed to %s topic" % (server, topic))

     try:
         while True:
             client.check_msg()
             sleep(1)

     finally:
         client.disconnect()

main(SERVER, TOPIC, USER, PASSWORD)
```

Here's a line-by-line description of the provided MicroPython script that controls a servo motor on the Raspberry Pi Pico by receiving angle values from an MQTT broker:

1. Importing Libraries: Required modules such as Pin, PWM, sleep, MQTTClient, and network are imported.

2. Pin and Frequency Definition: GPIO pin 18 and a frequency of 50 Hz (typical for servo motors) are defined.

3. Creating PWM Object: A PWM object is created on the defined GPIO pin.

4. Setting PWM Frequency: The frequency of the PWM signal is set to 50 Hz.

5. Defining set_angle Function: This function takes an angle as an argument and calculates the corresponding duty cycle.

- Calculating Duty Cycle: It ranges from 2.5% for 0 degrees to 12.5% for 180 degrees.
- Setting Duty Cycle: The duty cycle uses the full 16-bit resolution.

6. Wi-Fi Credentials Configuration: SSID and PASSWORD are defined for connecting to the local Wi-Fi network.

7. Wi-Fi Connection Setup: The script activates the WLAN interface, connects to the specified Wi-Fi network, and waits until connected.

8. Print Connection Information: Prints the successful connection message and IP configuration.

9. MQTT Configuration: The MQTT broker IP, client ID, topic, username, and password are defined.

10. Defining mqtt_connect function: This function creates and returns an MQTT client connected to the specified server.

11. Defining reconnect function: If the connection to the MQTT broker fails, this function will attempt to reconnect after 5 seconds.

12. Defining on_message Function: A callback function that takes the received MQTT message (angle), decodes it, prints it, and calls set_angle to adjust the servo.

13. Defining main() function: The main() function manages MQTT connection and subscription to the specified topic.
    - Setting Callback: The on_message function is the callback for incoming messages.
    - Connecting and Subscribing: The client connects to the MQTT broker and subscribes to the specified topic.
    - Message Checking Loop: In an infinite loop, the client checks for incoming messages and sleeps for 1 second between checks.
    - Disconnection Handling: The client is disconnected correctly if the loop exits (e.g., due to an exception).

14. Calling main() function: The main() function is called with the previously defined server, topic, username, and password, initiating the process.

The script continuously receives angle values from an MQTT broker. It sets the servo motor's position accordingly, providing integration with a Node-RED flow or other systems that can send MQTT messages.

You can use the techniques you learned in this chapter in other projects that require precise motor control. For example, you can adjust the angle of window blinds based on weather, time of day, or manual input through a Node-RED dashboard that receives environmental and other information from sensors and cloud services. Or you can design an automatic pet feeder that dispenses food at scheduled times or through manual activation from a mobile device.

5.2. DC Motor

In this project, you will learn how to control a DC motor using the Raspberry Pi Pico W and the TA6586 motor driver. The TA6586 is a motor driver integrated circuit (IC) capable of driving various motors, including DC motors, and it will handle the power requirements. At the same time, the Raspberry Pi Pico (or any other microcontroller) manages the control signals. This chip provides the functions to translate logic-level control signals into higher currents and voltage required to drive the motor's windings. One TA6586 chip can drive (or "control") a single DC motor.

Figure 2.5.2.5: The TA6586 motor controller DIP-8 IC.

TA6586 applications and pins

The TA6586 is typically found in applications such as:

- Cooling fans in computers and electronics
- Electric pumps
- Small household appliances
- Various types of small electrically driven motors

The TA6586 is available in a DIP-8 package. As the name suggests, this chip has eight pins and can be plugged into a regular breadboard. Below, you can see the roles of each pin:

1. BI: Back input.
2. FI: Forward input.
3. GND: Ground.
4. Vcc: Power supply.
5. FO: Forward Output (connect one wire from the motor here).

6. FO: Forward Output (same as pin 5).
7. BO: Back Output (connect the other wire from the motor here).
8. BO: Back Output (same as pin 7).

Pins 5 and 6 are electrically shorted, as are pins 7 and 8. This means you can connect one wire of the DC motor to either pins 5 and 6 and the other to pins 7 or 8.

This motor controller features several valuable features, such as a very low stand-by current (less than two micro-amperes), making it very economical when the motor is not working. It also has a wide range of compatible supply voltages, from as low as 3 V up to 14 V. It also has built-in safety features, such as a thermal shutdown function, which automatically turns off the device when it gets too hot for safe operation, and over-current and short-circuit protection functions.

The device's maximum safe voltage is 15 V, and the maximum current is 9 A.

Wiring Instructions
In this project, we will use the TA6586 to control and power the DC motor and the Raspberry Pi Pico to control the TA6586. We will use an external power supply because the DC motor requires more power than the Pico can provide. The Sunfounder kit contains a LiPo battery, a battery holder with wires, and a battery charger board. We will use these components to provide the necessary power to the TA6586 to supply the motor.

Follow the steps below to connect the components (you can also consult the wiring diagram below, which includes the pin numbers for the TA6586 chip and the charger module):

1. Plug the Raspberry Pi Pico into the breadboard, with the USB port towards the right side of the breadboard.

2. Plug the TA6586 chip on the breadboard. Take care to orient the chip so that the cutoff (semicircle) of the package is towards the Pico. Also, notice the circular mark on the package near the top left corner of the chip. This marking indicates pin 1 of the chip.
   - Connect pin 1 of the IC to GP15 of the Pico.
   - Connect pin 2 of the IC to GP14 of the Pico.
   - Connect pin 3 of the IC to the ground rail of the breadboard.
   - Connect pin 4 of the IC to the power rail of the breadboard.
   - Connect one motor wire to pin 5 of the IC (it does not matter which motor wire you connect).
   - Connect the other motor wire to pin 8 of the IC.

3. Continue with the battery and charger module. The Sunfounder kit contains a charger module and battery holder with wires and matching connectors. Connect the battery holder plug to the charger module plug. These plugs cannot be misconnected.

- Next, plug the charger module into the breadboard to align the charger module pin "V1" with the Pico pin "VBUS". Charger pin "V0" aligns with Pico pin "VSYS", and charger pin "GND" aligns with Pico pin "GND".
- Use a longer jumper wire and connect charger pin "V0" (the middle pin) to the breadboard power rail from where the motor controller draws its power.

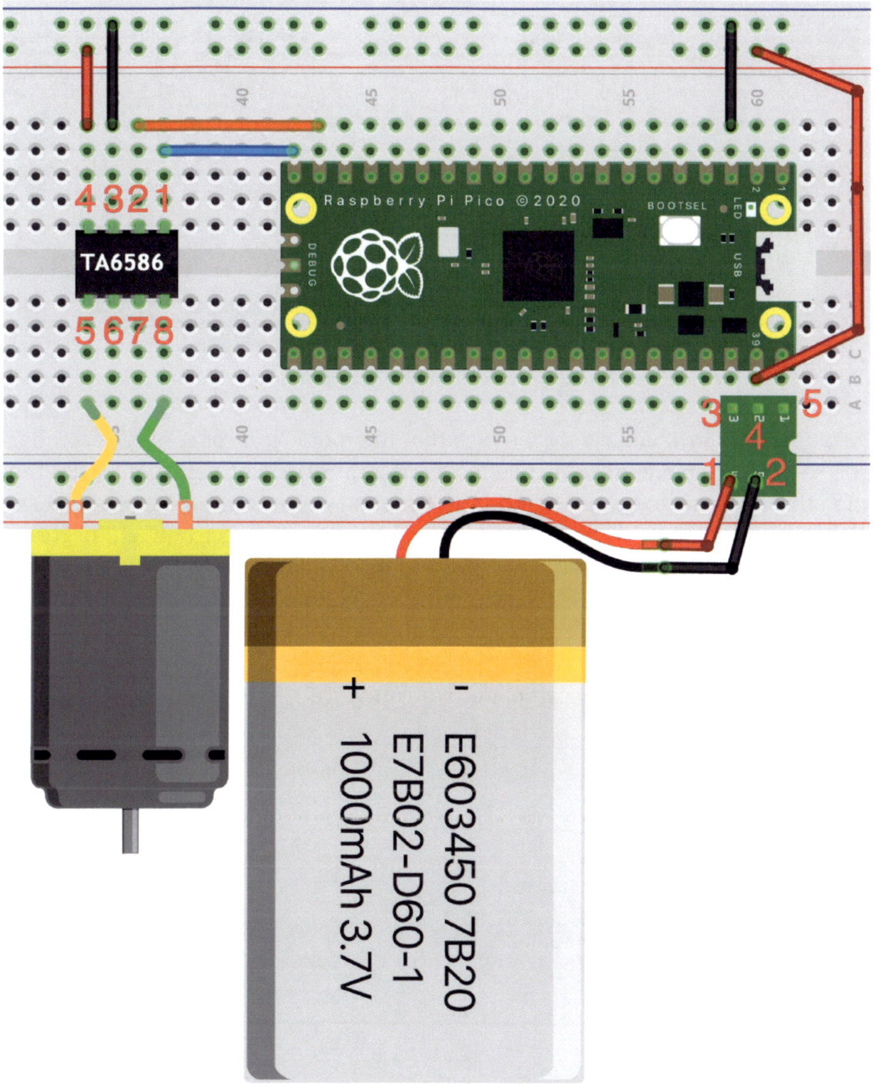

Figure 2.5.2.6: Wiring of the TA6586, DC motor, battery, charger, and Raspberry Pi Pico.

You can see my assembled circuit in the photograph below:

Figure 2.5.2.7: Assembled circuit.

In my experiments, however, rarely, I notice that the motor controller may become "stuck" and keep the motor spinning even though the program has finished running on the Pico. If or when this happens to you, resetting the Raspberry Pi Pico is an easy way to stop the motor from spinning. You can do this with a jumper wire. Connect one end of the wire to pin "Run" of the Pico (this is physical pin 30) and the other to one of the ground pins. This will reset the Pico, and you can restart the program in Thonny normally.

MicroPython code example

Now that the hardware setup is complete, we can write MicroPython code to control the DC motor.

```
import machine
import utime

motor1 = machine.Pin(14, machine.Pin.OUT)
motor2 = machine.Pin(15, machine.Pin.OUT)

def clockwise():
    motor1.high()
    motor2.low()

def anticlockwise():
    motor1.low()
    motor2.high()

def stopMotor():
    motor1.low()
```

```
    motor2.low()

while True:
    clockwise()
    utime.sleep(1)
    stopMotor()
    utime.sleep(1)
    anticlockwise()
    utime.sleep(1)
    stopMotor()
    utime.sleep(1)
```

In this script, we first initialise two GPIO pins (14 and 15) as output pins connected to pins 2 and 1 on the TA6586. We then define the "clockwise" and "anticlockwise" functions to control the motor's direction and spinning time. If the direction is "clockwise", pin 1 is set high, and pin 2 is set low. If the direction is "counterclockwise", pin 1 is set low, and pin 2 is set high. After setting the motor to spin in the desired direction, we let it spin for the specified time, then stop the motor by setting both pin 1 and pin 2 low. Finally, we call the "clockwise" function to make the motor spin clockwise for 5 seconds, pause for 2 seconds, and then call "anticlockwise" to spin in the opposite direction for 5 seconds.

Now, disconnect the USB cable from the computer, leaving only the battery to power the circuit. The Pico, motor controller and DC motor will continue to operate if it doesn't check that that battery is charged. The charger module has an LED status indicator for the battery. When this LED is lit, the battery is being charged. When it is off, it indicates that the battery is fully charged. You can charge the battery by leaving the USB cable connected to the computer for 2-3 hours. This will provide sufficient charge to the battery to make it possible to supply the circuit.

Controlling speed

Before continuing with the Node-RED extension of this experiment, let's look into the MicroPython code that we can use to control the motor's rotation speed. In the previous chapter, you learned how to use PWM to control the servo motor's shaft position. We can use PWM to control their speed when working with DC motors.

In the MicroPython script below, you can see the relevant code:

```
from machine import Pin, PWM
import time

motor1    = Pin(15, machine.Pin.OUT)
motor2pwm = Pin(14)

# Create a PWM object attached to motor2pwm pin
motor_pwm = PWM(motor2pwm)
```

```
# Set the PWM frequency (usually 50Hz to 2kHz for motor control)
motor_pwm.freq(50)

# Function to set motor speed (0 to 1023)
def set_speed(speed):
    if speed < 0:
        speed = 0
    elif speed > 1023:
        speed = 1023
    motor_pwm.duty_u16(speed * 64) # PWM duty is 16-bit,
                                   # so multiply by 64

# Example of using set_speed function to control motor speed
motor1.low()    # set spin direction
set_speed(1000) # Set speed close to max for this direction
time.sleep(2)

set_speed(0) # Stop
motor1.low()
time.sleep(5)

motor1.high()   # set oposite spin direction
set_speed(100)  # Set speed close to max for this direction
time.sleep(2)

set_speed(0) # Stop
motor1.low()
time.sleep(2)
```

In this script, just like in the examples in the servo motor chapter, we use PWM. We pass a 16-bit PWM value to the "duty_u16" function to control the motor's speed. As our DC motor can spin clockwise or anticlockwise, we use the value of GPIO 15 to control the direction. When this pin is "low", the motor will rotate in one direction; when it is "high", it will spin in the other direction.

We control the speed by creating a PWM object on GPIO 14. We can achieve maximum speed in one spin direction by writing "1023" to function "set_speed". We write the value "0" to the "set_speed" function to achieve maximum speed in the other direction. You might wonder why we use 0 to 1023 to set the PWM value instead of 0 to 1023 * 64 = 65,472. I have chosen to do this because it is easier for me to think from 0 to 1023 after many years of working with the Arduino UNO. However, if you prefer, you can go ahead and remove this multiplication and write the PWM value for the Raspberry Pi Pico in its native range (0 to 65,472).

Now that you know how to control the motor speed, let's apply all this knowledge in a Node-RED flow.

Node-RED flow

Let's turn our attention to Node-RED. We'll create a simple flow to control the DC motor on the Raspberry Pi Pico side. This flow contains two widgets that will allow us to control the motor direction of spin and speed from the dashboard. This flow will also provide an opportunity to use global variables.

A global variable in Node-RED is a variable that is accessible across the entire Node-RED flow. Unlike flow variables, which are limited to a specific flow, or context variables, which are confined to individual nodes, global variables can be accessed by any node in any flow within the Node-RED instance. Global variables are often used to store information that needs to be shared between different parts of a flow or between different flows. This can include configuration parameters, shared state information, or any other data multiple nodes need access.

An example use case for a global variable might involve storing a user setting or a system configuration parameter that multiple flows need to reference. For instance, you might set a global variable to hold the connection details for a database or an external API. Then, various nodes in different flows can retrieve that information when needed.

In the flow of this example, we'll use global variables to store the direction and speed values from the dashboard widgets so that they can be combined into a single comma-delimited string value in a function node. This makes it possible to use a single MQTT topic to convey direction and speed values to the Raspberry Pi Pico.

The flow will implement this dashboard:

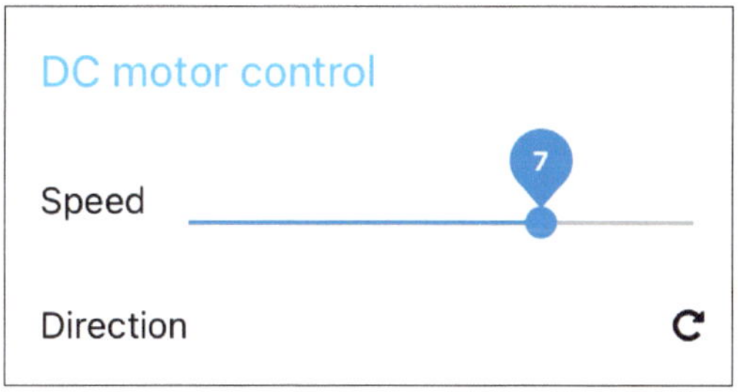

Figure 2.5.2.8: This dashboard allows direction and speed control.

In this dashboard, I use a slider for the speed and a switch for the direction. The slider has a range of 0 to 10. The MicroPython script will convert the speed value from this range to 0 to 65,472.

For the direction, I have made a minor customisation so that Font Awesome icons replace the regular switch icon to better represent this widget's purpose. I have used the "fa-rotate-right" and "fa-rotate-left" icons (see https://fontawesome.com/v4/icons/ for the details and search for "rotate"). When you click on the direction icon, the widget emits the new value and its icon changes to the alternate. You can see the implementation details of both widgets shortly.

Here is the Node-RED flow for this experiment:

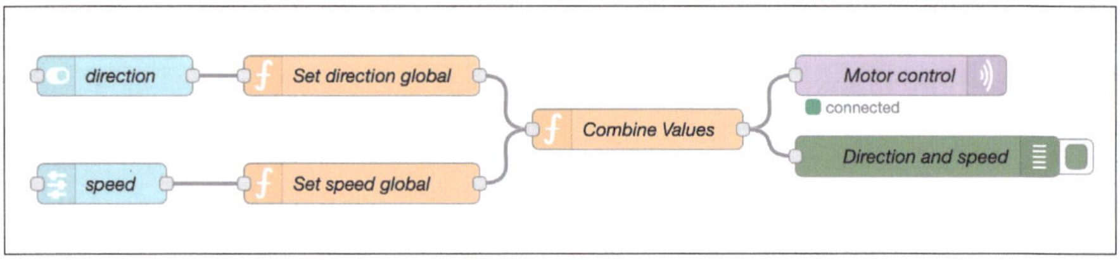

Figure 2.5.2.9: This flow can control a DC motor.

This flow is similar to those from previous experiments. One crucial difference, though, I mentioned earlier: the use of global variables.

In this flow, I want to demonstrate how to take input (alphanumerical) from two (or more) widgets, combine them into a single comma-delimited string, and send it to the MQTT broker from where the microcontroller can receive it. There are several techniques this can be achieved. The one used in this flow involves two steps:

1. Once the value is produced by a node (such as the switch node), it is sent to a function node. The function node uses the "global.set" function to set a global variable to the value received in the payload.

2. At another part of the flow where we want to use the value stored in the variable, we can use a second "function" node to read the variable using the "global.get" function.

Based on the above, let's look at the configuration of the nodes:

"Direction" node
This node implements the Switch widget that we can use to define the direction of spin. You can see its configuration in the screenshot below:

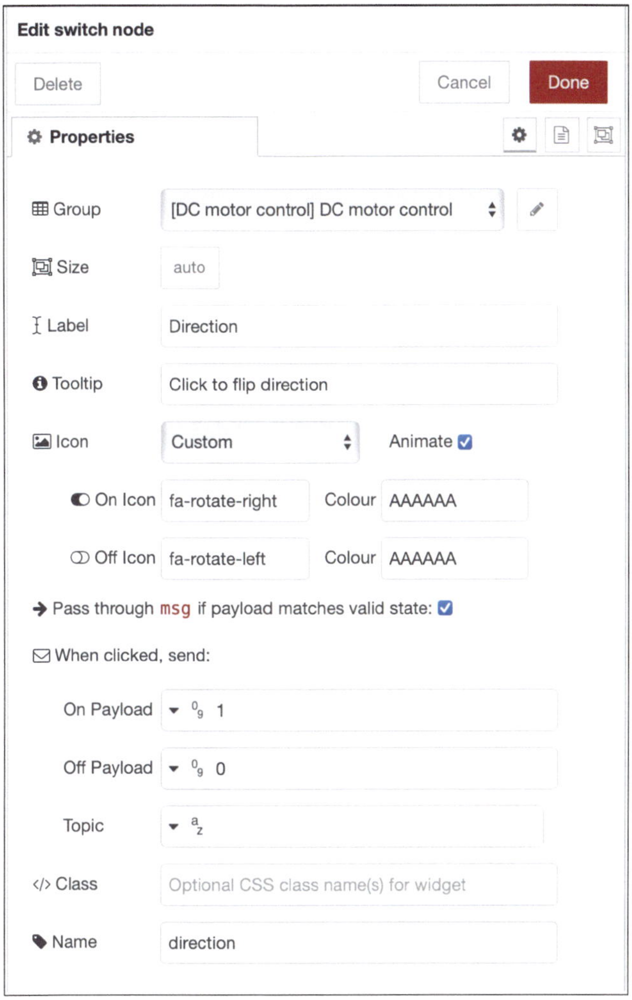

Figure 2.5.2.10: The configuration of the Direction switch node.

Notice that I have selected a custom icon and have specified a colour and Font Awesome icons for the On and Off widget states. For the Payload, I have specified the numeral "1" as the value for one direction and "0" for the other direction.

"Speed" node

The Speed node configuration is simple, and I provide the details below:

- Group: same as the "switch" node.
- Label: Speed
- Range:
  - min: 0
  - max: 10
  - step: 1

- Output: Only on release.
- Name: speed

"Set direction global" node

This is a function node that sets the "direction" global variable. You can see the JavaScript embedded in this node below:

```
global.set('direction', msg.payload);
return msg;
```

"Set speed global" node

This is a function node that sets the "speed" global variable. You can see the JavaScript embedded in this node below:

```
global.set('speed', msg.payload);
return msg;
```

"Combine Values" node

This node will get the direction and speed values from the global variables with the same name, combine them into a single comma-delimited string, and propagate this string to the next node in the flow (the "MQTT out" node). Here is the embedded Javascript:

```
var dir = global.get('direction') || 0;
var speed = global.get('speed') || 0;

var combinedString = dir.toString() + ',' + speed.toString();

msg.payload = combinedString;
return msg;
```

"Motor control" node

This is an MQTT-out node. It receives a payload from the "Combine Values" function node and sends it to the MQTT broker. Here are the details:

- Server: Your local MQTT server IP and port.
- Topic: motor_control
- Name: Motor Control

Go ahead, set up this flow in your Node-RED instance, and deploy it. When you are ready, continue with the MicroPython script.

MicroPython script

Here is the MicroPython script to match the Node-RED flow. A description of its operation follows.

```
from machine import Pin, PWM
from time import sleep
from umqtt.simple import MQTTClient
import network

motor1    = Pin(15, machine.Pin.OUT)
motor2pwm = Pin(14)

# Create a PWM object attached to motor2pwm pin
motor_pwm = PWM(motor2pwm)

# Set the PWM frequency (usually 50Hz to 2kHz for motor control)
motor_pwm.freq(50)

def map_value(value, from_min=0, from_max=10, to_min=0, to_max=65472):
    # Ensure the value is within the expected range
    value = max(min(value, from_max), from_min)

    # Perform the mapping
    return int((value - from_min) * (to_max - to_min) / (from_max - from_min) +
to_min)

# Function to set motor speed (expected input is from 0 to 10)
def set_speed(speed):
    if speed < 0:
        speed = 0
    elif speed > 10:
        speed = 10
    motor_pwm.duty_u16(map_value(speed))

# Configure your Wi-Fi credentials
   SSID      = '<your_wifi_network_SSID>'
   PASSWORD  = '<your_wifi_network_password>'

wlan = network.WLAN(network.STA_IF)
wlan.active(True)
wlan.connect(SSID,PASSWORD)

while not wlan.isconnected():
    pass

print('Connected to Wi-Fi')
print(wlan.ifconfig())

# Setup MQTT
SERVER          = '192.168.20.18'  # Replace with your MQTT broker IP
```

```
CLIENT_ID       = b'PICO'
TOPIC_TIMESTAMP = b'motor_control'
USER            = b'<your_MQTT_username>'  # Replace with your MQTT username
PASSWORD        = b'<your_MQTT_password>'  # Replace with your MQTT password

def mqtt_connect():
   client = MQTTClient(CLIENT_ID, SERVER, user=USER, password=PASSWORD,
keepalive=3600)
   client.connect()
   print('Connected to %s MQTT Broker'%(SERVER))
   return client

def reconnect():
   print('Failed to connect to the MQTT Broker. Reconnecting...')
   sleep(5)
   machine.reset()

def on_message(topic, msg):
    direction_speed = msg.decode()
    print("Received message:", direction_speed)
    components  = direction_speed.split(',')
    direction   = int(components[0])
    speed       = int(components[1])

    if direction == 0:
        motor1.low()
    else:
        motor1.high()
        speed = 10 - speed # reverse the speed to make
                           # it compatible with the direction

    set_speed(speed)

    sleep(2)
    # now, stop the motor
    motor1.low()
    set_speed(0)
    sleep(1)

def main(server, topic, username, passw):
    client = MQTTClient(CLIENT_ID, server, user=username, password=passw,
keepalive=3600)
    client.set_callback(on_message)
    client.connect()
    client.subscribe(topic)
    print("Connected to %s, subscribed to %s topic" % (server, topic))
```

```
    try:
        while True:
            client.check_msg()
            sleep(1)

    finally:
        client.disconnect()

main(SERVER, TOPIC_TIMESTAMP, USER, PASSWORD)
```

Let's look at this script in detail (although I will not look at the Wi-Fi and MQTT code segments you have learned about in previous scripts).

Basic Configuration

- Setting Up Motor Pins: motor1 is defined as a standard output pin to control the direction, and motor2pwm is defined as the PWM signal that will control the speed.

- Creating a PWM Object: motor_pwm is a PWM object attached to motor2pwm.

- Setting PWM Frequency: The frequency of the PWM signal is set to 50Hz, suitable for motor control.

Mapping function

- map_value function: Maps a given value from one range to another. In this case, it maps the input value (0 to 10) to a 16-bit range for PWM (0 to 65472). I prefer the input speed range of 0 to 10 because I find it more user-friendly for the user interface than the native PWM range of 0 to 65,472. This mapping function makes pushing the range conversion to the microcontroller easy.

Speed control

- set_speed function: Controls the speed of the motor. The speed is expected to be in the range of 0 to 10, and it's then mapped to the PWM range.

Motor control via MQTT

- on_message function: This function decodes an incoming comma-delimited MQTT message containing direction and speed. It controls the direction of the motor through motor1 and the speed through motor_pwm. The motor is stopped after a 2-second delay. Notice that the value for the speed depends on the direction of movement. To get consistent speed when the slider widgets are set to a particular position, the function considers the direction of spin, as specified by the value of the direction variable. If the direction value is "0", then the speed directly matches the value of the "speed" variable. However, if the direction value is "1", then the speed must be flipped to make it consistent, so this function will subtract the speed value from 10 and store the result back in the "speed" variable. With this calculation, the speed, as specified by the slider widget, will be the same regardless of the spin direction.

As you can see, controlling a DC motor safely is possible, considering the motor's power requirements. You should always ensure that your circuit is wired correctly and do not try to stop the rotation of the motor shaft, as this can cause a large amount of current to flow through the motor controller and trigger its thermal shutdown function.

Chapter 2.6 • Sensors

The projects in this chapter are designed to be aware of their surroundings using temperature, motion, and light sensors. Here, you'll explore a range of sensors that can measure environmental variables such as temperature and water levels. You'll learn to integrate these sensors into your Node-RED flows, Raspberry Pi Pico W circuits, and MicroPython scripts.

The first project uses the DHT11 sensor to monitor temperature and humidity. With the DHT11, you can build a weather station or a smart home system, so understanding how to use it will be invaluable.

Next, you'll work with the HC-SR04 ultrasonic sensor. This device can measure distance accurately, making it perfect for obstacle avoidance in robotics or parking sensors in smart garages.

You'll also get hands-on experience with a motion sensor, allowing your projects to detect movement in their environment. Motion sensors like the one you will learn about in this chapter can be used in security applications or interactive installations; this sensor adds a new layer of functionality.

Water level sensing is up next, a helpful skill if you're into gardening automation or flood warning systems. You'll learn to set up and integrate a water level sensor into your existing systems.

Then, you'll dive into the world of thermistors, specialised sensors offering even more precise temperature readings. Thermistors are perfect for projects requiring high accuracy, like a DIY incubator or a temperature-controlled fan.

Finally, you'll round out the chapter by exploring the analog light sensor. This sensor will enable your projects to respond to ambient light conditions, opening the door to applications like automatic night lights or sunrise alarm clocks.

Each project in this chapter comes with its own Node-RED flow, Raspberry Pi Pico W circuit, and MicroPython script. By the end of this chapter, you'll be well-equipped to make your projects more responsive and interactive.

6.1. Temperature with DHT11

The DHT11 is a simple digital temperature and humidity sensor. It uses a capacitive humidity sensor and a thermistor to measure the surrounding air and outputs a digital signal on the data pin. Through this digital pin, the microcontroller receives the temperature and humidity values in a format that is ready to use and does not require any processing. You can compare this to the thermistors (which you will learn about later in this chapter), which outputs a voltage instead of a final temperature value, requiring further processing on the microcontroller side.

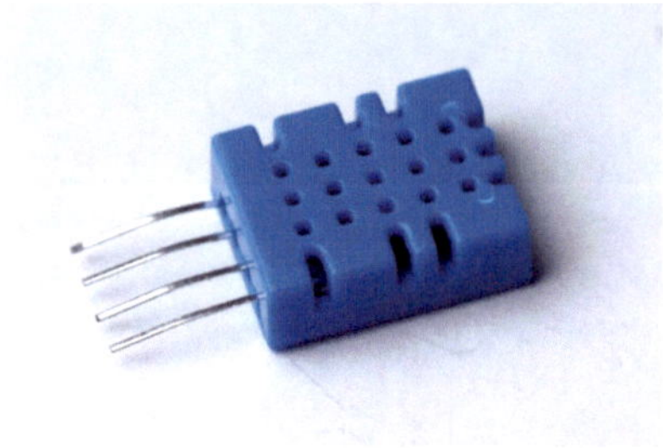

Figure 2.6.1.1: The DHT11 sensor.

The DHT11 sensor includes a resistive-type humidity measurement component and an NTC temperature measurement component. An NTC (Negative Temperature Coefficient) temperature measurement component is a type of thermistor used to measure temperature based on changes in electrical resistance with temperature variation. In NTC thermistors, the resistor resistance decreases as the temperature increases (hence the name "Negative Temperature Coefficient").

The operation of an NTC thermistor is based on the properties of its semiconductor material, commonly made from metal oxides like manganese, nickel, or cobalt. The semiconductor material has fewer free charge carriers at lower temperatures, resulting in higher resistance. As the temperature increases, thermal energy excites the atoms in the material, freeing up more charge carriers and reducing the resistance.

The relationship between resistance and temperature for an NTC thermistor is generally nonlinear and often approximated using the Steinhart-Hart or simpler Beta parameter equations. These equations allow for the precise calculation of temperature based on the measured resistance.

In a typical application, an NTC thermistor is integrated into a circuit and connected to a voltage divider or a current source. The changing resistance of the thermistor alters the voltage or current in the circuit, which can then be measured and converted into a temperature reading. This data can be fed into a microcontroller for further processing or directly used to control other components like fans or heaters.

The DHT11 has become one of the most popular sensors in IoT applications related to weather, home automation, and environmental sensing because of its low cost and reasonable performance.

About the DHT11 sensor module

The DHT11 sensor operates using a one-wire communication protocol, making it easy to interface with microcontrollers. The sensor transmits data in a 40-bit packet. This packet includes 5 bytes: integral and decimal parts of humidity, integral and decimal parts of temperature, and a checksum. The microcontroller measures the duration of each signal and decodes it to extract the humidity and temperature values. The checksum is used to verify data integrity. The microcontroller further converts the humidity and temperature data to meaningful values using calibration factors provided by the manufacturer.

Luckily, as the end users of the DHT11, we don't have to implement the communication protocol I described above. All this is handled by thoroughly tested and documented libraries and modules we can use in our programs.

Let's look at the main technical characteristics of the DHT11 sensor:

- Measurement Range: The DHT11 sensor can measure temperatures in the range of 0 °C to 50 °C (32 °F to 122 °F) with a precision of ±2 °C (±3.6°F). It can measure humidity in the range of 20% to 90% RH with a precision of ±5%.
- Supply Voltage: The sensor typically operates at 3.3 V to 5 V, making it compatible with various microcontroller platforms. It can work with the Arduino UNO at 5 V logic voltage and the Raspberry Pi Pico, which operates at 3.3 V.
- Communication: It employs a single-wire digital communication protocol, simplifying its connection to microcontrollers. We only need one wire for the data and two wires for power.
- Response Time: The DHT11 sensor takes around 2 seconds to provide updated temperature and humidity values, which makes it suitable for applications that do not require real-time data.
- Size and Form Factor: The sensor is compact, usually encased in a protective housing, and can be easily integrated into different projects.
- Cost-Effectiveness: One of the key advantages of the DHT11 sensor is its affordability, making it accessible for a wide range of applications.
- Reliability: While the DHT11 offers cost savings, it has slightly lower accuracy than more advanced sensors like the DHT22 or other specialised sensors.

Reading temperature and humidity using Raspberry Pi Pico and DHT11

In this experiment, you will connect a DHT11 sensor to a Raspberry Pi Pico and read temperature and humidity data every 5 seconds, printing the output to the console. Let's start with the connection of the sensor to the Raspberry Pi Pico. The wiring diagram below will help you with the process. Notice the pin numbers that annotate the DHT11 package. The sensor does not have pin markings on its package, so hold the sensor with its grill facing you. Pin 1 is the left-most pin.

Figure 2.6.1.2: The DHT11 sensor connected to the Raspberry Pi Pico.

Follow this process:

1. Connect the sensor's VCC pin (pin 1) to the Raspberry Pi Pico's 3.3V (OUT) pin.
2. Connect the sensor's GND pin (pin 4) to any GND pins on the Raspberry Pi Pico.
3. Connect the sensor's DATA pin (pin 2) to the Raspberry Pi Pico's GP18 pin (you can choose another pin if you want).
4. Leave sensor pin three unconnected (this pin is not used for any purpose).

Here is a photograph of my assembled circuit:

Figure 2.6.1.3: The DHT11 connected to my Raspberry Pi Pico.

Next, we'll move on to the software. The easiest way to read from the DHT11 sensor with MicroPython is to use the DHT module. The DHT module is part of the MicroPython firmware and does not require a separate install. You can read the official documentation for the DHT module one on the MicroPython website (see https://txplo.re/upython_dht). Note that the documentation refers to the ESP32 when discussing the DHT module. However, the module works similarly on the Raspberry Pi Pico and other supported MicroPython platforms.

Here is a simple MicroPython script to read and print the temperature and humidity data every 5 seconds:

```
mport time
from machine import Pin
from dht import DHT11

# Create a DHT11 instance
dht11 = DHT11(Pin(18))  # the DATA pin is connected to GP2 on the Raspberry Pi
Pico

while True:
    dht11.measure()  # make a new measurement
    temp = dht11.temperature()  # read temperature
    hum = dht11.humidity()  # read humidity
    print('Temperature:', temp, 'C', 'Humidity:', hum, '%')
    time.sleep(5)  # wait for 5 seconds
```

This script initialises the DHT11 sensor and reads the temperature and humidity measurements every 5 seconds, printing the results to the console. Let's look at the most important parts of this script:

- Within the infinite loop, the dht11.measure() method is called. This triggers the DHT11 sensor to perform a new temperature and humidity measurement. You can call the "measure" method whenever you want to get a fresh measurement from the sensor.
- The measured temperature and humidity values are obtained by calling the dht11.temperature() and dht11.humidity() methods, respectively.
- After printing the data, the script pauses execution for 5 seconds using time.sleep(5). This delay ensures the loop doesn't read data too frequently and allows the sensor to provide new measurements.

After uploading the script to the Raspberry Pi Pico, the temperature and humidity readings will be printed to the console.

Node-RED example

Let's add Node-RED to the mix. We'll create a Node-RED flow that implements a dashboard that uses widgets to display the temperature and humidity values obtained from the DHT11 sensor. To make this work, we'll write a MicroPython script that takes readings from the sensor and publishes the values to the MQTT broker. The script will utilise an MQTT topic for the temperature and another for humidity. The Node-RED flow will have a subscription to these topics and update the dashboard widgets as soon as there is a new published value.

MicroPython

Let's begin with the MicroPython script since the process starts with the Raspberry Pi Pico reading values from the sensor and publishing them to the MQTT broker. Here is the complete script, followed by a line-by-line description of its operation.

```
from machine import Pin, I2C
from umqtt.simple import MQTTClient
from time import sleep
import network
from dht import DHT11

# Configure your Wi-Fi credentials
   SSID      = '<your_wifi_network_SSID>'
   PASSWORD  = '<your_wifi_network_password>'

wlan = network.WLAN(network.STA_IF)
wlan.active(True)
wlan.connect(SSID,PASSWORD)

while not wlan.isconnected():
    pass

print('Connected to Wi-Fi')
print(wlan.ifconfig())

# Configure MQTT
SERVER     = '192.168.20.18'  # Replace with your MQTT broker IP
CLIENT_ID  = b'PICO'
TOPIC_TEMP = b'pico/dht11/temperature'
TOPIC_HUMD = b'pico/dht11/humidity'
USER       = b'<your_MQTT_username>'  # Replace with your MQTT username
PASSWORD   = b'<your_MQTT_password>'  # Replace with your MQTT password

def mqtt_connect():
   client = MQTTClient(CLIENT_ID, SERVER, user=USER, password=PASSWORD,
keepalive=3600)
   client.connect()
   print('Connected to %s MQTT Broker'%(SERVER))
   return client

def reconnect():
   print('Failed to connect to the MQTT Broker. Reconnecting...')
   sleep(5)
```

```
    machine.reset()

try:
    client = mqtt_connect()
except OSError as e:
    reconnect()

# Create a DHT11 instance
dht11 = DHT11(Pin(18))  # the DATA pin is connected to GP18 on the Pico

while True:
     dht11.measure()  # make a new measurement
     temp = dht11.temperature()  # read temperature
     hum  = dht11.humidity()  # read humidity
     print('Temperature:', temp, 'C', 'Humidity:', hum, '%')
     client.publish(TOPIC_TEMP, str(temp))
     client.publish(TOPIC_HUMD, str(hum))
     sleep(5)  # wait for 5 seconds
```

The script begins by importing the required modules and defining variables for Wi-Fi credentials, MQTT configuration, and other settings. Then, it establishes a Wi-Fi connection and waits until the connection is established. After connecting to Wi-Fi, it prints the Wi-Fi details.

Next, the script defines functions for connecting to an MQTT broker and handling reconnections in case of failures. It attempts to connect to the MQTT broker using the provided credentials.

A DHT11 instance is created, connected to GPIO pin 18. The script measures the temperature and humidity within an infinite loop using the DHT11 sensor. It then prints and publishes the obtained values to MQTT topics using the MQTT client. The script waits for 5 seconds before the next iteration.

The script's primary purpose is to read temperature and humidity data from the DHT11 sensor and publish it to an MQTT broker, allowing remote data monitoring through MQTT topics.

Node-RED

Now, we'll create a Node-RED flow that subscribes to the temperature and humidity MQTT topics and displays them on a dashboard. The main objective of this flow is to make this monitoring dashboard for the sensor data:

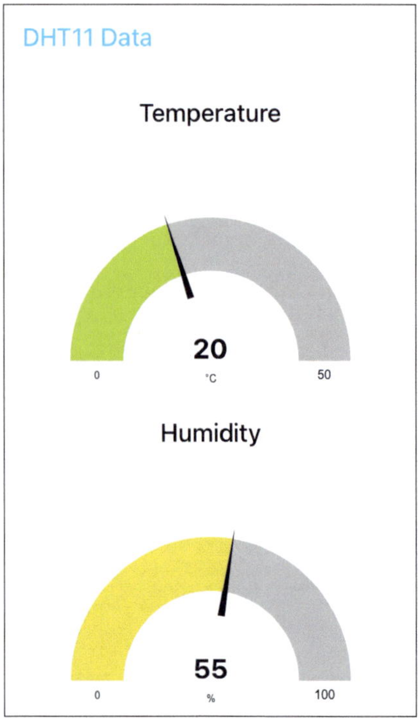

Figure 2.6.1.4: The Node-RED dashboard that displays the sensor data.

This dashboard uses two gauge widgets to display the sensor data.

Here is the Node-RED flow:

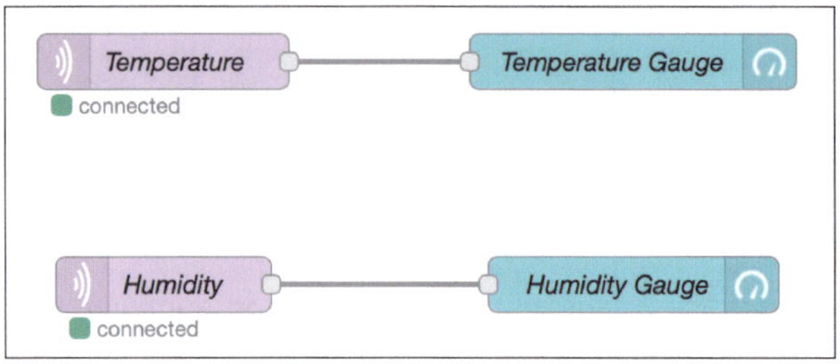

Figure 2.6.1.5: The Node-RED dashboard for this experiment.

This flow can't get any simpler! It contains two MQTT-in nodes that subscribe to the temperature and humidity topics and send their values to the gauge nodes for display in the dashboard.

Below, you can see the configuration of the Temperature MQTT-in node:

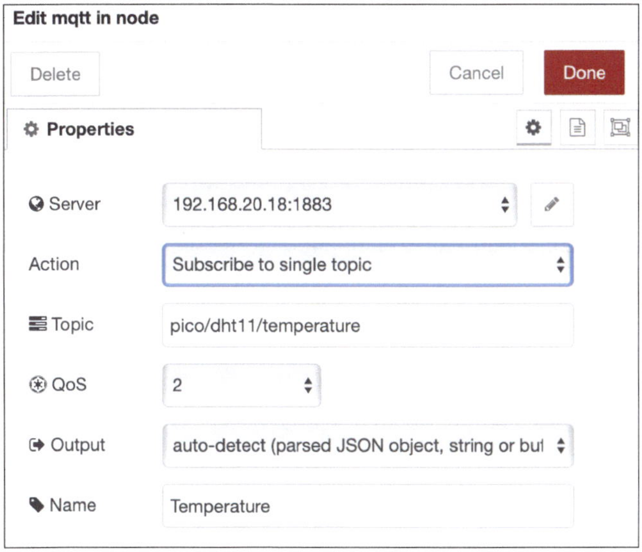

Figure 2.6.1.6: The MQTT-in node for the temperature topic.

The configuration for the humidity MQTT-in node is similar. Use these values:

- Server: same as your temperature MQTT-in node.
- Action: same as your temperature MQTT-in node ("subscribe to single topic").
- Topic: "pico/dht11/humidity"
- QoS: same as your temperature MQTT-in node ("2")
- Output: same as your temperature MQTT-in node ("auto-detect").
- Name: "Temperature"

When ready, deploy this flow and open the dashboard. Notice how the gauge widgets update their values every five seconds as the Raspberry Pi Pico publishes new values.

This Node-RED flow can be reused with minimal changes to display data from many other sensors. For example, instead of the DHT11, you can use a thermistor, an ultrasonic distance sensor or an analog light sensor. As you learned in this chapter, these sensors output numbers that can be displayed graphically in the Node-RED dashboard. You will learn how to use these sensors in later chapters.

6.2. HC-SR04 Ultrasonic Sensor

The HC-SR04 is a low-cost ultrasonic distance sensor in many robotics and automation projects. It measures distance by emitting an ultrasonic sound wave and then measuring the time it takes for the sound wave to bounce back, similar to how a bat or submarine would use echolocation.

The sensor has a measurement range of 2 cm to 400 cm with an accuracy of around 3 mm. It operates at a frequency of 40 kHz and can work at any temperature between 0 °C and 70 °C. It sends a 10-microsecond pulse to the trigger input, then sends eight cycles of ultrasonic burst at 40 kHz and waits for the reflected ultrasonic burst. When it receives the reflected wave, it sets the echo output to high for a duration that is the same as the time it took for the wave to return. This sensor better detects objects with a flat reflective surface because of the ultrasonic signal's characteristics.

Using the HC-SR04 with a Raspberry Pi Pico and MicroPython

In this section's first part, you will learn how to use the ultrasonic distance sensor with the Raspberry Pi Pico and MicroPython. Let's begin with the wiring of the sensor on the breadboard.

The sensor module has four pins, each marked with its role on the circuit board. You will see these markings on the front side of the sensor module, where the ultrasonic speaker and receiver are mounted. Follow these instructions and complete the wiring:

1. Connect the VCC (power) pin of the HC-SR04 sensor to the 5 V pin on the Raspberry Pi Pico. This is the preferred supply voltage for this sensor. However, I could also use this sensor by supplying it with power from the 3.3 V of the Pico "3V3(OUT)".

2. Connect the GND (ground) pin of the HC-SR04 sensor to any GND pin on the Pico.

3. Connect the Trig (trigger) pin of the HC-SR04 sensor to any GPIO pin on the Pico. I will use GP19 in this example.

4. Connect the Echo pin of the HC-SR04 sensor to any GPIO pin on the Pico. I will use GP18 in this example.

Here is the wiring diagram for this circuit:

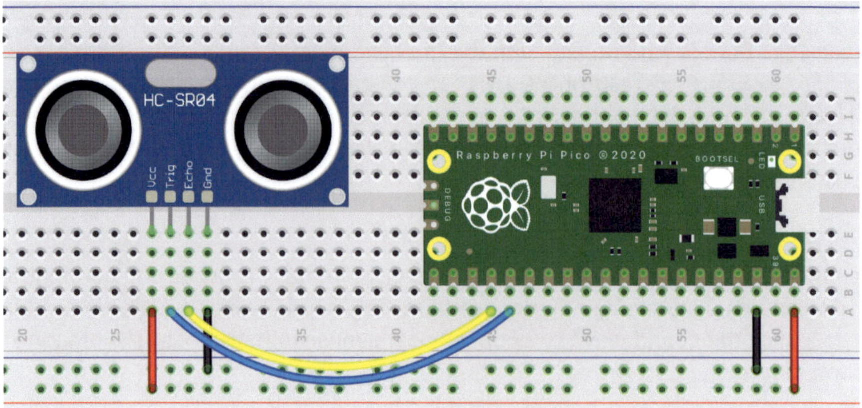

Figure 2.6.2.7: The HC-SR04 sensor connected to the Raspberry Pi Pico.

Below, you can see my assembled circuit for this experiment:

Figure 2.6.2.8: The HC-SR04 sensor connected to the Raspberry Pi Pico on the breadboard.

Here is a simple MicroPython script that prints the distance detected by the HC-SR04 sensor to the console every second.

```
import machine
import time

# Define the pins where the sensor is connected
trigger = machine.Pin(19, machine.Pin.OUT)
echo = machine.Pin(18, machine.Pin.IN)

def get_distance():
    # Send a 10-microsecond pulse.
    trigger.low()
    time.sleep_us(2)  # Wait for 2 microseconds low pulse
    trigger.high()
    time.sleep_us(10)  # Keep the trigger high for 10 microseconds
    trigger.low()

    # Wait for the pulse on the echo pin.
    while echo.value() == 0:
        pass
    start = time.ticks_us()

    # Measure the duration of the high pulse.
    while echo.value() == 1:
        pass
    finish = time.ticks_us()

    # Calculate the distance in centimeters and return it.
    return (finish - start) / 58.0

while True:
    print("Distance: ", get_distance(), "cm")
    time.sleep(1)
```

The script starts by importing the necessary modules, machine (for controlling pins) and time (for timing functions). It then defines the trigger and echo pins.

The get_distance() function measures the distance using the HC-SR04 sensor. Here is what happens inside the function:

1. A 10-microsecond pulse is sent to the trigger pin to trigger the ultrasonic pulse emission. The trigger pin is set low, then after a 2-microsecond delay, it's set high for 10 microseconds and then set low again.
2. The script waits for the echo pulse to start (the echo pin goes from low to high) using a while loop and the echo.value() function.
3. The time when the echo pulse starts is recorded using time.ticks_us().
4. Using another while loop, the script then measures the duration of the echo pulse (the total time the echo pin stays high).
5. The time when the echo pulse ends is recorded.
6. The distance is calculated by dividing the time duration by 58 (to convert microseconds to centimetres) and returned by the function.

In the main loop, the program calls the get_distance() function to measure the distance from an object before the sensor and then prints the result to the console. The script waits for 1 second before the subsequent measurement using time.sleep(1).

If you are curious about the figure "58" in point 6 above, I encourage you to read my article on the Tech Explorations website about this topic. You can find it at https://txplo.re/ultrasonic_sensor.

Node-RED example

Let's extend the functionality and have the Raspberry Pi Pico publish the range value to an MQTT topic, from where a Node-RED flow will fetch the value. To keep this example simple, this flow will show the distance value in the debug tab instead of setting up a dashboard. You can use the knowledge you have acquired from previous experiments to create a dashboard if you prefer that option.

MicroPython script

As the Raspberry Pi Pico is responsible for getting a distance measurement and publishing the value to the MQTT topic, we'll first look at the MicroPython script. Here is the script in its entirety, followed by a brief discussion.

```
from machine import Pin
from umqtt.simple import MQTTClient
import time
import network

# Configure your Wi-Fi credentials
   SSID     = '<your_wifi_network_SSID>'
   PASSWORD  = '<your_wifi_network_password>'
```

```
wlan = network.WLAN(network.STA_IF)
wlan.active(True)
wlan.connect(SSID,PASSWORD)

while not wlan.isconnected():
    pass

print('Connected to Wi-Fi')
print(wlan.ifconfig())

# Configure MQTT
SERVER     = '192.168.20.18'  # Replace with your MQTT broker IP
CLIENT_ID  = b'PICO'
TOPIC      = b'pico/range'
USER       = b'<your_MQTT_username>'  # Replace with your MQTT username
PASSWORD   = b'<your_MQTT_password>'  # Replace with your MQTT password

# Define the pins where the sensor is connected
trigger  = machine.Pin(19, machine.Pin.OUT)
echo     = machine.Pin(18, machine.Pin.IN)

def get_distance():
    # Send a 10-microsecond pulse.
    trigger.low()
    time.sleep_us(2)  # Wait for 2 microseconds low pulse
    trigger.high()
    time.sleep_us(10)  # Keep the trigger high for 10 microseconds
    trigger.low()

    # Wait for the pulse on the echo pin.
    while echo.value() == 0:
        pass
    start = time.ticks_us()

    # Measure the duration of the high pulse.
    while echo.value() == 1:
        pass
    finish = time.ticks_us()

    # Calculate the distance in centimeters and return it.
    return (finish - start) / 58.0

def mqtt_connect():
   client = MQTTClient(CLIENT_ID, SERVER, user=USER, password=PASSWORD,
   keepalive=3600)
```

```
    client.connect()
    print('Connected to %s MQTT Broker'%(SERVER))
    return client

def reconnect():
    print('Failed to connect to the MQTT Broker. Reconnecting...')
    time.sleep(5)
    machine.reset()

try:
    print("Connecting to the MQTT Broker...")
    client = mqtt_connect()
except OSError as e:
    reconnect()

# Send range to MQTT broker every second
while True:
    distance = get_distance()
    print("Distance: ", distance, "cm")
    client.publish(TOPIC, str(distance))
    time.sleep(1)
```

This script follows the pattern of the script from previous experiments. It imports required modules for the operation, configures Wi-Fi credentials, initialises the WLAN (Wi-Fi) interface and attempts to connect to the specified Wi-Fi network using the provided credentials. Upon successful Wi-Fi connection, the script configures MQTT settings and connects to the broker.

Then, the script uses code from the first MicroPython example in this chapter. It defines the GPIO pins (trigger and echo) where the ultrasonic distance sensor (HC-SR04) is connected and the function get_distance(), which calculates the distance using the HC-SR04 sensor.

The bulk of the work in this script happens in the infinite loop where it:

1. measures the distance using the get_distance() function;
2. prints out the measured distance in centimeters;
3. publishes the distance to the specified MQTT topic using the MQTT client;
4. sleeps for 1 second before repeating the loop.

Upload this script to your Raspberry Pi Pico (remember to set up the credentials for your WiFi network and MQTT broker). This completes the work on the Raspberry Pi Pico side.

Node-RED

In Node-RED, create a simple flow to receive and display the messages in the debug tab. You can see the flow and its debug tab output below:

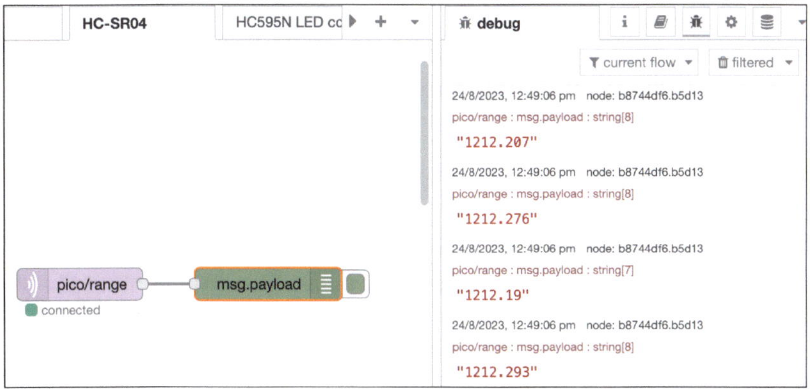

Figure 2.6.2.9: This flow receives distance values from the HC-SR04 sensor connected to the Raspebrry Pi Pico.

This flow only has one functional node, the MQTT-in node. Below, you can see its properties:

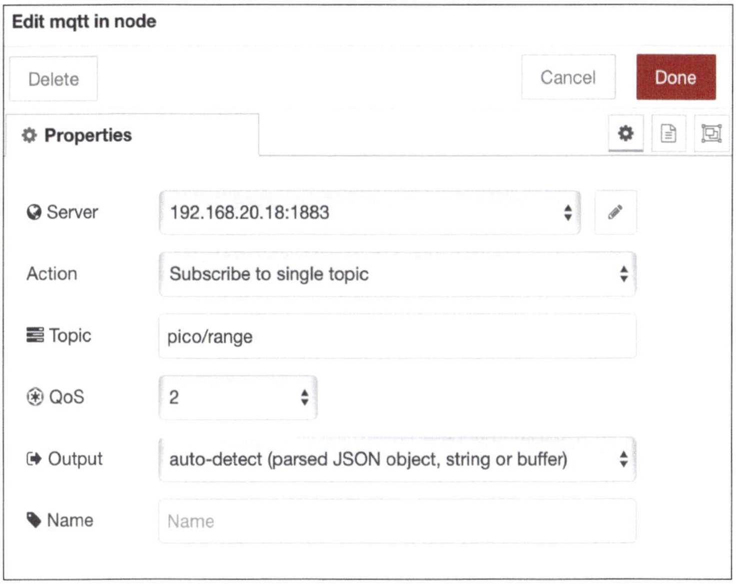

Figure 2.6.2.10: This properties for the MQTT-in node.

Set the topic to the string you used in the "TOPIC" variable of the MicroPython script. The rest of the properties are similar to those in previous experiments.

Once you deploy this flow, the "range" value will be printed in the Debug tab each time a new value is posted to the MQTT topic by the Pico.

6.3. Motion Sensor

In this project, you will learn to use the HC-SR501 Human Sensor Module, also known as the Passive Infrared (PIR) motion sensor. This device, coupled with our Raspberry Pi Pico W and the Node-RED platform, opens up a world of possibilities for interactive and responsive projects.

The heart of the HC-SR501 sensor is the PIR sensor, which detects the infrared light radiating from human bodies in motion. When the sensor picks up a sudden change in this infrared radiation, it sends a signal. Everything emits some level of infrared light, a type of light we can't see but can feel as heat. The sensor is designed to detect changes in the amount of infrared light it receives. When a warm object like a human or animal moves within the sensor's range, the amount of infrared light hitting the detector changes. This change is what the sensor identifies as motion.

Common uses for this sensor are wide-ranging and depend on your creativity. Here are a few examples to get your imagination going:

- Home Security System: The sensor can detect unauthorised movement and trigger alarms.
- Energy Saving: It can turn off lights in a room if there's no movement for a certain period.
- Interactive Installations: Artists and designers often use PIR sensors in installations to create responsive experiences.

HC-SR501 PIR sensor technical details

The HC-SR501 PIR sensor does come with two potentiometers and a jumper onboard, each with a specific function. You can see the top and bottom sides of the sensor below.

Figure 2.6.3.11: The HC-SR501 PIR sensor bottom (left) and top (right) with the lens removed.

In the photograph on the right side (above), you can see the sensor component of the module. To expose the sensor, you can remove the lens cover, taking care not to break its plastic pins. The sensor will not work correctly without the lens, so remember to put the lens back on the module before you start this experiment.

Potentiometers

You can turn these small orange knobs or dials with a screwdriver. They are used to adjust the behaviour of the sensor. You can experiment with the setting of these potentiometers until you reach an appropriate configuration for your application. For example, if you want the sensor to detect movement near your desk, you can set the sensitivity to around 1-2 meters. If you want the sensor to detect movement across your lab, you can max it out at 7 meters. There are no markings on the potentiometers to help you with the calibration, so you must do this by trial and error.

- Sensitivity Adjust: The first potentiometer (the one closest to the jumper, marked as "2" in the photogram above), usually labelled as Sensitivity or Distance Adjust, controls the sensing range of the sensor. By adjusting this potentiometer, you can set how far away the sensor will detect motion. The range can vary from approximately 3 meters to about 7 meters.

- Time Delay Adjust: The second potentiometer (marked as "1" in the photograph above) adjusts the time delay, often labelled Time. It controls how long the sensor output remains high after detecting motion. This can be adjusted from around 5 seconds to 5 minutes or more. This is useful for scenarios where you want the sensor to keep signalling movement longer after the initial detection.

Jumper

The jumper onboard has two positions and controls the trigger modes of the sensor. The jumper in the photograph above is marked as "3".

- Single Trigger Mode (non-repeatable): In this mode, indicated by placing the jumper in the H position, the sensor will send a high signal when it detects motion and then will not send another signal until the delay period has ended and motion has been detected again. I have configured my sensor to single trigger mode by placing the jumper across the two outer pins.

- Repeat Trigger Mode (repeatable): In this mode, indicated by placing the jumper in the L position, the sensor will send a high signal when it detects motion and keep the signal high as long as it continues to detect motion. Once motion is no longer detected, it will keep the signal high for the duration set by the time delay potentiometer and switch to low if no more motion is detected.

These adjustments allow you to fine-tune the behaviour of the HC-SR501 sensor based on your specific project requirements.

MicroPython example

Let's construct the circuit that will allow us to learn how to use this PIR sensor. Connecting the sensor to the Raspberry Pi Pico is straightforward:

- Connect the VCC pin (usually the rightmost pin) on the HC-SR501 to the 5 V pin on the Raspberry Pi Pico W. You can confirm the role of each pin by removing the lens from the module and looking at the pin marking on the module PCB. You can see the pin marking of my sensor in Figure 2.6.3.11 above.

- Connect the GND pin (usually the leftmost) on the HC-SR501 to any Ground (GND) pin on the Raspberry Pi Pico W.

- Finally, connect the OUT pin (the middle pin) on the HC-SR501 to one of the GPIO pins on the Raspberry Pi Pico W. I use GPIO 18 for this example.

You can see the wiring diagram below:

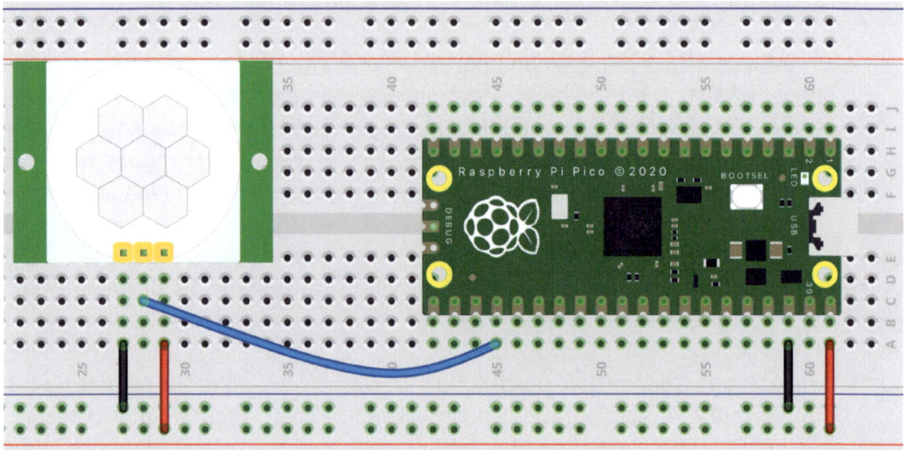

Figure 2.6.3.12: The HC-SR501 PIR sensor wiring diagram.

I have used a header extension to attach my PIR sensor to the breadboard because the module has protruding components that make it harder to connect directly. You can also use jumper wires if you don't have a header extension. You can see my circuit below:

Figure 2.6.3.13: My HC-SR501 PIR circuit on the breadboard.

Finally, here is the script that reads the state of the sensor and prints a message to the console:

```
from machine import Pin
import time

pir = Pin(18, Pin.IN)

while True:
    if pir.value():
        print("Motion detected!")
    else:
        print("No motion.")
    time.sleep(1)
```

This script does not require any external modules. It treats the PIR sensor as a simple on/off switch (or a button) and reads its state (or switch) once per second.

The script starts by importing the necessary modules: Pin from the machine package for controlling GPIO pins and time for timing functions. Then, it creates a Pin object named pir to represent the PIR sensor, which is connected to GPIO pin 18 (change the pin number if needed). Then, the script enters an infinite loop using the while True: statement. This loop will continuously run as long as the condition True remains true.

Inside the loop, the script checks the value of the PIR sensor using the pir.value() method. If the PIR sensor detects motion and its value is high (1), the condition if pir.value(): is satisfied, and the script prints "Motion detected!" using print("Motion detected!"). If the PIR sensor's value is low (0), indicating no motion, the condition else: is satisfied, and the script prints "No motion." using print("No motion."). After printing out the statue, the script pauses (sleep) for 1 second using time.sleep(1). This pause ensures that the loop does not execute too quickly and provides a delay between consecutive measurements.

Node-RED example

Let's add Node-RED to this experiment, as we did with the HC-SR04 sensor in the previous chapter. I will keep the MicroPython code and Node-RED flow as small as possible to keep this chapter concise. At the end of this part of the book, you will learn how to combine multiple sensors into a single circuit and create one dashboard that displays their values.

MicroPython script

Data flows from the Raspberry Pi Pico and the sensor attached to the Node-RED instance. So, let's first look at the Microptynon script. Here is the script:

```
from machine import Pin
import time
from umqtt.simple import MQTTClient
from time import sleep
```

```
import network

# Configure your Wi-Fi credentials
SSID      = '<your_wifi_network_SSID>'
PASSWORD  = '<your_wifi_network_password>'

print('Starting...')

wlan = network.WLAN(network.STA_IF)
wlan.active(True)
wlan.connect(SSID,PASSWORD)

print('Trying to connect to Wifi...')

while not wlan.isconnected():
    print('.', end='')
    time.sleep(1)
    pass

print('Connected to Wi-Fi')
print(wlan.ifconfig())

# Configure MQTT
SERVER      = '192.168.20.18'  # Replace with your MQTT broker IP
CLIENT_ID   = b'PICO'
TOPIC       = b'pir'
USER        = b'<your_MQTT_username>'  # Replace with your MQTT username
PASSWORD    = b'<your_MQTT_password>'  # Replace with your MQTT password

def mqtt_connect():
   client = MQTTClient(CLIENT_ID, SERVER, user=USER, password=PASSWORD,
   keepalive=3600)
   client.connect()
   print('Connected to %s MQTT Broker'%(SERVER))
   return client

def reconnect():
   print('Failed to connect to the MQTT Broker. Reconnecting...')
   sleep(5)
   machine.reset()

try:
   client = mqtt_connect()
except OSError as e:
   reconnect()
```

```
pir = Pin(18, Pin.IN)

while True:
    if pir.value():
        print("Motion detected!")
        client.publish(TOPIC, "Motion detected!")
    else:
        print("No motion.")
        client.publish(TOPIC, "No motion.")
    time.sleep(1)
```

Upload this script to your Raspberry Pi Pico (remember to set up the credentials for your WiFi network and MQTT broker). This completes the work on the Raspberry Pi Pico side.

Node-RED

In Node-RED, you will create a simple flow to receive and display the messages in the debug tab. You can see the flow and its debug tab output below:

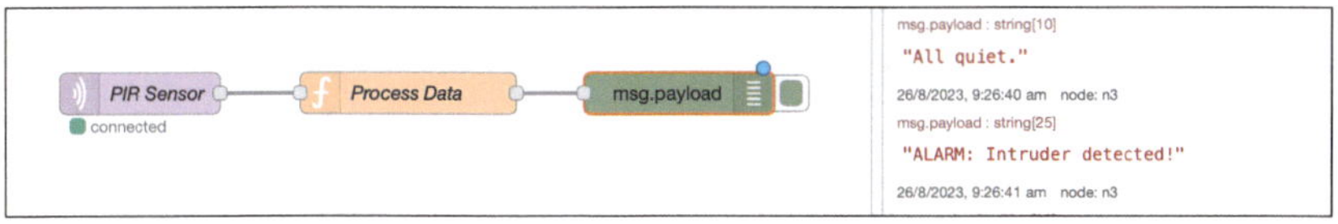

Figure 2.6.3.14: The Node-RED flow and some of its output in the Debug tab.

The PIR sensor MQTT-in node setup is straightforward, as you might expect. Ensure that the Topic parameter is set to "pir" (to match the "TOPIC" variable in the MicroPython script.

In the function node, Javascript transforms the value received from the MQTT topic to something more meaningful for the user looking at the Debug tab output. Here is the Javascript:

```
if (msg.payload === 'Motion detected!') {
    return { payload: 'ALARM: Intruder detected!' };
}
return { payload: 'All quiet.' };
```

When the MQTT input is the string "Motion detected!", the output from the node is an alarm message. Of course, you can customise this transformation to something that makes more sense for your application.

In the following two chapters, you will learn about the thermistor and the analog light sensor. These sensors are simple in operation, so their discussion will be brief. Learning about these last two sensors in the book will allow us to work on a more comprehensive project that combines several sensors into a single circuit and a Node-RED dashboard.

6.4. Water Level Sensor

In this project, you will learn how to measure the amount of water present in a container using a water level sensor. This simple sensor can monitor the water level in various environments, from home plants to fish tanks.

The water level sensor has ten exposed copper traces, split into two groups of five. One group is for power, and the other is for sensing. When these traces come into contact with water, they are bridged, completing a circuit. These copper traces act like a variable resistor. That means the resistance they provide changes based on the water level. The more immersed the sensor is, the better the conductivity and the lower the resistance. On the flip side, if the sensor is less immersed, the conductivity drops, and the resistance increases. The sensor processes this resistance into an output signal voltage that can be read by our Raspberry Pi Pico W's microcontroller, thus giving us a measure of the water level.

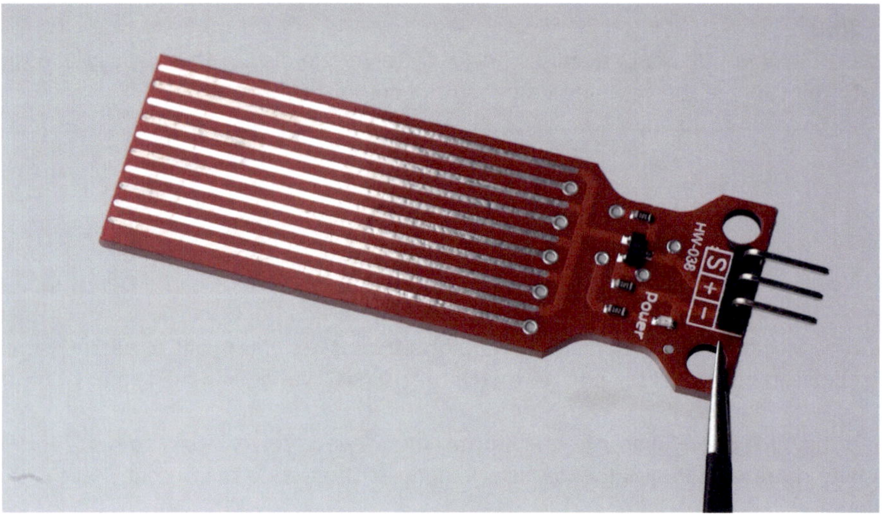

Figure 2.6.4.15: A resistive water level sensor.

It's important to note that the sensor cannot be fully submerged in water. Only the part with the ten traces should come into contact with water. Also, avoid supplying power to the sensor in a humid environment, as it could accelerate the corrosion of the probe and shorten the sensor's lifespan. Only power it up when you're taking readings.

Here are a few ideas of things you can do with a water level sensor:

1. Home Automation Systems: In smart homes, water level sensors are used to monitor the level of water in various appliances, like washing machines or dishwashers. This way, the system can alert the user or shut off the water supply if a leak or overflow is detected.
2. Irrigation Systems: In agriculture, these sensors can monitor the water level in irrigation systems. This allows farmers to ensure their crops get the right amount of water and not waste resources.

3. Aquariums: For those who keep fish or other aquatic pets, a water level sensor can help monitor the water level in the tank. This can help ensure the health and safety of the aquatic life.
4. Water Tanks: In industrial settings, water level sensors can monitor the water level in large storage tanks. This data can then be used to plan for refills and manage water usage efficiently.
5. Flood Detection: Water level sensors can be used in flood-prone areas to warn early about rising water levels.

Let's go ahead and set up an experiment with the water level sensor, the Raspberry Pi Pico W, and Node-RED.

Circuit

For this practical example, we'll create a simple circuit connecting the water level sensor to the Raspberry Pi Pico W. Here is the wiring diagram:

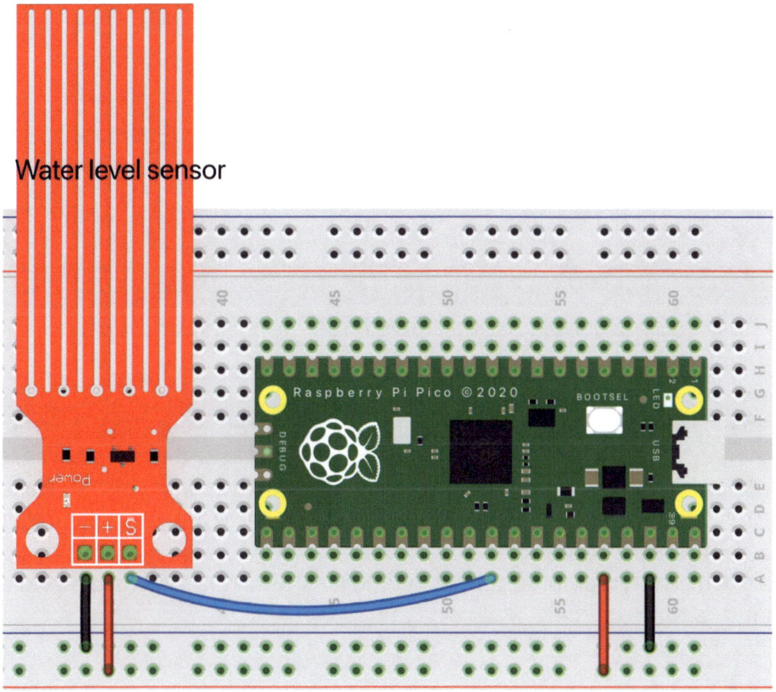

Figure 2.6.4.16: The water level sensor connected to the Raspberry Pi Pico

Here's a step-by-step guide to assemble the circuit:

1. First, connect the VCC pin of the water level sensor to the 3.3V pin on the Raspberry Pi Pico W. This will supply power to the sensor.

2. Next, connect the GND pin of the water level sensor to any ground (GND) pin on the Raspberry Pi Pico W. This completes the power circuit.

3. Finally, connect the signal (S) pin of the water level sensor to an ADC GPIO pin on the Raspberry Pi Pico W. I am using GPIO 26, but GPIO 27 and 28 will also work.

You can see my assembled circuit in the photograph below. Notice a small glass with water that I used during testing. The sensor is partially immersed in the water. Do not allow water in the sensor contacts, as this will cause a short circuit.

Figure 2.6.4.17: My assembled water level sensor circuit.

MicroPython code

This experiment will look at two programs to work with the water level sensor. A simple program that only reads the raw sensor value on the ADC pin and a more advanced version with a threshold for detecting significant water level changes could indicate leakage.

Version 1: simple ADC reading

Here is the first MicroPython program. Its purpose is to take a raw reading from the ADC pin and print it to the console.

```
import machine
import utime

sensor = machine.ADC(26)

while True:
    value = sensor.read_u16()
    print(value)
    utime.sleep_ms(200)
```

Here's what this code does:

1. It imports the machine and utime modules, which allow us to interact with the hardware and manage time-related functions, respectively.

2. It initialises the sensor as an ADC (Analog-to-Digital Converter) on GPIO pin 26.

3. Then, in a continuous loop, it reads the 16-bit unsigned integer value from the sensor and prints it out. This is the value that represents the water level. It repeats this every 200 milliseconds.

Insert the bottom end of the sensor in a cup of water. Be careful not to immerse the entire sensor in the water! Ensure that only the conductive traces at the bottom half of the sensor become wet. A short circuit will likely damage the Raspberry Pi Pico if the connectors become wet.

With the script running and the bottom of the sensor immersed in water, you will see values above 20000 appear in the console. The values I see in my console with the amount of water in the test glass, as shown in the photograph above, is around 28000.

Version 2: detecting significant changes in the water level
The second program example includes a threshold for detecting significant changes in water level:

```
import machine
import utime

sensor = machine.ADC(26)
threshold = 30000 # This value needs to be modified with the environment.

while True:
    value = sensor.read_u16()
    if value > threshold :
        print("Liquid leakage!")
    utime.sleep_ms(200)
```

This code works similarly to the previous example but adds a comparison against a threshold value. If the sensor's reading exceeds this threshold, it interprets that as a leak and sends a warning message. Note that the threshold value should be adjusted according to your specific environment.

Exploring the ADC Capabilities of the Raspberry Pi Pico
The Raspberry Pi Pico offers several features that make it an exceptional choice for various electronics projects. Among these is its Analog-to-Digital Converter (ADC), which we've used in our sensor experiments. Let's look at this feature and how it enhances our projects.

An ADC converts an analog signal into a digital one that our microcontroller can understand. In the context of our water level sensor experiment, the ADC converts the varying voltage levels that the sensor produces into digital values that the Raspberry Pi Pico can interpret.

The Pico's RP2040 microcontroller comes with an ADC, which has three input channels for GP26, GP27, and GP28. This means we can connect up to three analog sensors to the Pico at the same time. Each channel has a 12-bit resolution, which means it can represent the input voltage level with a value between 0 and 4095. The higher the value, the higher the input voltage.

It's important to note that Pico's ADC supports input voltages between 0 and 3.3V. Suppose your sensor produces a voltage higher than this. In that case, you must use a voltage divider or another method to bring the voltage into the acceptable range, as higher voltages could damage the Pico.

To use the ADC in MicroPython, we import the machine module and create an ADC object with the pin number as a parameter. We can then use the read_u16 function to get the current reading as a 16-bit value. For example, if our sensor is connected to GP28, we can read the value like this:

```
import machine

sensor = machine.ADC(26)
value = sensor.read_u16()
print(value)
```

In this code, the value will be between 0 and 65535 as the read_u16 function scales the 12-bit ADC value up to 16 bits. If you want the original 12-bit value, you can divide the result by 16.

The ADC capabilities of the Raspberry Pi Pico open up a world of possibilities for creating projects that interact with the physical world. With the ability to read analog sensors like the water level sensor, you can create complex and exciting projects such as automated plant watering systems, weather stations, etc. I encourage you to experiment with different analog sensors and see what you can create!

Node-RED Flow

Now, let's switch to Node-RED and create a flow that receives the sensor readings from our Raspberry Pi Pico W and displays them in a dashboard. We'll use a gauge for the dashboard to show the detected water level graphically.

Once the flow is running, the dashboard will look like this:

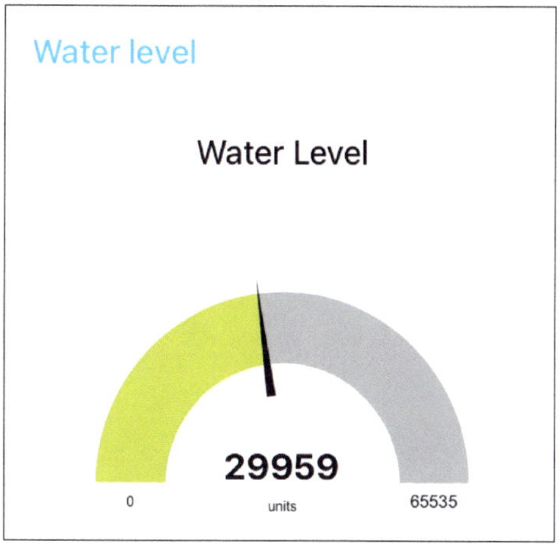

Figure 2.6.4.18: The dashboard indicates that the glass is approximately half-full!

The Node-RED flow comprises just two nodes: An MQTT-in node and a gauge node to display the measured value in the dashboard. Here's the flow in the editor:

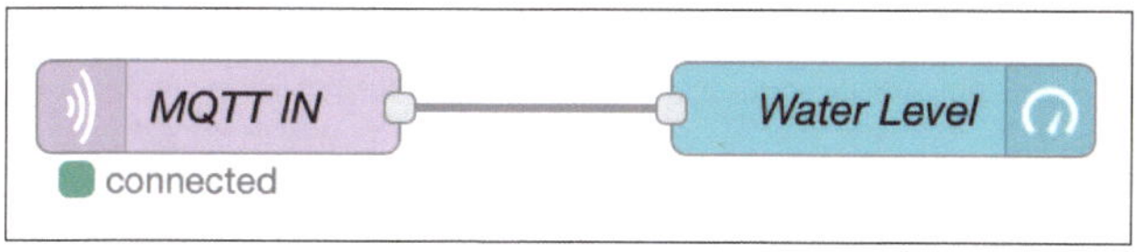

Figure 2.6.4.19: The water level Node-RED flow.

In this Node-RED flow, the MQTT-in node receives water level data from the topic "water level". This data is wired to a gauge on the Node-RED dashboard, which displays the current water level.

This flow is simple; you may adapt it to your use case. For example, add nodes for data processing, alerting, or controlling other devices based on the water level. Additionally, you need to ensure that your MicroPython code on the Raspberry Pi Pico is publishing the water level data to the correct MQTT topic.

When you set up the MQTT-in node, remember to:

1. Set the Server property to your MQTT server's IP address and port number.
2. Action: set it to "Subscribe to single topic"
3. Topic: set to "waterlevel" (to match the TOPIC variable in the MicroPython program that you will see below).

Here is the MicroPython program that will work with this Node-RED flow. The program adapts the first example in this chapter, with added code for the Wi-Fi connectivity and the MQTT broker.

```
import machine
import utime
import time
from umqtt.simple import MQTTClient
from time import sleep
import network

# Configure your Wi-Fi credentials
SSID      = '<your_wifi_network_SSID>'
PASSWORD  = '<your_wifi_network_password>'

print('Starting...')

wlan = network.WLAN(network.STA_IF)
wlan.active(True)
wlan.connect(SSID,PASSWORD)

print('Trying to connect to Wifi...')

while not wlan.isconnected():
    print('.', end='')
    time.sleep(1)
    pass

print('Connected to Wi-Fi')
print(wlan.ifconfig())

# Configure MQTT
SERVER      = '192.168.20.18'  # Replace with your MQTT broker IP
CLIENT_ID   = b'PICO'
TOPIC       = b'waterlevel'
USER        = b'<your_MQTT_username>'  # Replace with your MQTT username
PASSWORD    = b'<your_MQTT_password>'  # Replace with your MQTT password

def mqtt_connect():
   client = MQTTClient(CLIENT_ID, SERVER, user=USER, password=PASSWORD,
keepalive=3600)
   client.connect()
   print('Connected to %s MQTT Broker'%(SERVER))
   return client

def reconnect():
   print('Failed to connect to the MQTT Broker. Reconnecting...')
   sleep(5)
   machine.reset()
```

```
try:
   client = mqtt_connect()
except OSError as e:
   reconnect()

sensor = machine.ADC(26)

while True:
    value = sensor.read_u16()
    print(value)
    client.publish(TOPIC, str(value))
    utime.sleep_ms(200)
```

The following two projects will teach you about the thermistor and the analog light sensor. These sensors are very similar to the water level sensor because they also operate on variable resistance, where the sensor's resistance depends on the property it measures. Consequently, we will use the Raspberry Pi Pico's ADC capabilities and reuse the example MicroPython programs from this chapter.

6.5. Thermistor

In this project, you will learn about the electronic device called "thermistor". The thermistor is a type of resistor whose resistance changes with temperature. They're typically made of ceramic or polymer materials and are commonly used for accurate temperature measurement, compensation, and control.

A thermistor's operation principle is based on the change in electrical resistance of certain materials with temperature. This change in resistance can be measured and translated into a temperature reading. There are two types of thermistors: Negative Temperature Coefficient (NTC) and Positive Temperature Coefficient (PTC). NTC thermistors decrease in resistance as temperature increases, while PTC thermistors increase.

Figure 2.6.5.20: A type 103 thermistor.

How to calculate the temperature from a thermistor

The resistance-temperature relationship for thermistors is typically non-linear and can be approximated by the Steinhart-Hart equation, which is widely used for temperature calculation with NTC thermistors:

```
T = 1 / (A + B*ln(R) + C*(ln(R))^3)
```

Where T is the temperature in Kelvin, R is the thermistor's resistance, and A, B, and C are coefficients specific to each thermistor (usually provided by the manufacturer).

The thermistor bundled in the Sunfounder kit is type 103. Its specifications are in the sensor data sheet (see https://txplo.re/thermistor_103). The datasheet contains the A, B and C coefficients. The thermistor's nominal resistance is 10 kΩ.

Wiring and programming

Now, let's see how to connect a thermistor to our Raspberry Pi Pico and write a MicroPython code to read the temperature. You can see the wiring diagram below:

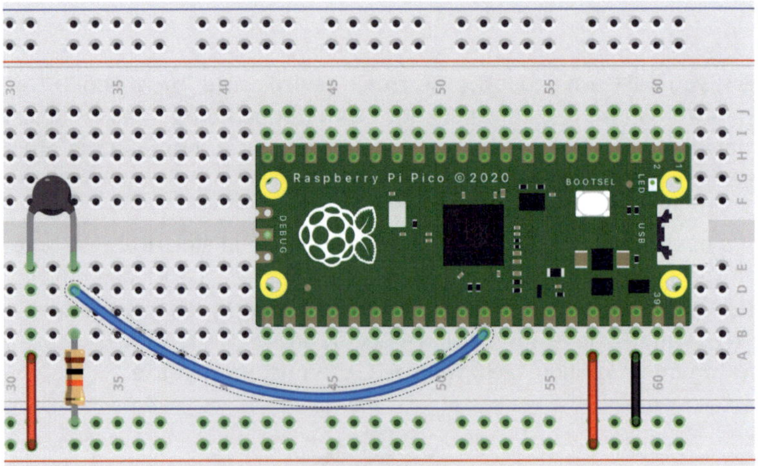

Figure 2.6.5.21: The thermistor connected to the Raspberry Pi Pico

Here are the steps to wire the circuit:

1. Connect one thermistor lead to the 3V3 (3.3 V) pin on the Raspberry Pi Pico.
2. Connect the other thermistor lead to a GPIO pin that supports ADC (Analog-to-Digital Conversion). I will use pin 26 (ADC0).
3. Connect this thermistor lead to a 10-kΩ resistor, and then connect the other end of the resistor to GND on the Pico. This will create a voltage divider circuit necessary to operate the thermistor properly.

Here is my circuit:

Figure 2.6.5.22: My assembled circuit.

With this wiring, the voltage at the ADC pin (in this circuit, pin 26) will change based on the thermistor's resistance, which varies with temperature.

The MicroPython code would look like this:

```
import machine
import math
import utime

# Define the ADC pin
thermistor = machine.ADC(26)

# Define the Steinhart-Hart coefficients
A = 0.001316
B = 0.0002128
C = 0.0000001003

# Define the resistance of the resistor in the voltage divider (in Ohms)
R2 = 10000

# Read the ADC value and convert it to a voltage
while True:
    adc_value = thermistor.read_u16()
    voltage = 3.3 * (adc_value / 65535)

    # Calculate the resistance of the thermistor from the voltage
    R1 = R2 * (3.3 - voltage) / voltage
```

```
    # Use the Steinhart-Hart equation to calculate the temperature
    lnR1 = math.log(R1)
    T_inv = A + B*lnR1 + C*(lnR1**3)
    T = 1 / T_inv

    # Convert the temperature to degrees Celsius
    Tc = T - 273.15

    print("Temperature: ", Tc, " degrees Celsius")
    utime.sleep(1)
```

This code will print the temperature in degrees Celsius every second. Please note that the values of A, B, and C in this code are just example values, and they should be replaced with the actual values provided by the manufacturer of your specific thermistor.

The thermistor provides an analog voltage proportional to temperature, and the ADC reads this voltage on the Raspberry Pi Pico.

Now that your Raspberry Pi can read the thermistor and produce the temperature output in degrees Celsius, you can send the data to the MQTT broker and use the Node-RED flow from the water level example to display the temperature in the dashboard. Remember to change the gauge widget's upper and lower limits to match the temperature range you are working with.

How to show the temperature in °F

To display the temperature in degrees Fahrenheit instead of Celsius, you must modify the final calculation in the MicroPython code. The conversion from Celsius to Fahrenheit is done by multiplying the Celsius temperature by 9/5 and then adding 32. In the program below, the formula that calculates the temperature in Fahrenheit is at the end of the code and looks like this:

$$T_f = (T_c \times 9.0/5.0) + 32.0$$

Here's the modified code:

```
import machine
from math import log
import utime

# Constants for the specific thermistor (10k NTC Thermistor with B=3380)
A = 0.001129148
B = 0.000234125
C = 0.0000000876741
```

```
# Pin connected to the thermistor
adc = machine.ADC(26)

# The resistance of the fixed resistor
R_fixed = 10000

while True:
    # Read the ADC value
    adc_value = adc.read_u16()

    # Convert the ADC reading into a voltage
    Vout = 3.3 * (adc_value / 65535)

    # Calculate the resistance of the thermistor
    R_thermistor = R_fixed * (3.3 - Vout) / Vout

    # Use the Steinhart-Hart equation to calculate the temperature in Kelvin
    logR = log(R_thermistor)
    T = 1.0 / (A + B * logR + C * logR * logR * logR)

    # Convert from Kelvin to Celsius
    Tc = T - 273.15

    # Convert from Celsius to Fahrenheit
    Tf = (Tc * 9.0/5.0) + 32.0

    print("Temperature: {:.2f} F".format(Tf))
    utime.sleep(1)
```

Remember to replace the constants A, B, and C with the appropriate values for your specific thermistor. The variable R_fixed should be set to the resistance of the fixed resistor in your voltage divider circuit. In this example, the thermistor is connected to ADC pin 26, but you can change this to match your actual circuit.

6.6. Analog Light Sensor

A light-sensitive resistor, also known as a photoresistor or photocell, is a type of resistor whose resistance changes based on the amount of light it is exposed to.

The principle of operation is based on photoconductivity, an optical phenomenon in which a material's electrical conductivity changes when it absorbs light. The photoresistor is made of a high-resistance semiconductor material. When photons fall on this material, they are absorbed by the semiconductor, which provides energy to the electrons, enabling them to jump the energy gap and reach the conduction band. This results in free charge carriers (electrons) that conduct electricity.

In simpler terms, when a photoresistor is exposed to light, the light photons provide enough energy to the material's electrons to move and conduct electricity, lowering the device's resistance. Conversely, when the light intensity decreases, the photoresistor's resistance increases as fewer free electrons conduct electricity.

Therefore, by measuring the resistance of a photoresistor, we can determine the intensity of light it's exposed to. This makes photoresistors useful in various applications, such as light sensing, light-activated switches, light meters, and more.

It's important to note that photoresistors are not linear devices - the relationship between the light intensity and resistance is not a straight line and can vary based on the specific design and material of the photoresistor.

Figure 2.6.6.23: A photoresist.

Wiring and programming
In a previous project, the one on the water level sensor, you learned how to use one of the ADC pins of the Raspberry Pi Pico to take analog readings. You can use the save MicroPython code with any sensors operating on the variable resistor principle. In the case of the water level sensor, the sensor's conductivity changes depending on the amount of moisture in the sensor's environment.

A photoresistor is a light-controlled variable resistor. The resistance of a photoresistor decreases with increasing incident light intensity. It can be used with a Raspberry Pi Pico, similar to the water level sensor since both are analog sensors. Instead of water level changes, the photoresistor will respond to changes in light intensity.

In this project, you will reuse the MicroPython code from the water level sensor experiment to work with the light sensor.

Wiring
Let's start with the wiring. Follow these step-by-step instructions. You can consult the wiring diagram for assistance.

Figure 2.6.6.24: The photoresistor connected to the Raspberry Pi Pico

You will need to connect the photoresistor to the Raspberry Pi Pico. This involves connecting one side of the photoresistor to a power source (like the 3.3V pin) and the other to an analog Pico pin. In this example, we'll use GPIO 26. You'll also need to connect a pull-down resistor from the analog pin to the ground to form a voltage divider circuit. The specific value of the resistor can vary, but 10 k-Ωs is a common choice as it closely matches the sensor's resistance in a moderately lit environment.

Here is my circuit:

Figure 2.6.6.25: My assembled circuit.

Once your circuit is set up, you must adjust the code to read the value from the correct pin. If, for example, you connect the photoresistor to GPIO26 (the same pin used in the water level sensor code), you won't need to change the pin number in the code.

Now, the value read from the sensor will correspond to the light level instead of the water level. A higher value will indicate that more photons have reached the sensor, meaning more ambient light is in the surrounding space. You may need to adjust the threshold value in the code depending on the specific behaviour you want.

Part 3: Raspberry Pi Pico, a Primer

Chapter 3.1 • Introduction to the Raspberry Pi Pico and Raspberry Pi Pico W

The Raspberry Pi Foundation is renowned for its tiny computers. It was founded in 2009 with a mission to promote the study of computer science. The Raspberry Pi Foundation launched its first computer in 2012, the hugely popular Raspberry Pi Model B, with more than 2 million units sold. I have one that has been working non-stop since 2013!

Over the years, the Raspberry Pi Foundation has designed and manufactured increasingly powerful computers, always following the formula established by the original Raspberry Pi Model B. Raspberry Pi computers have an open and "hackable" architecture. This means that the learner can access the core functionality of the microprocessor and other peripherals directly by connecting external components (such as sensors) to the pins of the Pi's header.

In 2021, the Raspberry Pi Foundation added another significant innovation to their illustrious lineup - the Raspberry Pi Pico and the Raspberry Pi Pico W. These microcontrollers stand apart from their predecessors due to their smaller size, flexibility, and the specific set of applications they target, and because they are not microcomputers but microcontrollers. As microcontrollers, they are in the same general category as the Arduino and ESP32 boards.

Think of a microcontroller as a very low-cost computing device with limited storage, memory and processing capabilities, designed to consume small amounts of power and be part of an embedded system to perform relatively simple and repetitive tasks.

In the case of the Raspberry Pi Pico, the device is based on the RP2040 chip, a dual-core Arm Cortex-M0+ processor, while the Raspberry Pi Pico W is its wireless variant. Both devices are designed to provide an affordable, easy-to-use path for beginners and hobbyists to dive into microcontrollers. The Pico has no video output, no operating system, just 264 KB of static RAM, and no flash memory for storage. As a comparison, the Raspberry Pi 4 comes with a BCM2711 system-on-a-chip processor, an integrated video controller that supports full high-definition video; it can run many different operating systems, including Ubuntu Linux and Windows, up to 8 GB of RAM and lots of options for storage, including hard disks.

Overview of Raspberry Pi Pico

The Raspberry Pi Pico and Pico W are very different to the traditional Raspberry Pi models because they focus more on embedded applications than full-fledged computing tasks. With a concentration on physical computing, these models come into their own when used in sensor reading, light control, and driving motors, making them perfect for IoT, wearable technology, and robotics applications.

Like the Arduino UNO, the Raspberry Pi Pico is a microcontroller board. Both devices are microcontroller boards designed to be low-cost, robust, and perfect for simple, repetitive tasks.

Like other Raspberry Pi devices, the Pico maintains the ethos of ease-of-use and community support that have been pillars of the Raspberry Pi project. However, they have a much smaller form factor and are significantly cheaper, making them perfect for prototyping or embedding in other devices. Adding the Pico W brings wireless capabilities to this formula, making the Pi Pico even more versatile.

Main Features of Raspberry Pi Pico

The Raspberry Pi Pico and Pico W have a few standout features that distinguish them from their competition.

1. **RP2040 Microcontroller Chip**: The RP2040, designed in-house by the Raspberry Pi Foundation, is a powerful dual-core Arm Cortex-M0+ processor that can run up to 133 MHz. Its performance is surprising, given its low power draw. This chip is now used by microcontroller boards designed by others, such as in the Arduino Nano RP2040 and the Adafruit Feather RP2040.

2. **Programmability**: The Raspberry Pi Pico is programmable using MicroPython and C/C++. This allows for many development options for beginners and advanced users. It is also possible to use a variety of development environments, such as the Arduino IDE, Thonny, and Visual Studio Code, to name a few.

3. **GPIO Pins**: The Raspberry Pi Pico and Raspberry Pi Pico W have 26 GPIO pins, including three analog inputs. This is the interface for the Pico to interact with the physical world, allowing users to connect many peripherals and sensors. With these pins, your projects can use the entire set of capabilities in the Raspberry Pi Pico, from simple GPIO on/off to high-speed SPI communications. You will learn more about The Raspberry Pi Pico's pins later in this chapter.

4. **Wireless Connectivity**: The Raspberry Pi Pico W adds Wi-Fi and Bluetooth functionality to the mix, which allows it to connect with other devices and the internet, making it even more versatile for IoT applications. With the arrival of the "W" model, the adoption of the Raspberry Pi Pico board exploded since now it was possible to use it for Internet of Things applications (or any application that requires communications) without additional components.

Detailed Specifications

Below is a summary of the Raspberry Pi Pico and Raspberry Pi Pico W specifications.

Raspberry Pi Pico

Here is an outline of the Pico's specifications:

- RP2040 microcontroller chip
- Dual-core Arm Cortex-M0+ processor, clocked at up to 133 MHz
- 264 KB on-chip SRAM
- 2 MB on-board QSPI Flash
- 26 GPIO pins. Through these pins, you can access:

  - Two SPI interfaces.
  - Two I^2C interfaces.
  - Two serial UART interfaces.
  - Three analog-to-digital converters (ADC) with 12-bit resolution.
  - 16 controllable Pulse Width Modulation (PWM) pins.
  - 26 multi-function GPIO pins.
- A temperature sensor
- An accurate clock and timer
- Castellated module for soldering directly to carrier boards
- Micro-USB B 1.1 port for power and data transfer. This port supports decide and host mode to connect peripherals such as a keyboard and mouse.

Raspberry Pi Pico W

In addition to the features of the Raspberry Pi Pico, the "W" variant offers the following:

- On-board Wi-Fi (2.4GHz b/g/n)
- WPA3
- Soft access point that supports up to four clients
- On-board Bluetooth 5.2 with support for Bluetooth LE and Bluetooth Classic.

If you are using the "W" variant (as we are in the projects of this book), it is essential to remember that the presence of the Wi-Fi module imposes some limitations on how some of the pins can be used. You can learn more about this in the official documentation (available at https://txplo.re/rpipico_doc). I have copied the relevant part below:

> Due to pin limitations, some of the wireless interface pins are shared. The CLK is shared with VSYS monitor, so only when there isn't an SPI transaction in progress can VSYS be read via the ADC. The Infineon CYW43439 DIN/DOUT and IRQ all share one pin on the RP2040. Only when an SPI transaction isn't in progress is it suitable to check for IRQs. The interface typically runs at 33 MHz.

The VSYS monitor on the Raspberry Pi Pico is an onboard power management system feature. It allows the RP2040 chip to monitor the voltage level on the VSYS pin. This monitoring capability enables the microcontroller to check the power supply's stability and respond to low-voltage conditions, potentially by entering a power-saving mode or signalling that the voltage is out of the acceptable range to the user or software. This function is essential for applications where power levels may fluctuate, and system stability is critical.

There is one more tip to remember about the performance of the Wi-Fi module, and it has to do with the positioning of the on-board antenna. Again, I copy from the documentation:

> For best wireless performance, the antenna should be in free space. For instance, putting metal under or close by the antenna can reduce its performance both in terms of gain and bandwidth. Adding grounded metal to the sides of the antenna can improve the antenna's bandwidth.

The Raspberry Pi Pico has gained popularity in the maker community for several reasons. First, it's affordable, especially compared with established peers, like the various Arduino and ESP32 boards. This means many people, from hobbyists to professionals, can experiment with it without a significant investment. The Pico also runs on the RP2040, a microcontroller chip developed by the Raspberry Pi team, which provides a good balance of power and energy efficiency.

Another big plus is its ease of use. The Raspberry Pi Pico can be programmed with MicroPython or C/C++, well-known languages among developers. MicroPython, in particular, is beginner-friendly, so it's not daunting for those new to electronics and programming.

Then there's its size. The Raspberry Pi Pico is relatively compact perfect for projects with space constraints. Despite its small size, it comes with a solid selection of GPIO pins, allowing various sensors and components to be connected.

Lastly, the Raspberry Pi community is a vibrant and supportive one. Many resources, tutorials, and projects are available for the Pico. This community support makes it much easier for anyone starting out or looking to tackle complex projects. All these factors combined make the Raspberry Pi Pico a hit in the Maker world.

Chapter 3.2 • Getting Started with Raspberry Pi Pico and Thonny

If you have never used the Raspberry Pi Pico, this section will help you get started. Specifically, this chapter will help you install the Thonny development environment, the software application we use to write programs for the Raspberry Pi Pico and manage various aspects of its operation. This chapter will also show you how to install the MicroPython interpreter on the Pico and upload your first MicroPython program. There is also a short discussion on alternatives to Thonny.

To follow the instructions in this chapter, you need the following hardware:

- A **Raspberry Pi Pico W** board. Using the Raspberry Pi Pico is also possible if you don't have the "W" variant. However, you cannot try out any example that uses Wi-Fi.

- A **micro-USB cable** for connecting your Raspberry Pi Pico to your computer. The cable needs to be capable of data transfer, not just charging.

No other hardware is needed for this chapter. You can read about hardware like LEDs and sensors in Part 2 of this book. In terms of software, you will need the following:

- A **MicroPython** interpreter. You will learn how to download an image of the interpreter from the MicroPython website and install it to your Raspberry Pi Pico.

- The **Thonny IDE**. This is an easy-to-use Integrated Development Environment (IDE) for Python with built-in MicroPython support on the Raspberry Pi Pico. You will learn how to download it from the Thonny website, install it on your computer, and use it to set up MicroPython on your Raspberry Pi Pico and upload your first MicroPython program.

Let's start with Thonny.

Setting up the Thonny development environment

Thonny is an Integrated Development Environment (IDE) designed for learning and teaching programming, particularly Python. With a simple and user-friendly interface, Thonny offers an excellent entry point for beginners while providing the features that seasoned developers expect. It comes with a Python 3 interpreter and offers step-through debugging, variable inspection, and a call stack viewer, enabling easier troubleshooting. It also has syntax highlighting, code completion, and inline error detection to assist in writing code efficiently. The Read-Eval-Print Loop (REPL) is integrated into the IDE, allowing quick code testing and iterative development.

What is even more essential for us, learning MicroPython with the Raspberry Pi Pico, is that Thonny offers built-in support for MicroPython, a lean implementation of Python for microcontrollers. With Thonny, you can quickly write, upload, and run MicroPython code on compatible devices. Thonny offers a simple process for "flashing" MicroPython firmware onto a microcontroller. It also provides a device shell within the IDE, allowing you to interact directly with the MicroPython interpreter on the microcontroller and file system management.

Thonny works on Windows, macOS and Linux. To install Thonny, follow the latest instructions on the Thonny GitHub repository. You will find the instructions for your operating system here: https://github.com/thonny/thonny/releases.

Install Micropypthon on your Raspberry Pi Pico W

Let's use Thonny to install MicroPython on the Raspberry Pi Pico W. You can use the same process to install MicroPython on a Raspberry Pi Pico, with a slight variation that I will mention inline below. Before starting the process, a note about the Pico's boot loader mode.

On the Raspberry Pi Pico W, notice the white button marked BOOTSEL. The BOOTSEL (Boot Select) button puts the device into bootloader mode. When in bootloader mode, the Pico appears as a mass storage device when connected to a computer via USB. This allows you to drag and drop firmware files onto the Pico or flash it with new firmware using development environments like Thonny.

To enter bootloader mode, hold the BOOTSEL button while plugging the Pico into your computer via a USB cable. You can release the button once connected and recognised as a mass storage device.

Figure 3.2.1: Press and hold the BOOTSEL button while connecting the board to the USB port to enter bootloader mode.

The BOOTSEL button is handy for:

- Initial setup: Installing MicroPython or other firmware for the first time.
- Firmware updates: Upgrading to a newer MicroPython or other compatible firmware version.
- Troubleshooting: When you need to wipe existing code or reset the device to a default state for debugging purposes.

Using the BOOTSEL button, you can easily manage the firmware and software running on your Raspberry Pi Pico.

Now that you know the bootloader mode let's flash the MicroPython interpreter on Pico's flash memory. Here is the process:

1. Set the Raspberry Pi Pico W into boot loader mode by holding down the BOOTSEL button while connecting the Pico to your computer via USB.

2. Start the Thonny IDE.

3. In Thonny, open the Thonny Options window. You can do this on all platforms by clicking Tools —> Options.

4. In the Thonny Options window, select the Interpreter tab.

5. In the Interpret tab, select your board from the drop-down titled "Which kind of interpreter should Thonny use for running your code?". Since we are using a Raspberry Pi Pico, select "MicroPython (Raspberry Pi Pico). Then, click "Install or update MicroPython" at the bottom right corner of the window.

6. A new window titled "Install of update MicroPython (UF2)" will appear. If your Raspberry Pi Pico W is in boot loader mode and connected to your computer via USB, it will be listed in the Target Volume dropdown. Continue to configure the installer with these settings:

   - Target volume: /Volumes/RPI-RP2 (the exact path depends on your OS and computer).
   - Family: RP2
   - MicroPython family: RP2
   - Variant: Raspberry Pi Pico W / Pico WH.
     - If you are working with Raspberry Pi Pico (no "W"), select "Raspberry Pi Pico / Pico H" from the list.

   - Version: 1.20.0 (or the latest stable version)

Here is a screenshot of the installer window at this point in the process:

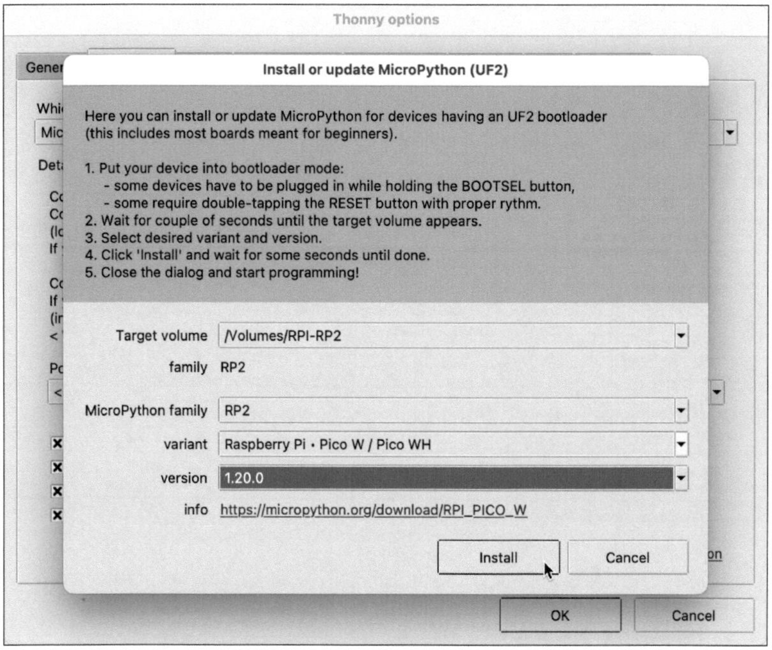

Figure 3.2.2: The MicroPython installer is ready to install the firmware on my Raspberry Pi Pico w.

1. Once you have configured the installed, click "Install" to start the process. A progress indicator shows a percentage completed for the download and write processes. Eventually, installed will display "Done!". Click "Close" to close the installer and "OK" to exit the options window.

2. Congratulations, you are ready to use MicroPython on the Raspberry Pi Pico W.

Quick test

After the firmware installation is complete, the Raspberry Pi Pico will automatically restart. The device in Thonny's bottom shell confirms that MicroPython is running on your Raspberry Pi Pico. In the screenshot below, you can see the Thonny editor and the Shell tab on the lower right side of the window. In the Shell, you can see the version of MicroPython running on the connected device (in this case, it is v1.20.0). In the bottom right side of the window, you can see the name and port path of the connected device.

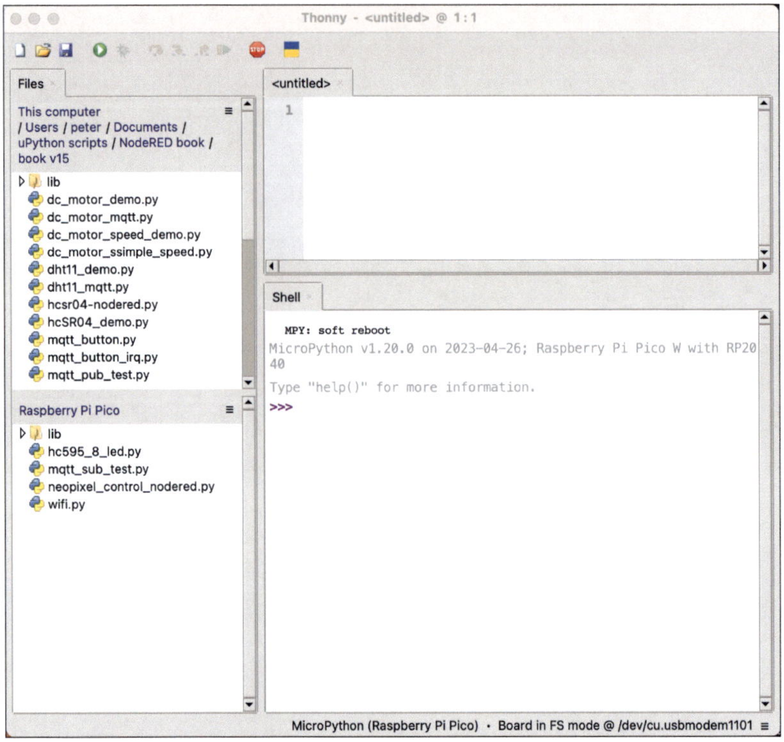

Figure 3.2.3: Thonny showing the MicroPython version in the Shell and connected device in the status bar.

Let's write a simple program and upload it to the Pico. This will confirm that the setup is correct. Here is what to do:

1. In the Thonny editor, write a simple MicroPython script, for example, "print("Hello, Pico")".

2. Press the Run button to execute the script.

3. The output will be displayed in Thonny's bottom shell, confirming that MicroPython is working on your Raspberry Pi Pico.

You can see this experiment in the screenshot below:

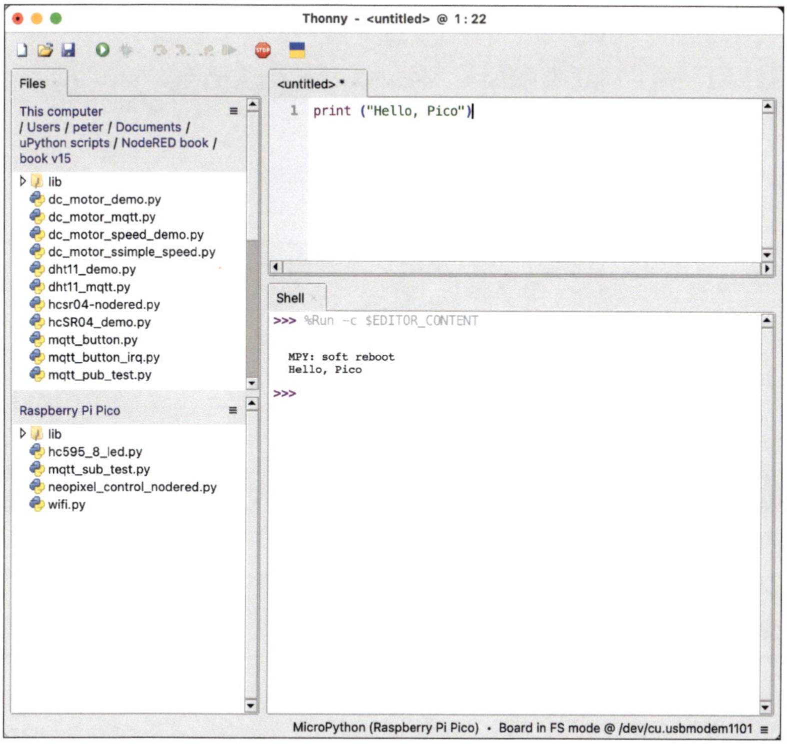

Figure 3.2.not set.4: Running a simple MicroPython program on the Pico.

As you can see, we could run this program on the Pico without saving it on the Pico's flash memory first. We wrote the program and tabbed editor and clicked the green "play" button.

If you want to save a program to reuse or edit it later, click on the blue diskette button. An aside: you have never seen an actual diskette. A diskette is a magnetic medium that is used to save files. Since the early 2000s, computers have no longer had a diskette drive as USB "thumb drives" and other media became preferable. Nevertheless, the diskette icon has survived to indicate the "save" function in GUIs.

When you click on the diskette (save) icon and the Pico is connected, Thonny will present a dialogue box asking you to select the save location for the file.

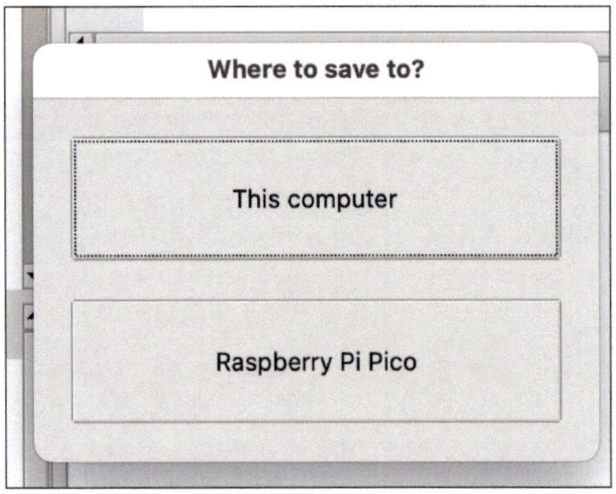

Figure 3.2.5: You can save your programs to your computer or the Pico's filesystem.

Remember, you can run a MicroPython program on your Raspberry Pi Pico without the program being stored on the Pico. It is convenient to store my working programs on the board. Click on "Raspberry Pi Pico", give it a name, like "hello.py", and click OK. Your program will be listed in the Raspberry Pi Pico part of the Files tab, as shown below.

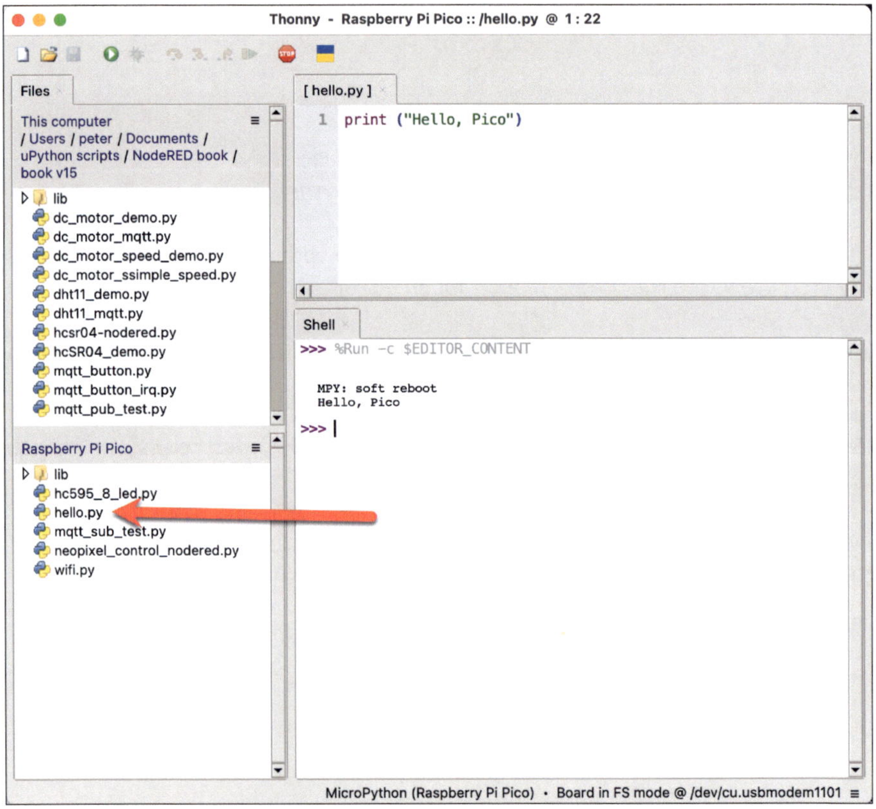

Figure 3.2.not set.6: You can save your programs to your computer or the Pico's filesystem.

The file listing in the Raspberry Pi Pico section of the Files tab shows the stored files and directories. Aside from the new "hello.py", you can see several other files that belong to projects elsewhere in this book.

Other development environment options

While Thonny is an excellent option for beginners due to its simplicity, there are other options if you prefer a different development environment. Here is a short list:

Mu Editor

Mu is a simple yet powerful code editor for Python development focusing on beginner programmers. It aims to make programming more accessible by providing an uncomplicated interface and essential tools. Mu offers syntax highlighting, error highlighting, and a built-in Python interpreter, making it a well-rounded choice for new and experienced developers.

Mu comes with multiple modes, including Python 3, Web, and even specialised modes for microcontroller programming, such as MicroPython and CircuitPython. This adaptability makes it particularly useful for various Python projects, including web development, data science, and embedded systems.

While both Mu and Thonny aim to make Python programming accessible, they cater to slightly different audiences and offer different features. Mu is designed to be straightforward for beginners, focusing on simplicity. Thonny, while still beginner-friendly, provides a broader range of features that can appeal to more experienced developers as well. The choice between the two would depend on your specific needs, level of expertise, and the nature of the project you're working on.

To learn more about Mu or download the editor, visit its official home page at https://codewith.mu/.

Visual Studio Code

Visual Studio Code offers a versatile and feature-rich platform for programming the Raspberry Pi Pico with MicroPython. As one of the most popular IDEs, Visual Studio Code brings an array of tools to the table, including version control, syntax highlighting, code completion, and an integrated terminal. The extensive library of extensions available in Visual Studio Code makes it highly customisable, allowing you to set up your development environment exactly how you want. While it doesn't provide out-of-the-box support for MicroPython, it can be configured to handle Raspberry Pi Pico development seamlessly.

On the other hand, Thonny is designed with simplicity in mind, primarily catering to beginners and those focusing exclusively on Python or MicroPython. Thonny offers a simpler user interface and straightforward tools for code analysis, debugging, and running programs. It has built-in support for MicroPython, allowing for direct installation of firmware on the Raspberry Pi Pico, and provides a more streamlined approach to microcontroller programming.

When it comes to debugging, Thonny has an edge with its built-in debugging tools that allow step-through debugging and variable inspection, making it easier for beginners to troubleshoot their code. Although featuring a robust debugging environment, Visual Studio Code requires an initial setup to get the debugger working specifically for MicroPython and Raspberry Pi Pico.

If you're a beginner or looking for a quick and hassle-free way to start programming your Raspberry Pi Pico with MicroPython, Thonny might be your better choice. But Visual Studio Code would likely be more suitable if you're an experienced developer seeking a more feature-rich environment or working on a larger, more complex project.

You can learn more about Visual Studio Code at https://code.visualstudio.com/.

Arduino IDE

The Arduino IDE is primarily known for C/C++ development on Arduino boards, but it can be configured to work with MicroPython and Raspberry Pi Pico. The IDE provides basic but essential features like code editing, serial monitoring, and uploading code to microcontrollers. Installing the right board support package lets you repurpose the Arduino IDE to write, compile, and upload MicroPython code to your Raspberry Pi Pico. However, it's worth noting that the Arduino IDE doesn't natively support Python, so it lacks some Python-specific features like code completion for Python libraries or Python-specific debugging.

Thonny, in contrast, is designed specifically for Python and MicroPython development. It offers a more beginner-friendly interface and a set of Python-specific tools, including a powerful debugger, a variables view, and the ability to manage MicroPython libraries directly. Thonny also has built-in support for Raspberry Pi Pico, allowing you to install the MicroPython firmware now from the IDE. This makes the process more straightforward for users either new to microcontroller programming or looking specifically for a Python-friendly development environment.

Thonny offers an integrated REPL (Read-Eval-Print Loop) for debugging and real-time code monitoring for Python and MicroPython. This is something that the Arduino IDE lacks, as its serial monitor only allows for simple text-based communication between the IDE and the connected device.

You can learn more about the Arduino IDE at https://www.arduino.cc/en/software.

uPyCraft

uPyCraft is a specialised IDE explicitly designed for MicroPython development on various microcontrollers, including the Raspberry Pi Pico. With its focus on MicroPython, uPyCraft offers tools optimised for quick development cycles, including built-in firmware burning and file management utilities. The interface is streamlined and intuitive, allowing for easy navigation, and it includes features like a REPL terminal and easy-to-access upload and download buttons. This makes uPyCraft highly specialised for MicroPython development but also somewhat limited regarding more generalised coding features.

Thonny, on the other hand, serves as a more general-purpose Python IDE that has been extended to support MicroPython development. Thonny offers a broader set of tools for code analysis, debugging, and editing, making it a good choice for standard Python development and more specialised MicroPython projects. It also has the advantage of being beginner-friendly with features like simple debugging tools, a variables view, and a rich Python shell. Thonny also provides built-in support for Raspberry Pi Pico, which makes setting up the development environment relatively straightforward.

Regarding debugging capabilities, both uPyCraft and Thonny offer integrated REPL terminals for real-time testing and debugging. Still, Thonny's debugger provides more comprehensive features like step-through debugging and variable inspection. This is particularly useful for beginners or those new to microcontroller programming who wish to understand how their code is executing.

Both uPyCraft and Thonny strongly support MicroPython and can be used effectively for Raspberry Pi Pico development. However, uPyCraft is more specialised and streamlined for MicroPython, making getting a MicroPython project off the ground quicker and easier. Thonny provides a more versatile and comprehensive development environment that can be used for a broader range of Python-related projects, and its debugging features make it particularly strong for educational purposes or those new to coding.

You can learn more about uPyCraft at the project home on Github:
https://github.com/DFRobot/uPyCraft_src

Geany
Geany is a lightweight and versatile IDE that supports various programming languages, including Python. Unlike Thonny, designed primarily for Python and MicroPython development, Geany offers a more generalised environment suitable for multiple languages. Geany's interface is straightforward and uncluttered, providing basic features like code highlighting, folding, and a built-in terminal. However, it lacks native support for MicroPython, and you'd need to set up external tools or scripts to upload your code to the Raspberry Pi Pico.

Thonny, designed explicitly for Python and MicroPython, has various Python-centric features such as a built-in Python shell, step-through debugging, and a variables view. It has built-in support for Raspberry Pi Pico, simplifying tasks like installing firmware and uploading code to the board. The interface is designed to be beginner-friendly, with easy-to-use debugging tools that are especially helpful for those new to programming or MicroPython.

Geany's lighter footprint makes it fast and responsive, making it a good choice for less-powerful computers or developers who prefer a more minimalistic environment. However, it requires manual configuration to enable features like code uploading to the Raspberry Pi Pico, which could be a hurdle for beginners. Thonny, on the other hand, provides a more out-of-the-box experience tailored for MicroPython development, with built-in features that simplify the entire development process from coding to debugging and code uploading.

In terms of debugging, Geany offers basic debugging features, but it lacks the integrated, beginner-friendly debugging experience that Thonny offers. Thonny provides valuable debugging tools such as step-through code execution and the ability to inspect variable values, which can be particularly beneficial for educational purposes and for developers new to MicroPython or programming in general.

You can learn more about Geany at https://www.geany.org/.

Remember, the best development environment is the one that suits your needs and preferences. Feel free to try different options and see what works best for you. With the Raspberry Pi Pico's flexibility and robust support, the possibilities for learning and creating are virtually limitless.

Chapter 3.3 • MicroPython and Raspberry Pi Pico

MicroPython is a lean and efficient Python 3 programming language implementation, optimised to run on microcontrollers like the Raspberry Pi Pico. MicroPython aims to be as compatible with desktop Python as possible, allowing Python programmers to dive into microcontroller programming with ease.

However, it's important to note that due to the resource constraints of microcontrollers, MicroPython does not include the entire standard library that comes with Python. Instead, it includes a small subset of the standard library and adds some modules specific to microcontrollers.

MicroPython on the Raspberry Pi Pico differs from other implementations, like the one for the ESP32. While the core language is the same, the libraries available vary based on the specific hardware of the microcontroller. For example, the machine module, which provides hardware-related functions, will have different classes and functions on the Raspberry Pi Pico and the ESP32, corresponding to their unique hardware features.

Basics of MicroPython on the Raspberry Pi Pico

Here are some basics to start with MicroPython on the Raspberry Pi Pico.

Running Code

MicroPython code can be run directly on the Raspberry Pi Pico. This can be done using an IDE like Thonny, which has a built-in REPL (Read-Evaluate-Print Loop) for running Python code. To run code, type it into the REPL and press Enter.

In the example of the screenshot below, you can see a simple MicroPython command I entered in the REPL inside the Shell tab.

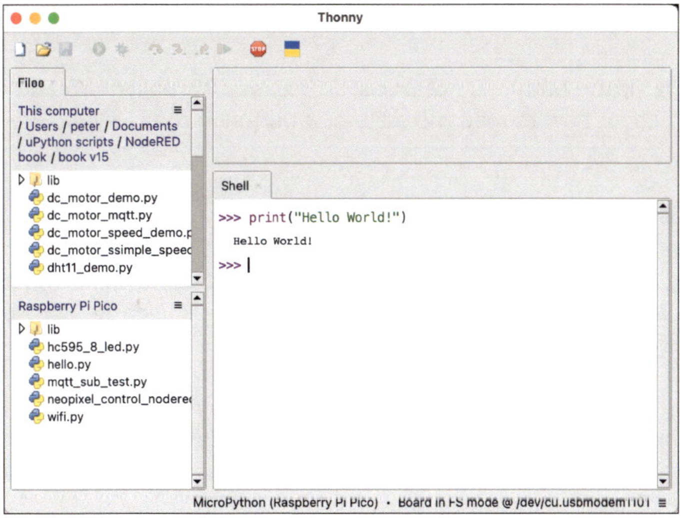

Figure 3.3.not set.1: You can run code directly in REPL.

Pin control

You can control the GPIO (General Purpose Input/Output) pins on the Raspberry Pi Pico using the machine module in MicroPython. Here is an example of how to control an LED connected to the Pico:

```
from machine import Pin

# Create a Pin object for pin GP14
led = Pin(14, Pin.OUT)

# Turn the LED on
led.value(1)

# Turn the LED off
led.value(0)
```

You can find more details about this code, including how to wire an LED, in Part 2 of this book.

Reading inputs

You can read inputs using the machine module. Here is an example of how to read the value of a button:

```
from machine import Pin

# Create a Pin object for pin GP15, set as input
button = Pin(15, Pin.IN, Pin.PULL_UP)

# Print the button state
print(button.value())
```

This will print "1" if the button is not pressed (because it's pulled up) and "0" if it is. You can find details about how to read the state of a button in the relevant experiment in Part 2 of this book.

Using libraries

MicroPython on the Raspberry Pi Pico has several useful libraries for controlling hardware and performing other tasks. Some of the projects in this book depend on libraries to make it easier for us to use the corresponding hardware. For example, the RFID project (Part 2) depends on a library named "mfrc522".

Unlike the Arduino IDE, Thonny does not provide a library manager. Instead, we copy a file that contains the library code to the Raspberry Pi Pico's filesystem in a directory titled "lib". The screenshot below shows the "lib" directory on my Raspberry Pi Pico, containing several required library files for this book's projects.

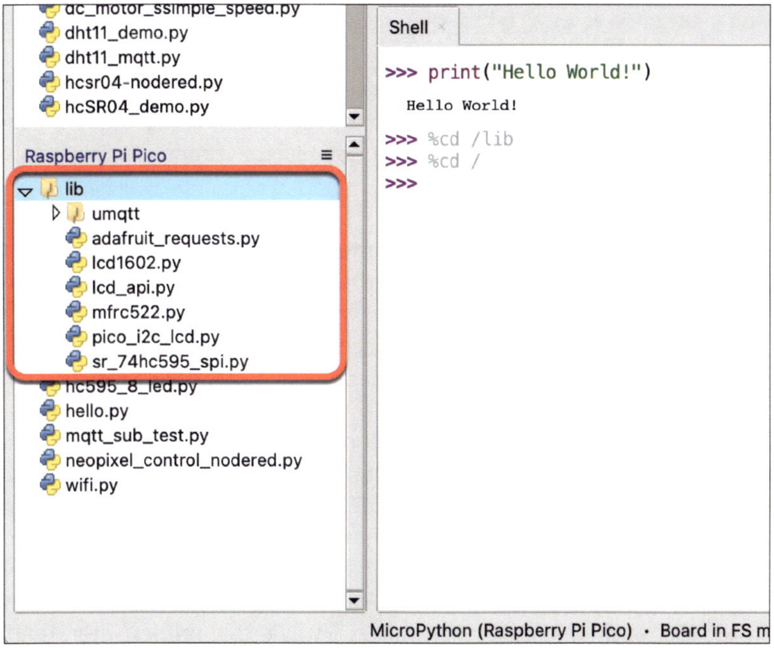

Figure 3.3.not set.2: Libraries are manually copied to the "lib" directory.

You can find documentation about the built-in libraries of MicroPython in this documentation page: https://docs.micropython.org/en/latest/library/index.html.

If you are looking for third-party libraries, an excellent source to look at first is the Python Package Index at https://pypi.org/. You can also use Google or other search engines to help you locate libraries that might not be catalogued in PyPi.

Chapter 3.4 • Micropython, a Primer

MicroPython emerged from a successful Kickstarter campaign in 2013. Damien George, a physicist passionate about computer science and electronics, created it. He conceived the idea of MicroPython to provide a compact and optimised Python interpreter for microcontroller usage.

George's vision was to make Python, a language known for its simplicity and readability, accessible for microcontrollers. Programming these small computing devices typically requires lower-level languages like C or C++. However, using Python could lower entry barriers, making embedded programming more accessible to hobbyists, educators, and those new to the field.

This part of the book is designed to help you learn more than enough MicroPython to complete all of the projects in this book quickly. Its chapters cover everything that you need to know (and more), including the language constructs (such as for control flow, creating functions, and making decisions), commands (such as for input and output, doing calculations, or creating data structures), and the basic use of modules. It also covers topics specific to using MicroPython on microcontrollers, such as reading and writing to files, working with GPIOs, and Wi-Fi communications. I wrote this primer with an emphasis on practicality, not theory. I use examples to demonstrate all topics of discussion. One of the most significant advantages of Python (and MicroPython) is that it is very readable; in most cases, the examples you will see, while they are executable code, can be understood by beginner programmers with little or no further explanation.

At the end of this chapter, you will also find advice for troubleshooting common problems, a list of best programming practices (most of them are universal programming best practices), a glossary of terms, and a list of references in case you want to delve deeper into these topics.

MicroPython overview

MicroPython is a lean and efficient implementation of Python 3. It's designed specifically for operation on microcontrollers and other constrained environments, but it doesn't cut corners when running Python code effectively.

What makes MicroPython stand out? Well, let's dive into some of its unique features:

- **Compactness**: MicroPython is designed to be as lean as possible, making it suitable for microcontrollers with limited resources. Its Python interpreter can fit within 256kB of ROM and 16kB of RAM.

- **Interactivity**: MicroPython offers an interactive prompt or REPL (Read-Evaluate-Print Loop). This allows you to interact with the microcontroller in real time, experimenting and debugging your code on the fly.

- **Python Compatibility**: Despite its trimmed-down nature, MicroPython strives to be as compatible with regular Python. Most Python scripts - especially those that avoid extensive libraries - can run under MicroPython without modification.

- **Hardware Access**: MicroPython includes a library to access low-level hardware, such as GPIO (General Purpose Input/Output) pins, ADCs (Analog to Digital Converters), and more. This provides you with direct control over various sensors and actuators.

- **Asynchronous Support**: MicroPython supports cooperative multitasking through Python's native coroutines and the async/await syntax. This is useful for managing multiple tasks without a full-blown operating system.

The importance of MicroPython in IoT and embedded development

Today, MicroPython is more relevant than ever. As our modern lifestyles become more dependent on Internet of Things (IoT) applications, we need accessible and efficient ways to program the "things" that form part of this Internet.

Here, MicroPython excels. It brings the simplicity and ease of Python to the world of microcontrollers. It enables developers to prototype quickly and hobbyists to learn embedded programming in a friendly and forgiving environment. Its balance of accessibility, performance, and low resource usage makes it a prime choice for modern embedded systems and IoT devices.

Furthermore, the broad adoption of MicroPython has resulted in an active community and a wealth of resources, including libraries, tutorials, and forums. This makes the journey into IoT development with MicroPython a well-supported one.

MicroPython is a powerful tool in the domain of microcontrollers and IoT. It combines Python's simplicity with the necessary features to interact with and control hardware. Whether you're a professional looking for a rapid prototyping tool, an educator seeking an accessible teaching resource, or a hobbyist venturing into the world of embedded systems, MicroPython is an excellent companion on your journey.

4.1. An Introduction to MicroPython

MicroPython is a lean and efficient Python programming language implementation designed to run on microcontrollers and other constrained systems. It includes a small subset of the Python standard library and is optimised to run on small devices. MicroPython is open-source software developed and maintained by a community of enthusiasts. Let's take a high-level view of the language and examine some of the most popular microcontrollers supporting it.

Important language constructs and commands

Just like Python, MicroPython also supports various programming constructs. If you are familiar with Python or other programming languages, the following constructs and commands will seem familiar:

1. **Variables and Types**: Variables in MicroPython can be integer, float, string, list, tuple, and dictionary types.

2. **Control Structures**: If-Else statements, For and While loops for control flow.

3. **Functions:** Functions are defined using the "def" keyword.

4. **Classes and Objects**: MicroPython supports object-oriented programming. You can define classes and create instances of those classes.

5. **Modules**: You can create and import modules (a Python code file) in MicroPython.

6. **Exception Handling**: MicroPython supports Error and Exception handling through "try", "except", and "finally" blocks.

7. **File I/O**: Reading from and writing to files is supported.

8. **Hardware Specific Features**: MicroPython provides libraries to easily control hardware such as GPIOs (General Purpose Input Output), ADC (Analog to Digital Converters), PWM (Pulse Width Modulation), and more.

Later in this chapter, you will learn more about each item in the list. Refer to the MicroPython Documentation pages for comprehensive information: https://docs.micropython.org/en/latest/.

Devices that support MicroPython and cross-platform capability

MicroPython can be run on several microcontroller boards and systems. MicroPython can work on Microcontrollers that share the same ARM Cortex-M architecture (such as many STM32 boards, the RP2040 boards, and SAMD21 boards). There are also "forks" of the MicroPython mainline firmware modified to work on other systems, such as the Micro:bit, and the ESP8266/ESP32.

Below is a list that contains only a tiny subset of boards that can work with MicroPython:

1. Pyboard: The original hardware that was built to run MicroPython. This board uses the STM32F4 microcontroller.

2. ESP8266 and ESP32 modules: Widely used in IoT projects. These boards are based on Tensilica microprocessors (the ESP32 uses the Xtensa LX6, while the ESP8266 uses the Tensilica Diamond Standard 106Micro).

3. BBC micro:bit: A tiny programmable educational computer with many peripherals on board and a very user-friendly interface. This board is using the Nordic nRF52833 microprocessor.

4. Raspberry Pi Pico: A microcontroller board from the Raspberry Pi Foundation. Enough said.

5. Adafruit Circuit Playground Express: A multi-featured board with sensors and LEDs. This board uses the ATSAMD21 ARM Cortex M0 processor.

6. WiPy, LoPy, FiPy boards: MicroPython boards by Pycom with networking features like WiFi, LoRa, Sigfox, etc.

7. Many more. You can find a complete list of compatible boards on the MicroPython website here: http://www.micropython.org/download/

MicroPython's portability is one of its key advantages, allowing developers to write code that can run on a wide range of microcontrollers that support the language. This cross-platform capability is mainly due to the architecture of MicroPython itself, which is designed to be a compact and optimised implementation of Python 3. Consequently, it can be deployed on microcontrollers with varying hardware capabilities, from lower-end models with minimal memory to more advanced boards with extensive features.

The foundational layer of MicroPython, often called the "MicroPython kernel", handles the core functionalities such as memory management, file operations, and basic input/output. This kernel can be ported to different microcontrollers, providing a consistent programming interface across various platforms. Once the kernel is ported, features like GPIO control, analog and digital I/O, communication protocols like UART, SPI, and I^2C, and other hardware-specific functionalities are typically added as modules. These modules provide a "Pythonic" interface to the underlying hardware, making it easier for developers to write portable code.

While MicroPython code can often run unmodified across different microcontrollers, some limitations should be considered. First, not all hardware features will be available on all boards. For example, a Wi-Fi module program may not be compatible with a microcontroller that lacks Wi-Fi support. Second, memory limitations may restrict the complexity or size of the programs that can be run. Finally, different microcontrollers may have varying processing speeds, impacting the performance of time-sensitive applications.

Despite these limitations, MicroPython's ability to run on diverse microcontrollers gives developers considerable flexibility. It simplifies the prototyping process and allows easier migration between different hardware platforms, ultimately enabling a more efficient and cost-effective development cycle.

Example Projects Using MicroPython

MicroPython is being used in a wide range of projects. As you embark on a journey to learn MicroPython and use it in your projects, perhaps this list of suggested projects can spark your imagination:

1. **IoT Projects**: Control and monitor IoT devices using MicroPython. For example, an ESP8266/ESP32 can be programmed using MicroPython to read sensor data and send it to a server for logging, processing, and control.

2. **Robotics**: MicroPython can be used in robotics for controlling motors and reading sensor data. The BBC micro:bit has been used widely in educational robotics projects.

3. **Wearables**: MicroPython can be used to program wearable devices. You can use it to control LEDs on clothes or read sensor data from wearable sensors.

4. **Home Automation**: With networking-enabled MicroPython boards, you can create home automation projects like controlling your lights, curtains, and other home appliances.

MicroPython is a powerful tool for microcontroller programming. Its simplicity makes it an excellent choice for beginners and professionals alike who want to quickly prototype and build hardware projects.

4.2. MicroPython Language Constructs

In this section, you will learn about some of the most important topics in MicroPython: Data Types and Variables, Control Flow Constructs, Functions and Modules, and Exception Handling.

Understanding Data Types and Variables is crucial as they serve as the building blocks of any software. They're the first step in writing effective MicroPython code that performs as expected.

Next, Control Flow Constructs like decisions (if, elif, else) and loops (for, while) are essential for guiding how your code executes, enabling you to build logical and efficient scripts.

In addition, Functions and Modules play a pivotal role in enhancing code reusability and manageability. Functions encapsulate code snippets for reuse, while modules help you organise those functions into distinct, manageable files.

Last but not least, Exception Handling is vital for robust programming. Errors and exceptions are inevitable in any software application; knowing how to handle them gracefully ensures that your programs are robust and user-friendly.

Understanding these key areas will equip you with the tools to develop efficient, scalable, and maintainable MicroPython applications.

Data types and variables

Just like Python, MicroPython supports a variety of data types. These include integers, floating-point numbers, strings, lists, dictionaries, etc. You declare variables in MicroPython the same way you do in Python. In the program below, variables are declared, and then their values (or, in the case of collections, one of their values) are retrieved and printed.

```
from array import array

# Integer
x = 10
print(x)

# Floating-point number
y = 20.5
print(y)

# String
name = "MicroPython"
print(name)

# Declare an array of integers
my_array = array('i', [1, 2, 3, 4, 5])
print(my_array[2])

# List
numbers = [1, 2, 3, 4, 5]
print(numbers[0])

# Dictionary
person = {"name": "John", "age": 30}
print(person["name"])

# Declare a tuple
my_tuple = (1, "apple", 3.14)
print(my_tuple[1])  # Output will be "apple"
```

Let's look at these data types closer.

In MicroPython, Integers are whole numbers that can be positive or negative. They are often used for tasks that require counting or ordering, such as iterating through loops. Integer arithmetic in MicroPython behaves the same way in standard Python, supporting everyday operations like addition, subtraction, multiplication, and division. Learn more here: https://txplo.re/upython_int.

Floating-point numbers represent real numbers and are written with a decimal point. They are handy for representing values that cannot be expressed as integers, such as fractions or irrational numbers. Floating-point numbers in MicroPython support the same set of operations as integers, but remember that they are approximations, which can sometimes lead to rounding errors. Learn more here: https://txplo.re/upython_float.

Strings are sequences of characters and are used for storing text. In MicroPython, you can create a string by enclosing characters in single or double quotes. Strings are immutable, meaning their elements cannot be changed directly once created. Operations like slicing, concatenation, and various built-in methods can be used to manipulate strings effectively. Learn more here: https://txplo.re/upython_str.

Arrays in MicroPython are similar to lists but are more memory-efficient and offer less functionality. They store elements of the same data type and are particularly useful in applications constrained by memory and computational power, such as microcontrollers. The array module in MicroPython enables you to create arrays and perform basic operations on them. In the example above, we use the "array" constructor function to create a new array. The first parameter of this function, "i", indicates to the constructor that we want to store integers in this array. The second parameter contains the integers to be stored. Notice that, unlike the other datatypes we cover here, the array data type is defined in the module with the same name, which must be imported before use. Learn more here: https://txplo.re/upython_array.

Lists are built-in to MicroPython. Lists are ordered collections of items that can contain elements of varying data types, including other lists. They are defined by enclosing their elements in square brackets. Lists are mutable, meaning their elements can be changed directly. They are incredibly versatile and commonly used for tasks that require storing and manipulating an ordered set of items. Learn more here: https://txplo.re/upython_list.

Dictionaries are unordered collections of key-value pairs. They are useful for storing and organising data in a way that makes it easy to look up values associated with a particular key. In MicroPython, dictionaries are defined using curly braces and support various methods for adding, removing, and modifying key-value pairs. Unlike lists and arrays, dictionaries are optimised for retrieving values when the key is known. Learn more here: https://txplo.re/upython_dictionary.

A tuple is an ordered, immutable collection of elements. Once a tuple is created, its size cannot be changed, nor can its elements be modified. This makes tuples useful for representing collections of items that should not be altered, such as configurations or sets of constants. Tuples are defined by enclosing their elements in parentheses and separating them with commas. They can contain elements of different data types, including other tuples, allowing for complex data structures like nested ones.

Accessing elements within a tuple is done via indexing, similar to lists and arrays. However, unlike lists and arrays, the immutability of tuples means that operations such as appending or removing elements are not supported. Despite this limitation, tuples offer advantages in terms of computational efficiency and are generally faster than lists for read-only operations.

Another critical feature of tuples is that they can be unpacked. This means that the values in a tuple can be assigned to a set of variables in a single statement, making it easier to work with individual elements. Moreover, because of their immutable nature, tuples can be used as keys in dictionaries, something not possible with lists. Learn more here: https://txplo.re/upython_tuple.

Unlike languages like C or Java, Python and MicroPython are dynamically typed. This means you don't have to declare the type of a variable when you create it. The interpreter infers the type based on the value you assign.

Control flow constructs

Control flow in MicroPython is handled using conditionals (if, elif, else) and loops (for, while). Here is an example of a conditional statement:

```
temperature = 25

if temperature > 30:
    print("It's hot!")
elif temperature < 10:
    print("It's cold!")
else:
    print("The weather is nice.")

And here's how loops work:

# For loop
for i in range(5):
    print(i)

# While loop
count = 0
while count < 5:
    print(count)
    count += 1
```

In MicroPython, control flow constructs guide the execution path of the program, enabling developers to introduce logic, repetition, and decision-making capabilities. The if, elif, and else statements are used for conditional execution. With these constructs, you can specify a condition to test, and depending on whether the condition evaluates to True or False, the corresponding block of code will execute. For example, you might use if to check whether a number is positive, negative, or zero, running different code for each scenario.

Loops are another fundamental control flow construct. The while loop continues execution as long as a given condition remains True. This loop type is useful when the number of iterations is not known in advance. Conversely, the for loop iterates through a sequence of items, executing its code block for each item. The for loop is commonly used when the number of iterations is predetermined, such as when traversing an array or list.

Another essential feature of MicroPython control flow is the break and continue statements. The break statement exits the innermost loop, terminating its execution immediately, while the continue statement skips the current iteration and proceeds to the next one. These statements are beneficial for managing exceptional conditions or specific cases within loops.

Additionally, MicroPython supports nested loops and conditional statements, where one loop or conditional statement can be placed inside another. This allows for complex logic and computation, albeit with the caution that nested structures should be used judiciously to maintain code readability and efficiency. Understanding these control flow constructs is critical for writing programs that are functional and efficient in MicroPython.

Functions and Modules

Functions in MicroPython are defined using the def keyword:

```
def greet(name):
    print(f"Hello, {name}!")

greet("MicroPython")
```

You can also create modules in MicroPython, simply Python scripts you can import into other scripts. To create a module, save your Python code in a file and then use the import keyword to import it. For instance, if you have a file greetings.py with the greet function defined in it, you can import and use it like so:

```
import greetings
greetings.greet("MicroPython")
```

In MicroPython, functions are blocks of reusable code designed to perform a specific task. They enhance code reusability and organisation, making structuring and managing complex programs easier. Functions in MicroPython can be defined using the def keyword followed by the function name and a pair of parentheses. Inside these parentheses, you can specify parameters that allow the function to accept inputs. The code block within the function is executed when the function is called. Functions can also return values using the return statement, making them useful for computations that need to be used in multiple program parts.

Modules in MicroPython serve as containers for functions, variables, and runnable code. They enable you to organise your code into separate files, which can be imported and used in other MicroPython scripts. This leads to more maintainable, readable, and modular code. To create a module, you save your MicroPython code as a .py file. This file can then be imported into other MicroPython scripts using the import statement. MicroPython also has standard modules that provide additional functionalities such as mathematical operations, file handling, and data manipulation. Learn more about modules in MicroPython here: https://txplo.re/upython_module.

Both functions and modules are indispensable tools for writing scalable and efficient MicroPython programs. Functions offer a way to encapsulate code for reuse, while modules provide the means to organise those functions into clean, easy-to-navigate files. Understanding these topics is essential for anyone building robust, modular MicroPython applications.

Exception Handling

Exception handling in MicroPython follows the same structure as Python, using try, except, finally blocks. This allows your program to respond appropriately to errors or exceptions during execution.

Here's an example:

```
try:
    print(5 / 0)
except ZeroDivisionError:
    print("You can't divide by zero!")
finally:
    print("This is the end of the program.")
```

In the above code, attempting to divide by zero raises a ZeroDivisionError. The except block catches this exception and prints an appropriate message. The final block is executed regardless of whether an exception was raised, providing a place to put any cleanup code.

Exception handling in MicroPython is a mechanism for gracefully managing runtime errors, thereby enhancing the robustness and reliability of a program. As mentioned, exception handling is implemented using the try, except, finally, and optionally else clauses. When a code block is enclosed within a try statement, MicroPython runs that code and watches for exceptions. If an exception occurs, the program flow is immediately diverted to the corresponding except block, where corrective measures or error messages can be executed.

The finally clause is optional, but if used, the code within this block is executed regardless of whether an exception is raised or not. This ensures that resources like files or network connections are closed or released, maintaining the system's integrity. The else clause is also optional and is executed if the try block does not raise any exceptions. This is useful for code that should only run if the try block is successful and can help improve code readability.

Exception handling also allows for catching specific error types, which means you can write tailored responses depending on the nature of the error. For example, a ValueError can be caught separately from a TypeError, allowing for precise debugging information and error-handling strategies. In addition to the built-in exceptions, MicroPython allows for the definition of custom exceptions through class inheritance, offering even more granularity in error management.

Understanding exception handling in MicroPython is crucial for building robust applications. It allows developers to anticipate and prepare for potential issues, offering a controlled environment for managing errors, thus improving the overall user experience and system reliability.

4.3. MicroPython Frequently Used Commands

MicroPython shares most of its commands with Python, allowing you to leverage the power of Python on your microcontroller. Let's delve into some fundamental commands and touch on some advanced ones. This list is not meant to be exhaustive. To see detailed information on all of the available commands in MicroPython, you can look at the Python Language Reference documentation available online at https://docs.python.org/3/reference/. Also, you should review the differences between MicroPython and Python (also known as "CPython") on the MicroPython documentation site at https://docs.micropython.org/en/latest/genrst.

To make this review more straightforward to understand, I have organised the comments into categories: Basic, Strings, GPIOS, Time and network operations.

Basic Commands

In this group, I include commands for displaying text and variables, reading user input, converting data from one type to another, and the essential math operatives.

Print(): Display text or variable values

The print() function outputs data to the console. It's your go-to command for debugging or displaying information; we use it extensively in this book. Here is an example:

```
print("Hello, MicroPython!")

# Define a variable
temperature = 25

# Use the print command to display the variable
print("The current temperature is:", temperature, "degrees.")
```

The "print" example in the code above prints a fixed string to the console. The second "print" example prints a string that contains a variable ("temperature"). As you can see, the "print" command is versatile and allows you to mix fixed strings with variables.

Input(): Read user input

The input() function reads a line of text from the user:

```
name = input("Enter your name: ")
print(f"Hello, {name}!")
```

Even though we will not use this command in this book, it is worth knowing because it is helpful in any situation where your programs require console input from the user. In the example above, the program asks the user to enter their name at the prompt, and the user input is stored in the "name" variable. The code in the second line prints out the content of the variable. In the print() command, the "f" before the string stands for "formatted string." It's a way to embed variables or expressions within a string using curly braces {}. The variable inside the curly braces gets evaluated at runtime, and its value is inserted into the string.

In this specific example, f"Hello, {name}!" evaluates to a string that includes the value of the name within it. The "f" lets you include variables like names directly within the string.

Type conversions

In the previous chapter, you learned about the most important data types available in MicroPython. It is often the case that your program will need to do conversions between data types. A frequent such conversion is from a string to a number. You can see several examples below:

```
x = int("10")       # Converts a string to an integer
y = float("10.5")  # Converts a string to a float
z = str(10)         # Converts a number to a string
a = list("hello")  # Converts a string to a list
```

You will see similar conversions in many experiments in this book. Typically, when you use a command like "input" or have data sent from Node-RED to the Raspberry Pi Pico (or between systems of different types), the data is sent as text strings. If the MicroPython program needs to do operations that require numbers (not strings), then the original string must first be converted (or "cast") to a number.

In the examples above, you can see commands such as "int()" and "float()" that can achieve this. These commands accept a string and convert it to an integer or floating-point number. The opposite is also possible. You can use "str()" to convert a numeral to a string. You can also convert a string to a list with "list()".

Math Operations

MicroPython supports basic math operations like addition (+), subtraction (-), multiplication (*), division (/), and modulo (%):

```
x = 10 + 5  # 15
y = 10 - 5  # 5
z = 10 * 5  # 50
a = 10 / 5  # 2.0
b = 10 % 3  # 1
```

In the examples above, you can see addition, subtraction, multiplication, normal and modulo division. Modulo division yields the remainder of the normal division between two numbers.

String commands

Strings are a fundamental data type in Python with various built-in manipulation methods. Understanding these commands can help you efficiently work with text data. In this section, we will cover some of the most commonly used string commands, including upper(), lower(), split(), replace(), and find().

upper(): Convert String to Uppercase

The upper() method converts all the characters in a string to uppercase. Here is an example:

```
text = "hello"
uppercase_text = text.upper()
print(uppercase_text)  # Output: "HELLO"
```

In this example, text.upper() converts all characters in the string text to uppercase.

lower(): Convert String to Lowercase

The lower() method converts all the characters in a string to lowercase. Here is an example:

```
text = "WORLD"
lowercase_text = text.lower()
print(lowercase_text)  # Output: "world"
```

Here, text.lower() converts all the characters in the string text to lowercase.

split(): Split String into List

The split() method splits a string into a list of substrings based on a specified delimiter. Here is an example:

```
text = "apple,banana,cherry"
split_text = text.split(",")
print(split_text)  # Output: ['apple', 'banana', 'cherry']
```

The string text is split into a list using the comma as the delimiter.

replace(): Replace Substring

The replace() method replaces occurrences of a substring within a string with another substring. Here is an example:

```
text = "I like coffee"
replaced_text = text.replace("coffee", "tea")
print(replaced_text)  # Output: "I like tea"
```

In this case, all substring "coffee" occurrences are replaced with "tea".

find(): Find Substring

The find() method searches for a substring within a string and returns the index of the first occurrence. If the substring is not found, it returns -1. Here is an example:

```
text = "Python is fun"
index = text.find("fun")
print(index)  # Output: 10
```

The find() method locates the substring "fun" starting at index 10 in the string text.

Text processing is one of the core strengths of Python and, as an extension, of MicroPython. The commands in the list above are helpful in most text processing in Python programs.

GPIO commands

General-Purpose Input/Output (GPIO) pins are essential to MicroPython's hardware interface features. These pins can be configured for input and output operations, allowing you to read from sensors, control lights, drive motors, and more. In this section, we'll explore key GPIO commands in MicroPython, specifically Pin(), on(), off(), and value().

Pin(): Configure GPIO Pin

The Pin() function initialises a GPIO pin and sets it as an input or output. Here is an example:

```
from machine import Pin

# Configure GPIO pin 2 as an output
led = Pin(2, Pin.OUT)
```

In this example, we import the Pin class from the machine module and configure GPIO pin 2 as an output pin.

on(): Turn on Pin

The on() method sets the output level of a configured pin to high (usually 3.3V or 5V, depending on the board).

```
# Turn on the LED
led.on()
```

Here, the on() method turns on the LED by setting GPIO pin 2 to a high level.

off(): Turn off Pin

The off() method sets the output level of a configured pin to low (0V).

```
# Turn off the LED
led.off()
```

The off() method turns off the LED by setting GPIO pin 2 to a low level.

value(): Read Pin Value

The value() method reads the current level (high or low) of a configured GPIO pin. This can be useful for reading the state of a button or other input devices.

```
# Configure GPIO pin 4 as an input
button = Pin(4, Pin.IN)

# Read the value of the button
button_state = button.value()
print(button_state)  # Output will be either 0 or 1
```

In this example, we configure GPIO pin 4 as an input and use the value() method to read its current state. The output will be "0" if the button is not pressed (low level) and "1" if it is pressed (high level).

Understanding these basic GPIO commands provides a strong foundation for building various MicroPython-based hardware projects. From simple LED control to interfacing with complex sensors, effective use of GPIO commands allows you to extend the functionality of your MicroPython devices.

Time Operations

Managing time is essential in many MicroPython applications, whether you're looking to delay execution, timestamp data, or handle time-based events. The MicroPython utime library provides several time-related commands to help achieve these objectives. In this section, we will focus on the sleep(), time(), and localtime() commands to understand their functionality and use cases.

sleep(): pause execution

The sleep() function pauses the program's execution for a specified number of seconds. This is often used for adding delays. Here is an example:

```
import utime

# Pause execution for 2 seconds
utime.sleep(2)
```

In this example, `utime.sleep(2)` pauses the program execution for 2 seconds before moving on to the following line of code.

time(): Get Current Time

The time() function returns the current time in seconds since the epoch (usually January 1, 1970).

```
# Get the current time in seconds
current_time = utime.time()
print("Current time:", current_time)
```

Here, utime.time() returns the current time in seconds since the epoch, and it's stored in the variable current_time. Here is an example:

localtime(): Convert to Local Time

The localtime() function converts a time expressed in seconds since the epoch to a structured time in local time.

```
# Convert current time to local time
local_time = utime.localtime(current_time)
print("Local time:", local_time)
```

In this case, utime.localtime(current_time) takes the time in seconds stored in current_time and converts it to a tuple representing the local time. The tuple contains the year, month, day, hour, minute, and second.

Network operations

Networking is a critical aspect of IoT projects, enabling devices to communicate with each other and access the internet. MicroPython provides a set of network commands to manage wireless connections effortlessly. In this section, we will delve into the core WLAN (Wireless Local Area Network) commands, namely WLAN(), connect(), isconnected(), and disconnect().

WLAN(): Configure Wireless Network

The WLAN() function initializes the wireless interface, allowing you to configure its mode (Station, Access Point, or both). Here is an example:

```
from network import WLAN

# Initialize the WLAN interface in station mode
wlan = WLAN(mode=WLAN.STA)
```

In this example, the WLAN() function from the network module is used to initialise the WLAN interface in station mode (WLAN.STA).

connect(): Connect to Network

The connect() method attempts to connect the device to a specified wireless network using an SSID (Service Set Identifier) and password. Here is an example:

```
# Connect to a network
wlan.connect('Your_SSID', auth=(WLAN.WPA2, 'Your_Password'))
```

Here, wlan.connect() attempts to connect to a wireless network with the specified SSID and authentication credentials.

isconnected(): Check Connection Status

The isconnected() method checks if the device is successfully connected to a network. Here is an example:

```
# Check if connected
if wlan.isconnected():
    print("Connected to the network.")
```

In this example, wlan.isconnected() returns a boolean value, indicating whether the device is connected to the network.

disconnect(): Disconnect from Network

The disconnect() method disconnects the device from the currently connected wireless network. Here is an example:

```
# Disconnect from the network
wlan.disconnect()
```

The wlan.disconnect() command disconnects the device from the currently connected network, freeing up the interface.

MicroPython supports many more advanced commands and techniques, especially with hardware peripherals and network communication.

4.4. MicroPython Modules

MicroPython comes with built-in modules (also called "libraries") that provide essential functionality for interacting with hardware and performing complex tasks. Some modules are inherited from Python, while others are specific to MicroPython. Here are a few:

- machine: This module is specific to MicroPython and provides functions to interact with the hardware of your microcontroller, like pins, SPI and I²C interfaces, and more.
- micropython: This module provides access to MicroPython-specific functions.
- time: It provides functions for handling time.

- math: It provides mathematical operations beyond basic arithmetic like power, logarithm, and trigonometric functions.

Let's use the time module as an example:

```
import time

start = time.ticks_ms()  # Get the current millisecond count
time.sleep(1)  # Wait for one second
end = time.ticks_ms()  # Get the current millisecond count again

print(end - start)  # This will print approximately 1000
```

Let's break down this code into its components.

- The code starts by importing the time module, which provides various time-related functions. Once the module is imported, the program can use its functions.
- The function `time.ticks_ms()` returns the number of milliseconds since the device started running. This value is stored in the variable start.
- The `time.sleep(1)` function pauses the code execution for 1 second. The argument 1 indicates that the sleep time is 1 second.
- After the 1-second pause, `time.ticks_ms()` is called again to get the current millisecond count. This value is stored in the variable end.
- Finally, the elapsed time during the sleep period is calculated by subtracting the start from the end. This difference is then printed out and will be approximately 1000 milliseconds or 1 second.

This code provides a code example that is frequently seen in Python programs that need to measure elapsed time and implement a simple delay.

Importing and using modules

Importing and using modules in MicroPython is as simple as it is in Python:

```
import machine

pin = machine.Pin(10, machine.Pin.OUT)
pin.value(1)
```

In this example, we're importing the machine module, creating an instance of the Pin class, setting GPIO 10 as an output pin, and turning it on.

Building and implementing custom modules

You can create your modules by writing functions or classes in a Python file and then importing that file into your scripts. This is an excellent way to practice code reuse, reduce the risk of defective code (as you can reuse code you have already written and tested) and improve productivity.

For example, let's create a module called led_blink.py:

```
# led_blink.py
from machine import Pin
import time

def blink(pin_number):
    led = Pin(pin_number, Pin.OUT)

    for i in range(10):
        led.value(not led.value())
        time.sleep(0.5)

And use it in a different script:

import led_blink

led_blink.blink(10)
```

In the later code example, after importing the led_blink module, the program can call the module functions by name. The two programs must be in the same directory for the import function. You can also store your MicroPython modules in the " lib " directory. This is my preferred method as it helps keep primary programs and modules ("libraries") separate and tidy.

Most commonly used external modules

You can also add numerous external modules to MicroPython to extend its functionality. Here are some you might find helpful:

- urequests: A MicroPython version of Python's popular requests module for making HTTP requests.
- ujson: A MicroPython version of Python's json module for handling JSON data.
- usocket: A module for creating and handling sockets, essential for network communications.
- mqtt: A protocol for lightweight, publish-subscribe-based messaging commonly used in IoT applications.

Using these modules involves downloading their source code and copying it to your microcontroller's file system. Then, you can import them as you would any other module.

For example, to use urequests to make an HTTP GET request, you'd do the following:

```
import urequests

response = urequests.get('http://httpbin.org/get')
print(response.text)
```

By using and creating modules, you can write more structured and modular code and leverage the work of others to perform complex tasks more efficiently. It's a crucial skill in your MicroPython toolbelt. In this book, you will use several built-in and external libraries.

Exploring online resources for MicroPython modules

Finding the suitable module for your needs can sometimes be challenging. Several online resources can help you find, learn, and explore MicroPython modules.

- **MicroPython Documentation**: The official MicroPython documentation is the first place to look. It provides comprehensive information about the built-in modules and their usage.

- **GitHub**: Many external MicroPython modules are hosted on GitHub. Simply searching "MicroPython modules" on GitHub can reveal a treasure trove of available modules.

- **MicroPython Forum**: The MicroPython forum is an active community where you can ask questions and share your experiences. It's also an excellent place to discover new modules and libraries.

- **Awesome MicroPython**: Awesome MicroPython is a curated list of MicroPython resources on GitHub. It includes a variety of libraries and tools that can be extremely useful in your projects.

- **PyPI**: While not all Python modules are compatible with MicroPython due to resource constraints, some are. PyPI, the Python Package Index, may contain packages that are MicroPython-compatible.

- **Online tutorials and blogs**: Websites like Instructables, Hackaday, and Medium often feature projects and tutorials that use various MicroPython modules. Reading these can help you understand how the modules work in real life.

Remember that while MicroPython is designed for microcontrollers, it tries to be as compatible with Python as possible. This means that learning resources intended for Python often apply to MicroPython, particularly when understanding how to structure your code and use modules effectively.

4.5. MicroPython project examples

This chapter contains MicroPython code that belongs to three projects of increasing complexity. The first project aims to make an LED connected to a Raspberry Pi Pico pin blink. In the second project, you connect a sensor to a Pico, take measurement and humidity readings, and write the data to a text file on the Pico's filesystem. The third project brings IoT into the fold and sends the sensor data to an MQTT broker from where another device or Internet service can read and process them.

I challenge you to dive into these three projects, read the code, and try to understand how they work. Focus on the program structure, the commands, and the parameters passed into the various functions. Start to get a feeling of how MicroPython looks and works. I provide brief explanations for each project. Remember that the documentation (for MicroPython and Python) is always available anytime you want to look for details and additional information.

Use these resources:

- MicroPython documentation: https://docs.micropython.org/.
- Python documentation: https://docs.python.org/3/.

Let's begin.

Beginner: Blink an LED

Starting simple is always a good approach, so our first project is blinking an LED. This classic exercise is to embedded systems what "Hello, World!" is to general programming - a simple, rewarding way to ensure your system is functioning correctly.

Before diving into the code, it's essential to understand a bit about LEDs and GPIO pins. LEDs, or Light Emitting Diodes, are simple components that emit light when an electric current flows through them. GPIO pins (General Purpose Input/Output) on your microcontroller are the 'gateway' for controlling these and other hardware components.

Here's how you can blink an LED connected to GPIO 2 on a Raspberry Pi Pico board (this code will work on any board that supports MicroPython, such as the ESP32 and the Micro:bit):

```
from machine import Pin
import time

led = Pin(2, Pin.OUT)  # Define Pin 2 as output

while True:
    led.value(not led.value())  # Toggle LED on and off
    time.sleep(0.5)  # Pause for half a second
```

This code continuously switches the LED on and off, with a delay of 0.5 seconds. It's a simple yet effective way to test your environment, hardware, and the physical connections on your microcontroller.

Intermediate: simple temperature logger

Now that you've got your bearings, let's move up to a slightly more complex project. A temperature logger will introduce you to sensor reading, file operations, and real-world data logging.

Suppose we're using a DHT22 sensor for temperature and humidity measurement connected to GPIO 4 on our microcontroller:

```
from machine import Pin
import dht
import time

# Create a DHT object
d = dht.DHT22(Pin(4))

while True:
    d.measure()  # Take a measurement
    temp = d.temperature()  # Fetch the temperature
    humid = d.humidity()  # Fetch the humidity

    # Write the temperature and humidity to a file
    with open('temp_log.txt', 'a') as f:
        f.write('%s,%s\n' % (temp, humid))

    time.sleep(60)  # Log every minute
```

This code continually measures the temperature and humidity and logs them to a file named temp_log.txt. With the logged data, you can perform various analyses later, like observing daily or seasonal trends, checking peak values, etc.

Advanced: IoT weather station

We'll move our game to IoT or the Internet of Things for the advanced project. We aim to build an IoT weather station that reads environmental data and sends it to a cloud-based IoT platform. We'll use the MQTT (Message Queuing Telemetry Transport) protocol, a lightweight messaging protocol designed for constrained devices and high-latency networks, making it perfect for IoT applications.

Our weather station will use a DHT22 sensor to measure temperature and humidity. It will then send this data to the Mosquitto MQTT broker running on a local machine or cloud server.

To connect to an MQTT broker, we'll need the umqtt.simple module. You'll need to download and transfer it to your board if it's not installed. You can find the module at this location: https://txplo.re/umqtt.

Here is the MicroPython code that accomplishes this:

```
from machine import Pin
import dht

import time
from umqtt.simple import MQTTClient

# Define MQTT parameters
SERVER    = 'your_mqtt_server'
CLIENT_ID = 'your_client_id'
TOPIC     = b'temp_humidity'

# Create a DHT object
d = dht.DHT22(Pin(4))

def connect_and_send():
    client = MQTTClient(CLIENT_ID, SERVER)
    client.connect()

    while True:
        d.measure()  # Take a measurement
        temp   = d.temperature()  # Fetch the temperature
        humid  = d.humidity()  # Fetch the humidity
        msg    = (b'{0},{1}'.format(temp, humid))

        client.publish(TOPIC, msg)
        time.sleep(60)  # Send every minute

    client.disconnect()

try:
    connect_and_send()
except OSError as e:
    restart_and_reconnect()
```

In this code, the connect_and_send function establishes a connection to the MQTT broker, reads temperature and humidity values, and publishes them to a specified MQTT topic. As the MQTT protocol works with byte strings, we pack the temperature and humidity into a byte string.

On the other side of this system, you can set up an MQTT client to subscribe to the same topic and process the published messages, perhaps storing them in a database, visualising them in real-time, or triggering actions based on specific conditions.

This project combines multiple advanced concepts, including sensor interfacing, networking, data conversion, error handling, and working with third-party libraries. It's a solid foundation for many IoT applications you might explore in the future. Please note that the code in this example does not include Wi-Fi connectivity. Without it, the Pico board cannot connect to a network and, therefore, access the MQTT broker. You can learn about the Wifi-relevant code in the opening chapters of Part 2.

4.6. Troubleshooting and Best Practices

You have ventured quite far into the MicroPython landscape. Let's gear up to tackle this journey's occasional bumps and hitches. This chapter will walk you through common errors you might encounter while coding in MicroPython, effective coding practices, and some handy debugging techniques.

Let's start with common errors in MicroPython, their causes, and how you can fix them. The errors listed below are the most common ones.

ImportError: no module named 'xyz'

This error occurs when MicroPython can't locate the module you're trying to import. The module isn't built into the firmware, or you haven't uploaded the file to your board. To resolve this, ensure the module is in your board's file system or firmware.

You can find a list of libraries with MicroPython at the documentation site here: https://txplo.re/upython_libraries.

IndentationError: expected an indented block

This is a common issue, particularly for people new to Python and MicroPython. MicroPython, like Python, uses indentation to define code blocks. This error means you've missed an indentation somewhere. Review your code carefully and correct the indentation.

I prefer using simple white spaces (the kind produced when you press the space bar) to create indentations instead of tabs. Simple spaces work well across different editors and operating systems. On the contrary, the amount of space that tabs produce is not standardised and can vary. Using simple white spaces (space bar) will significantly reduce the risk of intention errors.

MemoryError: memory allocation failed, allocating xxx bytes

Microcontrollers have limited memory. This error occurs when your code uses more memory than the device can provide. To fix this, optimise your code, try using smaller data structures, or consider a board with more memory.

For example, you can use arrays instead of lists and integers instead of floating-point numbers to reduce your program's memory footprint.

OSError: [Errno 2] ENOENT

This error signifies that MicroPython couldn't find a file or directory you've referenced. Check your file paths and names.

TypeError: object of type 'xyz' has no len()

You're trying to use the len() function on an object that doesn't support it. Review the code to ensure you use len() on an appropriate object (like a list or string). You will encounter similar errors if you try to use a function on an object that does not support it. The error message will give clues to help you resolve such errors.

SyntaxError: invalid syntax

This message flags a syntax error in your code, meaning you've written something MicroPython can't understand. Check for missing or extra brackets, quotes, or colons. Make sure your indentation is consistent.

NameError: name 'x' is not defined

You're trying to use a variable or function named 'x', but it hasn't been defined yet. To fix this error, make sure that the variable or function you're trying to use has been defined before you try to use it.

Debugging techniques in MicroPython

Debugging is essential to any programming journey and refers to removing defects (bugs) from your program. Often, a bug in your program does not produce an error message like the ones you learned about in the previous section of this chapter. Bugs can be as subtle as a calculation returning an incorrect result (meaning an error in the formula).

To fix bugs that are not due to a straightforward programming error, you can empty the following techniques:

1. **Use print statements**: The most straightforward way to debug is by inserting print statements in your code. Print variable values or messages to understand the flow of your program. I use this method across almost all programming examples in the book.

2. **REPL**: This tool lets you interact with your board over a web interface. You can execute commands in real-time, which is excellent for testing and debugging. You can use the REPL to try out code that is causing problems. Try variations of the code and let the REPL give you instant feedback that does not require the program's execution. Once you have determined the code works, please insert it into the program.

3. **Tracebacks**: MicroPython's error messages include a traceback that shows the line number and function where the error occurred. Use this information to pinpoint the source of your problem. Here is an example of a traceback appearing in the Shell, indicating a syntax error in line 11:

Traceback (most recent call last):

```
  File "<stdin>", line 11
SyntaxError: invalid syntax
```

1. **mpy-cross**: If you're running out of memory, compile your Python scripts into bytecode files with the mpy-cross compiler. This can make your code run faster and use less memory, making it easier to debug resource constraints. This technique is advanced, and you will unlikely need it until your programs get larger. If you are curious to learn more, see this page: https://pypi.org/project/mpy-cross/.

2. **Hardware debugger**: Some boards support hardware debugging, which can provide detailed insight into the execution of your code. You might need additional hardware and software for this. The Raspberry Pi Pico has a hardware debugger called the "Raspberry Pi Debug Probe" that allows you to debug a program running on any ARM-based microcontroller (including the Pico). You can learn more about it on the debug probe's product page at https://www.raspberrypi.com/products/debug-probe/.

As you are starting your journey of learning MicroPython, please focus on using print statements, REPL, and understanding of tracebacks as your primary way of debugging your programs. Once you feel your programs have grown in size and sophistication, look at the hardware debugger as your next step in building quality programs. The mpy-cross tool is helpful in cases where you need to pre-compile MicroPython programs and is a tool for special circumstances.

Effective coding practices in MicroPython

Let's move on to good coding habits that will make your MicroPython journey smoother. Learn and apply these habits, and you will see your productivity increase and your bug rate decrease.

Here's the shortlist:

1. **Keep it simple and efficient**: Microcontrollers are limited in power and memory—Prioritise simplicity and efficiency in your code. Avoid complex data structures or operations if simpler ones can do the job.

2. **Manage your resources**: Use peripherals and memory wisely. Always close files and network connections when you're done with them. Release GPIOs and other peripherals when not in use.

3. **Handle exceptions**: Add try and except blocks to manage runtime errors gracefully.

4. **Use clear and consistent naming**: Descriptive names for variables, functions, and modules can make your code more readable.

5. **Comment your code**: Include comments to explain complex parts of your code. This can be beneficial when you revisit your code after some time or when sharing it with others.

In the list above, items 1 and 2 are most important for a resource-constrained device such as a microcontroller. Of course, keeping your programs simple, efficient and light on resource demand is an advantage regardless of the size of your computer. However, general-purpose computers are far more forgiving in these areas than microcontrollers.

Important programming conventions in MicroPython (and Python)

When starting with MicroPython or Python, the syntax and functionality are usually your main focus. However, as you become more comfortable with the basics, you'll find that how you write your code becomes as important as what your code is doing.

Following standard programming conventions helps make your code more understandable, maintainable, and accessible to other programmers. In this section, you will learn about the essential conventions like naming guidelines, code layout, commenting, and more, breaking them down in a way that's easy for beginners to grasp. By the end, you'll understand why these conventions are essential and how to incorporate them into your coding projects.

Naming Conventions

Variables

In Python and MicroPython, variables should be named in a way that makes their purpose clear. Use lowercase letters and separate words with underscores. For example:

- temperature_readings
- current_speed

Functions

Function names should also be lowercase and separated by underscores. This helps make it clear what the function does. For example:

- calculate_average()
- read_sensor()

Constants

Constants are variables whose values shouldn't change. By convention, these are written in all uppercase letters with underscores separating words. For example:

- MAX_SPEED
- PI_VALUE

Code Layout

Indentation

Python and MicroPython use indentation (whitespace at the beginning of a line) to define code blocks. Typically, four spaces are used for each level of indentation. Make sure to be consistent in the amount of indentation you use.

Line length

Limiting your lines to 79 characters for code and 72 for comments and docstrings is a good practice. This makes your code easier to read and understand.

Comments

Single-line comments

Use the hash symbol # followed by a space and your comment for single-line comments. These comments explain what the following line or block of code does. For example:

```
# This is a comment.
```

Multi-line comments

For multiple lines of comments, you can use triple quotes, either single ''' or double """, though they are technically docstrings. These comments often describe what a function does or briefly overview a code block.

Docstrings

Docstrings provide a way to document your code. They are written using triple quotes and placed right after defining a function, class, or module. For example:

```
def add_numbers(a, b):
    """This function adds two numbers and returns the sum."""
    return a + b
```

Import Statements

When importing modules or libraries, the general convention is to place all import statements at the top of your file. Python's standard library imports should come first, followed by third-party libraries and custom modules.

```
# Correct
import os
import sys
from math import sqrt
```

Spacing

It would be best if you avoided extraneous whitespace in the following situations:

- Immediately inside parentheses, brackets, or braces.
- Before a comma, semicolon, or colon.

Using underscores
In Python, a single leading underscore `_variable` indicates that a variable is for internal use and should not be accessed directly. A double underscore `__variable` mangles the attribute names, making creating subclasses that accidentally override private attributes and methods harder.
Following these conventions makes your MicroPython and Python code easier to read, maintain, and share with others. This is especially crucial in large projects or when working in a team setting.

Like any programming language, MicroPython has its quirks. But with knowledge of common errors, good coding practices, and effective debugging techniques, you can overcome the challenges and focus on building exciting projects.

4.7. Glossary of MicroPython Terms

Understanding the terms used in MicroPython can be crucial to making sense of the language and its operation. Here, we'll review some of the most common words and briefly explain each. Most items in this list should be familiar to you by now.

1. **MicroPython**: A lean and efficient Python 3 programming language implementation designed to run on microcontrollers and constrained environments. Further reading MicroPython Documentation: https://docs.micropython.org/en/latest/

2. **REPL (Read-Eval-Print Loop)**: The interactive MicroPython prompt you can access over USB or the network. It lets you execute Python commands in real time and see the result immediately. Further reading REPL Tutorial: https://docs.micropython.org/en/latest/wipy/tutorial/repl.html.

3. **Firmware**: The MicroPython software that runs on your microcontroller. It includes the Python interpreter, runtime library, and any built-in modules. You can find firmware for your board here: https://micropython.org/download/

4. **Board**: The microcontroller hardware that runs the MicroPython firmware. This could be a Raspberry Pi Pico (W), ESP32, ESP8266, PyBoard, or any other compatible device. You can see a list of all supported boards here: http://micropython.org/download/.

5. **GPIO (General Purpose Input/Output)**: The pins on your board that can be used for input or output. They can read the state of a button, drive an LED, and interface with other hardware. Learn about the Pin class in the documentation: https://docs.micropython.org/en/latest/library/machine.Pin.html.

6. **Pulse Width Modulation (PWM)**: A technique to control the perceived intensity of LEDs, speed of motors, or generate audio signals. Learn about PWM with MicroPython here: https://docs.micropython.org/en/latest/library/machine.PWM.html.

7. **ADC (Analog-to-Digital Converter)**: A hardware feature that converts analog voltage to a digital value. It's used to read analog sensors like temperature or light sensors. Learn about ADC in MicroPython here: https://docs.micropython.org/en/latest/library/machine.ADC.html.

8. **I2C (Inter-Integrated Circuit)**: A communication protocol used to connect low-speed peripherals to your board. Learn about I²C in MicroPython here: https://docs.micropython.org/en/latest/library/machine.I2C.html.

9. **SPI (Serial Peripheral Interface)**: Another communication protocol used for high-speed connections. Learn about SPI in MicroPython here: https://docs.micropython.org/en/latest/library/machine.SPI.html.

10. **Module**: A file containing Python definitions and statements. The file name is the module name with the .py suffix.

11. **Exception**: An error detected during execution. MicroPython has various exceptions, like TypeError or OSError, that you can catch and handle. To learn more about Python exceptions, see this documentation page: https://docs.python.org/3/library/exceptions.html.

Remember, this is just a shortlist of the most common terms in MicroPython programming. MicroPython has rich features and capabilities that go far beyond this glossary. As you continue exploring, you'll encounter more terms that will expand your MicroPython vocabulary and deepen your understanding of the language.

4.8. References and Further Reading

While this book aims to cover the key aspects of MicroPython, the world of embedded programming is vast and constantly evolving. This page provides a list of references and further resources you may find helpful in your continued exploration of MicroPython.

1. **MicroPython Official Documentation**: This is the official MicroPython documentation. It covers the core concepts and the standard library and provides specific details for individual boards—a must-visit for any MicroPython enthusiast. Go to https://docs.micropython.org.

2. **MicroPython Forum**: This is an active forum where MicroPython users and developers discuss issues, share projects, and help each other. It's a great place to get answers to your MicroPython questions. Go to https://github.com/orgs/micropython/discussions.

3. **GitHub Repository of MicroPython**: You can find the source code of MicroPython on GitHub. It's a great resource if you want to understand the internals of MicroPython, contribute to the project, or stay updated with the latest changes. Go to https://github.com/micropython/.

4. **Awesome MicroPython**: A curated list of excellent MicroPython libraries, frameworks, software, and resources. It's regularly updated and provides a wide variety of resources. Go to https://github.com/mcauser/awesome-micropython.

5. **OpenWeatherMap**: This website provides a free API for accessing weather data. It's often used in IoT projects, many of which can be implemented in MicroPython. Go to https://openweathermap.org/api.

6. **MQTT Essentials**: A comprehensive guide to the MQTT protocol used in many IoT projects. It's helpful for understanding and implementing MQTT communication in MicroPython. Go to https://www.hivemq.com/mqtt-essentials/.

7. **Raspberry Pi Pico Python SDK**: The official SDK from the Raspberry Pi Foundation, a good reference for doing advanced programming on the Raspberry Pi Pico with MicroPython. Go to https://datasheets.raspberrypi.com/pico/raspberry-pi-pico-python-sdk.pdf.

Remember that learning is an ongoing process. The more you read, experiment, and collaborate with the community, the more you develop your MicroPython skills.

Chapter 3.5 • Programming Raspberry Pi Pico with MicroPython

General Purpose Input/Output (GPIO) pins on a Raspberry Pi Pico are physical pins that allow you to connect to the outside world, providing an interface with various electronic components such as LEDs, buttons, sensors, and more. These pins can be programmed to act as an input or output and provide a way to read from or write to external devices.

In MicroPython, the machine module is used to control the GPIO pins. A pin is represented by a Pin object, which can be imported from the machine module. Here is the code example to do this in a program:

```
from machine import Pin
```

To create a Pin object, you need to specify the pin number and mode, which can be Pin.OUT for output or Pin.IN for input. Here is an example of both:

```
led = Pin(14, Pin.OUT) # Output pin
button = Pin(15, Pin.IN, Pin.PULL_UP) # Input pin with a pull-up resistor
```

Next, let's look at simple examples of using GPIOs on the Raspberry Pi Pico.

Digital inputs and outputs

Digital inputs and outputs are the most basic form of interaction with the physical world through the Pico. Digital means the pin can only be in one of two states: HIGH or LOW, often represented by "1" and "0". We can use a Pin object's value() method to get or set its state for digital output. For instance, if we have an LED connected to pin GP14, we can turn it on and off like this:

```
led = Pin(14, Pin.OUT)
led.value(1) # Turn on LED
led.value(0) # Turn off LED
```

For a digital input, such as a button, we use the value() method to read its state. Assuming the button is connected to pin GP15:

```
button = Pin(15, Pin.IN, Pin.PULL_UP)
print(button.value()) # Prints 1 if the button is not pressed, 0 if it is
```

Analog input

Analog inputs and outputs can handle a range of values, not just HIGH and LOW. This is useful for components like potentiometers or analog sensors. The Raspberry Pi Pico has three pins (GP26, GP27, GP28) that can read analog values. To read an analog input, we use the ADC (Analog-to-Digital Converter) class from the machine module:

```
from machine import ADC

potentiometer = ADC(Pin(26)) # Potentiometer connected to GP26
print(potentiometer.read_u16()) # Read the value
```

The Raspberry Pi Pico uses PWM (Pulse Width Modulation) for analog output, which we'll discuss next.

PWM output

PWM is a technique for getting analog results using digital means. There is some confusion on the difference between PWM output and true analog output, so before going any further, let's take a closer look at this topic.

What (exactly) is PWM?

PWM is a technique where the microcontroller quickly switches a digital pin on and off to simulate an analog-like behaviour. The "width" of the time the signal is "on" (the pulse width) determines how much power is delivered to the device you're controlling. If the pin is on half the time and off the other half, you'll get 50% power.

For instance, you can use PWM to dim an LED or control the speed of a motor. However, because it's a digital technique that mimics an analog signal, some sensitive analog components may not work well with PWM due to its inherently "choppy" nature.

The most important word in the previous two sentences is "mimic". With PWM, the microcontroller tries to achieve the effect of a true analog signal using a digital (ON/OFF) signal. We do this because it is much easier (and cheaper) for a microcontroller to create digital signals rather than true analog signals.

What is "true analog"?

True analog output, on the other hand, provides a continuous voltage level, unlike PWM, which is essentially a series of on-and-off pulses. When a device expects an analog input, it often means it's looking for a continuous, smooth change in voltage, which only a true analog output can provide.

In a true analog output system, you might use a digital-to-analog converter (DAC) to convert a digital value to a continuous voltage level. This is often used in sensitive applications like audio signal processing, where a smooth signal is crucial.

Key differences between PWM and true analog

Let's summarise the differences between PWM signals and true analog signals in a list.

1. **Nature of Signal**: PWM is a digital representation, rapidly switching between high and low states, whereas the true analog output is a continuous voltage level.

2. **Complexity**: PWM is generally more straightforward and requires fewer resources. True analog output often requires specialised hardware like a DAC.

3. **Compatibility**: Not all devices can tolerate PWM's "choppy" nature, but any device that expects an analog signal will work with true analog output.

4. **Resolution**: DACs used for true analog output often have a higher resolution than PWM, providing a more accurate representation of the desired analog value.

5. **Applications**: PWM is often good enough for LEDs and motors but may not suffice for sensitive components like audio processors, which would typically use true analog output.

If you need a simple and resource-efficient way to mimic an analog signal, PWM is often the way to go. However, true analog output is generally a better choice for applications requiring a smooth, high-resolution analog signal.

PWM programming in MicroPython
MicroPython provides the PWM class to use PWM on the Pico. Here's an example of how to dim an LED:

```
from machine import PWM

led = PWM(Pin(14)) # LED connected to GP14

# Set the PWM frequency to 1000 Hz
led.freq(1000)

# Set the duty cycle to 50%, which dims the LED
led.duty_u16(32768)
```

Programming the Raspberry Pi Pico with MicroPython provides a powerful and flexible way to interact with the physical world. Whether you're reading sensors, controlling LEDs, or driving motors, all this is possible with the Raspberry Pi Pico and MicroPython.

Chapter 3.6 • Serial Communications with the Raspberry Pi Pico

As virtually all microcontrollers, the Raspberry Pi Pico and Pico W also provide UART (Universal Asynchronous Receiver-Transmitter) channels that can be used to provide connectivity with nearby devices. The Pico and Pico W have two UART channels.

A UART channel allows bi-directional asynchronous serial communication and is one of the most common (and oldest) communications technologies.

UART (Universal Asynchronous Receiver/Transmitter) is a hardware protocol for asynchronous serial communication between devices. UART can interact with other microcontrollers, sensors, or even a computer terminal in the Raspberry Pi Pico. This section aims to provide a fundamental understanding of using UART communications on a Raspberry Pi Pico with MicroPython.

I decided to include this chapter in this book because UART, despite its age, is a beneficial communications technology. While Wi-Fi and Bluetooth are available in almost all modern microcontrollers, good-old UART is often a better option for your project. This may be because Bluetooth is unavailable on the other device, a WiFi hotspot might be unreliable or inoperative, or it is overkill considering the resources they require compared to UART. Therefore, knowing how to use UART in your project is helpful in your toolbox.

If you want to implement the instructions and code below, you will need two Raspberry Pi Picos. One will be the transmitter, and the other will be the receiver.

Setting up UART with machine.UART

The UART class from the machine module is used to set up UART communication. You'll need to specify the UART channel, baud rate, and, optionally, the pins for transmitting (TX) and receiving (RX).

Here is an example:

```
from machine import UART, Pin

# Initialize UART1
uart1 = UART(1, baudrate=9600, tx=Pin(4), rx=Pin(5))
```

This code first imports the UART and Pin classes from the machine module. We initialise a UART object named uart1 on UART channel 1 with a baud rate of 9600 bits per second. The transmitting (TX) is set to GPIO pin four, and the receiving (RX) is set to GPIO pin 5.

Sending data with write()

The write() method is used to send data over UART. Here is an example:

```
# Sending a string "Hello"
uart1.write("Hello")
```

The uart1.write("Hello") command sends the string "Hello" through UART1.

Receiving data with read()

The read() method reads a specified number of bytes from the UART buffer. Here is an example:

```
# Reading 5 bytes from the UART buffer
received_data = uart1.read(5)
```

The command uart1.read(5) reads 5 bytes of data from UART1 and stores it in the variable received_data.

Full UART communication example

Let's consolidate the example code into a single program that establishes serial communications between two Picos. Here is the complete program:

```
from machine import UART, Pin
import time

# Initialize UART1
uart1 = UART(1, baudrate=9600, tx=Pin(4), rx=Pin(5))

# Wait for a second
time.sleep(1)

# Send data
uart1.write("Hello")

# Wait to receive data
time.sleep(1)

# Read received data
received_data = uart1.read(5)
print("Received:", received_data)
```

You can upload this code to both Picos. You should customise the message that each one sends to the other. For example, Pico 1 should send the message "Hello from Pico 1" and Pico 2 the message "Hello from Pico 2". Using jumper wires to connect the TX pin of Pico 1 to the RX pin of Raspberry Pi Pico 2 and the RX pin of Pico 1 to the TX pin of Pico 2.

This code, UART1, is initialised with a baud rate of 9600, TX pin as GPIO 4, and RX pin as GPIO 5. A 1-second delay is introduced to ensure everything is set up. Then, the string "Hello" is sent over UART1. Another 1-second delay is added to wait for the incoming data.

And finally, 5 bytes are read from UART1, and the received data is printed.

Understanding UART communication on the Raspberry Pi Pico opens doors to many exciting projects, especially those that require communications between a microcontroller and devices that only support UART, such as GPS receivers.

Chapter 3.7 • SPI and I²C Serial Communications

In this section, you will learn about using MicroPython to communicate with other devices (such as sensors or other microcontrollers) over the Serial Peripheral Interface (SPI) and Inter-Integrated Circuit (I²C).

SPI and I²C — introduction and comparison

When you think of SPI (Serial Peripheral Interface), think fast — at least compared to the wired alternatives for connecting to peripherals. It's like the sports car of communication protocols. At a minimum, SPI connections require a minimum of four and two power wires. Here is an overview of the pins involved:

- SDO – Serial Data Out. (Required) This was formerly known as "MOSI" or "Master out Slave in". This is the device's output signal that sends data to a receiver.
- SDI – Serial Data In. (Required) This was formerly known as "MISO" or "Master In Slave Out". This is the input signal on a device that receives data.
- CS – Chip Select. (Required) This was formerly known as "SS" or "Slave Select". The controller activates this signal to start communication with a peripheral.
- SCK – Serial Clock. (Required) This is the clock signal generated by the microcontroller to keep communications synchronised.
- PICO (peripheral in / controller out). This is used for devices that can be either a controller or a peripheral. It is the signal on which the device sends output when acting as the controller and receives input when acting as the peripheral.
- POCI (peripheral out/controller in). This is also used for devices that can be either a controller or a peripheral. It is the signal on which the device receives input when acting as the controller and sends output when acting as the peripheral.
- SDIO – Serial Data In/Out. This is a bi-directional serial signal.

An important benefit from all those pins is the speed and the ability to talk and listen simultaneously, known as full-duplex communication. It's great for situations where you need quick data transfer, like sending pictures or audio files.

Conversely, we have I²C (Inter-Integrated Circuit), the humble yet versatile buddy. It's not as fast as SPI, but it's easier on the connections. You only need two wires (plus two for power), regardless of how many devices you have. One wire is for data (SDA), and the other is for the clock (SCL). While SPI requires one device to play the controller role, in I²C, any device can initiate communications with any other device. The I²C interface is generally better suited for tasks that don't require high-speed data transfers, like reading sensor data or sending text to a display.

A comparison between I²C and SPI

When comparing SPI and I²C, speed is an area where SPI has the advantage. Under ideal conditions, speeds up to 8 Mbits (or even more) are possible for SPI, while I²C can achieve around 400 kbit/s (a high-speed mode advertises up to 3.4 Mbit/s). SPI is fast but might need more resources (wires, in this case), and I²C is slower and more careful but with fewer requirements. SPI usually requires a controller device to control the traffic. I²C, on the other hand, does not require a single controller, and any device on the channel can initiate communications.

SPI might be the way to go if you've got a project where speed is of the essence, and you have pins to spare. Storage devices like SD card readers/writers and EEPROM memories work better with SPI. If you're looking for something more relaxed and tight on available pins, I²C is your friend. Sensors are more commonly users of the I²C option.

So, whether you're building a simple weather station or an advanced robot, knowing when to use SPI or I²C can make your project more efficient and effective.

SPI with MicroPython on the Raspberry Pi Pico

In MicroPython, we can use the SPI class from the machine module for SPI communication. Here is a code example:

```
from machine import SPI, Pin

# SPI(0) corresponds to SPI0 with default pins (SCLK=GP18, MOSI=GP19, MISO=GP16)
spi = SPI(0)

# Initialize SPI with baudrate of 10,000
spi.init(baudrate=10000)

To write data, use the write method:

data = bytearray([1, 2, 3])  # Data to send
spi.write(data)

To read data, use the read method:

data = spi.read(10)  # Read 10 bytes
```

I²C with MicroPython on the Raspberry Pi Pico

In MicroPython, the I²C class from the machine module is used for I²C communication. Here is an example:

```
from machine import I2C, Pin

# I2C(0) corresponds to I2C0 with default pins (SDA=GP8, SCL=GP9)
i2c = I2C(0)
```

```
# Initialize I2C with a baud rate of 100,000
i2c.init(freq=100000)
```

You can write to an I^2C device with the writeto method:

```
device_addr = 0x3C  # Address of the I2C device
data        = bytearray([1, 2, 3])  # Data to send
i2c.writeto(device_addr, data)
```

And you can read from an I^2C device with the readfrom method:

```
device_addr = 0x3C  # Address of the I2C device
data        = i2c.readfrom(device_addr, 10)  # Read 10 bytes
```

Here is what you should remember about SPI and I^2C: SPI is commonly used in applications where speed is a priority, such as SD card readers, TFT displays, and high-speed sensors like some gyroscopes. On the other hand, I^2C is often chosen for less time-sensitive tasks where you want to connect multiple devices with just two wires, like when using temperature sensors, OLED displays, or real-time clocks.

Chapter 3.8 • Wi-Fi and Bluetooth with the Raspberry Pi Pico

In this section, you will learn the basics of wireless communications with the Raspberry Pi Pico W and MicroPython. The Raspberry Pi Pico W offers Wi-Fi and Bluetooth wireless capabilities.

As you might expect, using both is relatively easy with MicroPython. WiFi is your go-to option when you want your Pico W to send or receive more data or communicate over longer distances or across the Internet. For example, you could use WiFi to fetch weather data from an online API and display it on a connected screen. You can also use it to create a small web server to control home appliances from your phone's browser.

Bluetooth is more suited for short-range, lower-data-rate applications and is easier on the battery. For instance, you can use Bluetooth to connect a Pico W to a smartphone app you've created to control a robot or a small car. The smartphone sends simple commands over Bluetooth, and the Pico W executes them, driving motors or flashing LEDs.

Networking is a crucial aspect of the Internet of Things (IoT). IoT devices often need to send or receive data over a network to communicate with other devices or online services. This can be done using various technologies, including Wi-Fi, Bluetooth, Zigbee, LoRa, etc.

Wi-Fi and Bluetooth are particularly popular due to their ubiquity in consumer devices. Wi-Fi provides a way to connect to local networks and the Internet, while Bluetooth provides a way to communicate directly with other devices over a short range.

Wi-Fi Networking with MicroPython on the Raspberry Pi Pico W

Wi-Fi on the Pico W is straightforward with MicroPython, a programming language for microcontrollers and beginners. Here's a simple MicroPython script to connect your Raspberry Pi Pico W to a Wi-Fi network. This code imports the required modules, sets up your Wi-Fi credentials, and connects to the network.

```
import network
import time

# Initialize WiFi
wifi = network.WLAN(network.STA_IF)
wifi.active(True)

# Set your WiFi credentials
SSID     = "your-ssid"
PASSWORD = "your-password"

# Connect to WiFi
wifi.connect(SSID, PASSWORD)
```

```
# Wait for the connection to be established
while not wifi.isconnected():
    time.sleep(1)

print("Successfully connected to WiFi!")
```

Let's look at each line of code and what it does. This is code you will see in all examples where Wi-Fi is used in this book.

- import network, time: These lines import the necessary modules for network and time functionalities.
- wifi = network.WLAN(network.STA_IF): Here, we create a WLAN (Wireless Local Area Network) object in station mode (STA_IF) to connect to an existing network.
- wifi.active(True): This line activates the WiFi interface.
- SSID and PASSWORD: Replace "your-ssid" and "your-password" with the name and password of your WiFi network.
- wifi.connect(SSID, PASSWORD): This line attempts to connect to the WiFi network using the provided SSID and password.
- while not wifi.isconnected(): time.sleep(1): The loop waits until the Pico W successfully connects to the WiFi network. It checks every second.
- print("Successfully connected to WiFi!"): Once connected, this message is printed.

Once you've successfully connected your Raspberry Pi Pico W to a WiFi network, you can take it further by fetching internet content. For example, let's retrieve the current temperature in Sydney, Australia, from openweathermap.org. We will use the API and make a HTTPS request to do this. You can find the API details I use in this example here: https://openweathermap.org/api/geocoding-api.

Accessing data from an API like OpenWeather involves making an HTTPS request to a specific URL and interpreting the returned JSON data. Below is a MicroPython example that fetches the current temperature from OpenWeather's API for a given city. Replace "YOUR-API-KEY" with your OpenWeather API key (this is a free service).

Here is the complete program (a line-by-line description follows):

```
import network
import time
import usocket
import ussl
import ujson

# Initialize WiFi
wifi = network.WLAN(network.STA_IF)
wifi.active(True)
```

```
# Set your WiFi credentials
SSID     = "your-ssid"
PASSWORD = "your-password"

# Connect to WiFi
wifi.connect(SSID, PASSWORD)

# Wait for connection
while not wifi.isconnected():
    time.sleep(1)

print("Successfully connected to WiFi!")

# Fetch weather data from OpenWeather API over HTTPS
addr_info = usocket.getaddrinfo("api.openweathermap.org", 443)
addr      = addr_info[0][-1]

s         = usocket.socket()
s.connect(addr)

# Wrap socket with SSL
s         = ussl.wrap_socket(s)

# Construct and send the HTTPS GET request
api_key   = "YOUR-API-KEY"
city      = "Sydney"
request   = "GET /data/2.5/weather?q={}&appid={}&units=metric HTTP/1.1\r\nHost:
api.openweathermap.org\r\n\r\n".format(city, api_key)
s.write(request.encode('utf-8'))

# Receive and decode the response
response  = b""
while True:
    data  = s.read(100)
    if data:
        response += data
    else:
        break

# Extract temperature from JSON response
response_str = str(response, 'utf-8')
start_idx    = response_str.find('{"temp":') + 8
end_idx      = response_str.find(',"feels_like"')
temperature  = response_str[start_idx:end_idx]
```

```
print(f"The current temperature in {city} is {temperature}°C.")
```

Let's see how this code works:

- import ujson: We use MicroPython's JSON handling library to process the JSON response from OpenWeather.
- api_key = "YOUR-API-KEY": Replace this with your OpenWeather API key.
- city = "Sydney": We are fetching the weather data for Sydney. You can change this to any city you prefer.
- request = "GET /data/2.5/weather?q={}&appid={}&units=metric...: This line creates the API request, including the city and API key.
- response = b"": We initialize an empty bytes object to store the incoming response.
- response_str = str(response, 'utf-8'): We convert the received bytes to a UTF-8 encoded string.
- start_idx = response_str.find('{"temp":') + 8...: We manually extract the temperature from the JSON response. A more advanced approach would involve parsing the JSON, but this simple method works for our example.

The last print statement will display the current temperature in Sydney or any other city you specify. With this, you've successfully fetched real-time temperature data using your Raspberry Pi Pico W and MicroPython. This is a foundational step towards building weather monitoring or home automation projects.

Bluetooth with MicroPython on the Raspberry Pi Pico

Bluetooth is a wireless technology that enables short-range communications between devices. The built-in Bluetooth capability on the Raspberry Pi Pico W allows you to connect with various Bluetooth-enabled devices such as keyboards, speakers, or even other microcontrollers. A typical use case for Bluetooth on the Pico W is to use the Pico as a Bluetooth peripheral to a smartphone. The Pico, using BLE, can connect to the smartphone and send data from sensors, buttons, etc.

Unlike Wi-Fi, the Bluetooth functionality is not in the mainstream of the MicroPython firmware. This means that using Bluetooth with MicroPython boards is not as straightforward as with Wi-Fi.

On the Raspberry Pi Pico W, you must install a particular version of the MicroPython firmware that includes the necessary Wi-Fi and Bluetooth packages. You will find this firmware on the Raspberry Pi Pico documentation page here: https://www.raspberrypi.com/documentation/microcontrollers/micropython.html

Download the firmware titled "Raspberry Pi Pico W with Wi-Fi and Bluetooth LE support", and install it on your Pico W. Because this is a custom firmware, you cannot use Thonny. Still, you will have to do the installation manually. Follow these instructions:

1. Download the firmware.

2. hold the BOOTSEL button down on the Pico W and connect the board to your computer.

3. drag the firmware file on the Raspberry Pi Pico device on the computer.

4. When the firmware file completes the copy, the Pico will automatically eject, and you can use Thonny to program.

Raspberry Pi publishes a guide on how to connect to the Internet with the Raspberry Pi Pico W. You can find this guide here:
https://datasheets.raspberrypi.com/picow/connecting-to-the-internet-with-pico-w.pdf.

In Chapter 6, you will find information about using Bluetooth on the Raspberry Pi Pico W using MicroPython. The information and examples below come directly from that chapter, and I encourage you to read it from the source. When I write these lines, the Bluetooth implementation in MicroPython is under active development, and many of the details may change. Therefore, watch that documentation if you want to use Bluetooth with MicroPython and the Raspberry Pi Pico W.

Let's look at an example from the Raspberry Pi book. In this example, we will program the Pico W as a peripheral that sends numbers to a smartphone. On the smartphone side, you can use a BLE client, such as BLE Scanner from Bluepixel Technologies. Once connected, the Pico W will send numbers to the app, and the app will display them as in the example screenshot below:

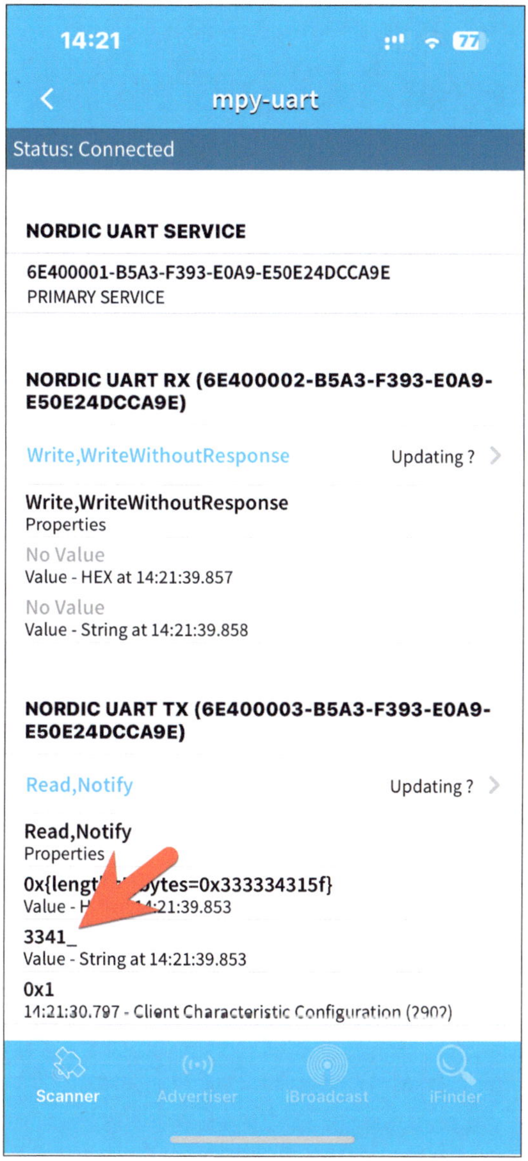

Figure 3.8.1: My smartphone is connected to the Raspberry Pi Pico W. The arrow shows a number that the Pico W sent to the phone.

Let's look at the code. First, you must create a new file titled "ble_advertising.py". This file handles the advertising functionality the Pico W can use to let other devices know it exists. Here is the code (you can copy the code from the MicroPython Github repository at https://github.com/micropython/micropython/blob/master/examples/bluetooth/ble_advertising.py):

```
# Helpers for generating BLE advertising payloads.

from micropython import const
import struct
import bluetooth

# Advertising payloads are repeated packets of the following form:
#    1 byte data length (N + 1)
#    1 byte type (see constants below)
#    N bytes type-specific data

_ADV_TYPE_FLAGS = const(0x01)
_ADV_TYPE_NAME = const(0x09)
_ADV_TYPE_UUID16_COMPLETE = const(0x3)
_ADV_TYPE_UUID32_COMPLETE = const(0x5)
_ADV_TYPE_UUID128_COMPLETE = const(0x7)
_ADV_TYPE_UUID16_MORE = const(0x2)
_ADV_TYPE_UUID32_MORE = const(0x4)
_ADV_TYPE_UUID128_MORE = const(0x6)
_ADV_TYPE_APPEARANCE = const(0x19)


# Generate a payload to be passed to gap_advertise(adv_data=...).
def advertising_payload(limited_disc=False, br_edr=False, name=None,
services=None, appearance=0):
    payload = bytearray()

    def _append(adv_type, value):
        nonlocal payload
        payload += struct.pack("BB", len(value) + 1, adv_type) + value

    _append(
        _ADV_TYPE_FLAGS,
        struct.pack("B", (0x01 if limited_disc else 0x02) + (0x18 if br_edr else
        0x04)),
    )

    if name:
        _append(_ADV_TYPE_NAME, name)

    if services:
        for uuid in services:
            b = bytes(uuid)
            if len(b) == 2:
                _append(_ADV_TYPE_UUID16_COMPLETE, b)
```

```
            elif len(b) == 4:
                _append(_ADV_TYPE_UUID32_COMPLETE, b)
            elif len(b) == 16:
                _append(_ADV_TYPE_UUID128_COMPLETE, b)

    # See org.bluetooth.characteristic.gap.appearance.xml
    if appearance:
        _append(_ADV_TYPE_APPEARANCE, struct.pack("<h", appearance))

    return payload


def decode_field(payload, adv_type):
    i = 0
    result = []
    while i + 1 < len(payload):
        if payload[i + 1] == adv_type:
            result.append(payload[i + 2 : i + payload[i] + 1])
        i += 1 + payload[i]
    return result


def decode_name(payload):
    n = decode_field(payload, _ADV_TYPE_NAME)
    return str(n[0], "utf-8") if n else ""


def decode_services(payload):
    services = []
    for u in decode_field(payload, _ADV_TYPE_UUID16_COMPLETE):
        services.append(bluetooth.UUID(struct.unpack("<h", u)[0]))
    for u in decode_field(payload, _ADV_TYPE_UUID32_COMPLETE):
        services.append(bluetooth.UUID(struct.unpack("<d", u)[0]))
    for u in decode_field(payload, _ADV_TYPE_UUID128_COMPLETE):
        services.append(bluetooth.UUID(u))
    return services


def demo():
    payload = advertising_payload(
        name="micropython",
        services=[bluetooth.UUID(0x181A), bluetooth.
UUID("6E400001-B5A3-F393-E0A9-E50E24DCCA9E")],
    )
    print(payload)
    print(decode_name(payload))
```

```
    print(decode_services(payload))


if __name__ == "__main__":
    demo()
```

Save this file to your Pico W and continue with the main program. You can find the main program, titled "ble_simple_peripheral.py", in the same GitHub repository folder as the advertiser program (see https://github.com/micropython/micropython/blob/master/examples/bluetooth/ble_simple_peripheral.py). I copy ble_simple_peripheral.py below:

```
# This example demonstrates a UART periperhal.

import bluetooth
import random
import struct
import time
from ble_advertising import advertising_payload

from micropython import const

_IRQ_CENTRAL_CONNECT = const(1)
_IRQ_CENTRAL_DISCONNECT = const(2)
_IRQ_GATTS_WRITE = const(3)

_FLAG_READ = const(0x0002)
_FLAG_WRITE_NO_RESPONSE = const(0x0004)
_FLAG_WRITE = const(0x0008)
_FLAG_NOTIFY = const(0x0010)

_UART_UUID = bluetooth.UUID("6E400001-B5A3-F393-E0A9-E50E24DCCA9E")
_UART_TX = (
    bluetooth.UUID("6E400003-B5A3-F393-E0A9-E50E24DCCA9E"),
    _FLAG_READ | _FLAG_NOTIFY,
)
_UART_RX = (
    bluetooth.UUID("6E400002-B5A3-F393-E0A9-E50E24DCCA9E"),
    _FLAG_WRITE | _FLAG_WRITE_NO_RESPONSE,
)
_UART_SERVICE = (
    _UART_UUID,
    (_UART_TX, _UART_RX),
)
```

```
class BLESimplePeripheral:
    def __init__(self, ble, name="mpy-uart"):
        self._ble = ble
        self._ble.active(True)
        self._ble.irq(self._irq)
        ((self._handle_tx, self._handle_rx),) = self._ble.
gatts_register_services((_UART_SERVICE,))
        self._connections = set()
        self._write_callback = None
        self._payload = advertising_payload(name=name, services=[_UART_UUID])
        self._advertise()

    def _irq(self, event, data):
        # Track connections so we can send notifications.
        if event == _IRQ_CENTRAL_CONNECT:
            conn_handle, _, _ = data
            print("New connection", conn_handle)
            self._connections.add(conn_handle)
        elif event == _IRQ_CENTRAL_DISCONNECT:
            conn_handle, _, _ = data
            print("Disconnected", conn_handle)
            self._connections.remove(conn_handle)
            # Start advertising again to allow a new connection.
            self._advertise()
        elif event == _IRQ_GATTS_WRITE:
            conn_handle, value_handle = data
            value = self._ble.gatts_read(value_handle)
            if value_handle == self._handle_rx and self._write_callback:
                self._write_callback(value)

    def send(self, data):
        for conn_handle in self._connections:
            self._ble.gatts_notify(conn_handle, self._handle_tx, data)

    def is_connected(self):
        return len(self._connections) > 0

    def _advertise(self, interval_us=500000):
        print("Starting advertising")
        self._ble.gap_advertise(interval_us, adv_data=self._payload)

    def on_write(self, callback):
        self._write_callback = callback

def demo():
    ble = bluetooth.BLE()
```

```
    p = BLESimplePeripheral(ble)

    def on_rx(v):
        print("RX", v)

    p.on_write(on_rx)

    i = 0
    while True:
        if p.is_connected():
            # Short burst of queued notifications.
            for _ in range(3):
                data = str(i) + "_"
                print("TX", data)
                p.send(data)
                i += 1
        time.sleep_ms(1000)


if __name__ == "__main__":
    demo()
```

The only difference between my copy above and the original from Github is that I have increased the sleep time to 1000 ms in the infinite loop at the program's end to reduce the transmission rate. This program will wait for the Pico W to connect to a host device (in my case, this is my smartphone), and then it will send an increasing number every 1 second.

I will not discuss the implementation details now as this is out of the scope of this book. However, if you are interested, the best resource for further study is the Raspberry Pi book I mentioned earlier in this section.

Chapter 3.9 • Interfacing with Sensors and Actuators

In this section, you will learn the basics of using sensors and actuators with the Raspberry Pi Pico and MicroPython. You'll learn to read data from various sensors controlling motors and other actuators.

Sensors example using MicroPython on the Raspberry Pi Pico: DHT11
Sensors provide a means to measure various physical quantities such as temperature, humidity, light, pressure, etc., and convert them into a form the microcontroller can understand. Let's take an example of a DHT11 sensor that measures temperature and humidity. This is a sensor you will also use in one of the experiments in this book.

In MicroPython, we use the dht module for DHT11 sensor. This module is built into the MicroPython firmware, so it is unnecessary to install it separately.

```
from machine import Pin
import dht

# Create a DHT object on GP22
sensor = dht.DHT11(Pin(22))

# Measure temperature and humidity
sensor.measure()

# Get temperature and humidity
temp = sensor.temperature()
hum = sensor.humidity()

print('Temperature:', temp, 'C')
print('Humidity:', hum, '%')
```

For other types of sensors, you may need to use different modules or communicate using SPI, I²C, or UART as appropriate.

Motors and other Actuators with MicroPython on the Raspberry Pi Pico
Actuators like LED, motors, servos, and relays provide a means to effect physical environmental changes. Here's an example of controlling a servo using PWM:

```
from machine import Pin, PWM

# Create a PWM object on GP14
servo = PWM(Pin(14))

# Set the PWM frequency to 50 Hz
servo.freq(50)
```

```
# Set the duty cycle to move the servo to the center position
# This may need to be adjusted depending on your servo
servo.duty_u16(3277)
```

If you replace the servo motor with an LED and use the same code example, the LED will fade on/off.

DC motors and stepper motors can be controlled using PWM and digital outputs, respectively, though you must provide adequate power.

Building Sensor-based Projects with MicroPython on the Raspberry Pi Pico

With the ability to interface with sensors and actuators, you can build various projects. Let's take a simple example of a thermostat:

```
from machine import Pin, PWM
import dht, time

# Create a DHT object on GP22 and a PWM object on GP14
sensor = dht.DHT11(Pin(22))
fan = PWM(Pin(14))

# Set the PWM frequency to control the fan speed
fan.freq(1000)

while True:
    # Measure temperature
    sensor.measure()
    temp = sensor.temperature()

    # If the temperature is above 25 C, turn on the fan
    if temp > 25:
        fan.duty_u16(65535)  # Fan at full speed
    else:
        fan.duty_u16(0)  # Fan off

    # Wait for a while before the next measurement
    time.sleep(2)
```

This project uses a DHT11 sensor to measure temperature and a fan as an actuator. When the temperature exceeds a certain threshold, the fan is turned on to cool down the environment. This is just a simple example. You can build various projects with the Raspberry Pi Pico and MicroPython by combining different sensors and actuators.

Part 4: More Node Red Topics

Chapter 4.1 • Other Useful Nodes and Features

In Part 1 of this book, you learned to use some of the most fundamental Node-RED nodes, including "debug" and "function". You used these nodes in almost all of the projects and examples in this book.

However, Node-RED has thousands of nodes that you can use. Most of them are made and contributed by Node-RED users and are not part of the Node-RED core set of nodes.

In this chapter, you will learn about several more nodes that I find helpful in my daily work with Node-RED. I will also highlight additional nodes that are useful for you to know about without getting into the details of their operations.

1.1. The "catch" Node

The catch node in Node-RED is used for error handling. It captures runtime errors or exceptions in a flow, allowing you to manage them effectively. When an error occurs in any node that the catch node is configured to watch, the catch node triggers and sends a message object containing details about the error.

The catch node in Node-RED is equivalent to similar commands in languages like Python and Ruby. For example, you can use the try, except, finally blocks for exception handling in Python. Here's a simple example:

```
try:
    result = 10 / 0
except ZeroDivisionError:
    print("Cannot divide by zero.")
finally:
    print("Execution completed.")
```

In this example, an attempt is made to divide ten by zero within the try block. This produces a ZeroDivisionError, caught in the except block, and the message "Cannot divide by zero." is printed. The "finally" block is executed regardless of whether an exception was raised, printing "Execution completed."

In Ruby, exception handling can be done using begin, rescue, and ensure blocks. Here's a straightforward example:

```
begin
    result = 10 / 0
rescue ZeroDivisionError
    puts "Cannot divide by zero."
ensure
    puts "Execution completed."
end
```

Like the Python example, this Ruby code attempts to divide ten by zero in the beginning block. This raises a ZeroDivisionError, caught in the rescue block. The message "Cannot divide by zero." is printed. The ensure block is always executed, regardless of an exception, and prints "Execution completed."

Let's look at how exception handling works in Node-RED with "catch".

How the "catch" node works

When you add a catch node to your flow, you can configure it to listen for errors from all nodes in the same tab or from specific nodes. Once an error is caught, the catch node generates a new message object that contains information about the error. This message object usually has properties like msg.payload for the error message, msg.topic for categorising the error, and msg.error containing the complete error object with extra details such as the error code and stack trace.

In the screenshot below, you can see the properties window for the "catch" node. The configuration is simple and only requires you to set the scope to "all nodes" or "selected nodes".

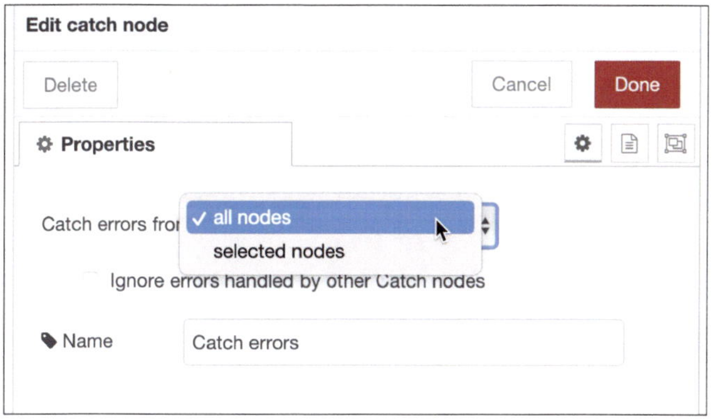

Figure 4.1.1.1: The "catch" node properties allow you to set the scope.

If you choose "selected nodes", all nodes in the current flow will be shown in a list from where you can select the ones you want to monitor for errors.

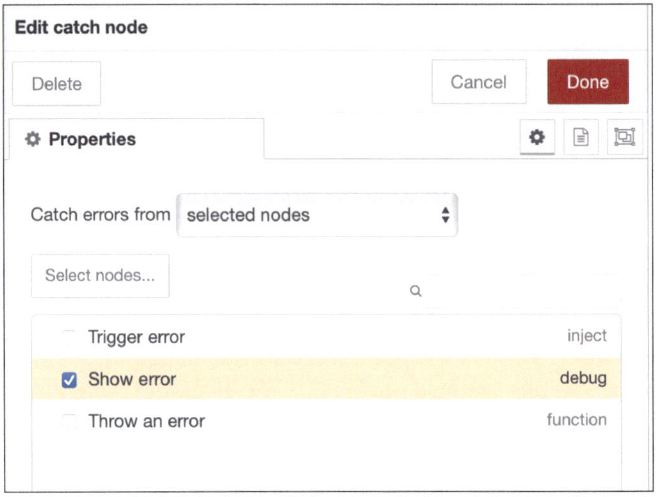

Figure 4.1.1.2: You can select specific nodes to monitor for errors.

An example showing the catch node operating within a flow

Let's create a simple flow to demonstrate how the catch node works. The flow will consist of an inject node, a function node that deliberately produces an error, and a catch node to capture that error.

Here are the nodes that make up this example:

1. **Inject Node**: Use this node to start the flow. You can set it to inject a timestamp or any other simple payload.

2. **Function Node**: Insert a function node right after the inject node. Inside the function node, write some code that will produce an error. For instance, you could use the following JavaScript code to throw an error:

```
throw new Error("This is a deliberate error");
```

1. **Catch Node**: Add and configure a catch node to catch errors from the function node.

2. **Debug Node**: Connect the catch node to a debug node. This will display the error message in the debug sidebar when the flow runs and an error is caught.

Connect these nodes sequentially: Inject Node → Function Node → Catch Node -> Debug Node. You can see the flow below, as well as the debug node output (I have expanded the JSON object):

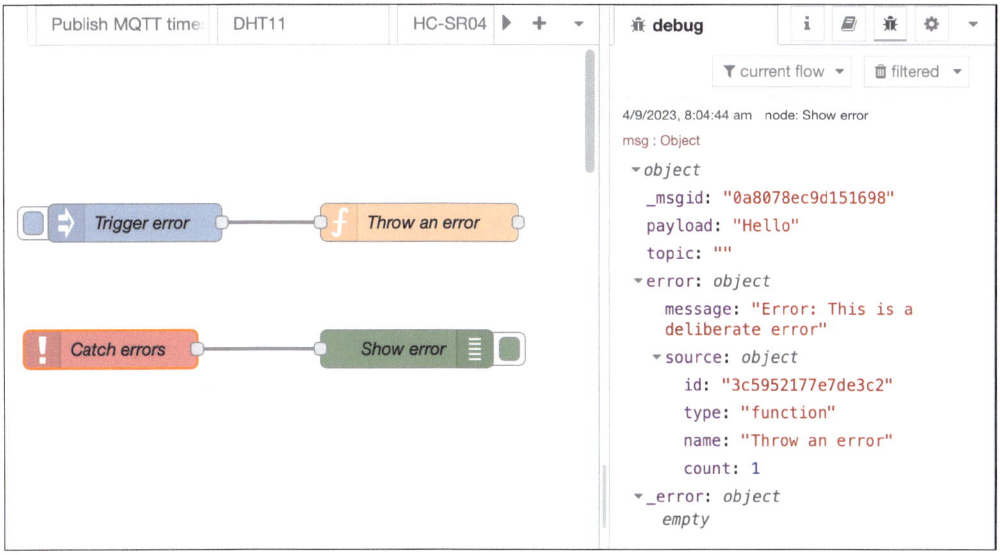

Figure 4.1.1.3: The "catch" node waits for errors in the "function" node.

After deploying this flow, click the inject node's button to initiate the sequence. The function node will throw an error, and the catch node will capture this error. The debug node will then display the error message, showing exactly what went wrong. Your flows can have multiple "catch" nodes. It can have one "catch" node listening for errors across all nodes in the flow or multiple "catch" nodes, each listening for errors in specific nodes.

1.2. The "linkout" and "linkin" nodes

Managing complex workflows and keeping them organised in Node-RED can sometimes be challenging. This is where the Link nodes, namely the "link in" and "link out" nodes, can be handy. These nodes allow you to create virtual wires between different parts of a flow or even across others, making your flows easier to understand and manage. In this chapter, we will dive deeper into the usage and benefits of these nodes.

Purpose of the Link Nodes

The "link out" node and "link in" node act as wireless connectors. Instead of creating a wired connection between two nodes, you can use a pair of these nodes to establish a virtual connection. The "link out" node will transmit any messages it receives to any "link in" nodes linked to it, regardless of where those nodes are located on the same tab or even on a different tab.

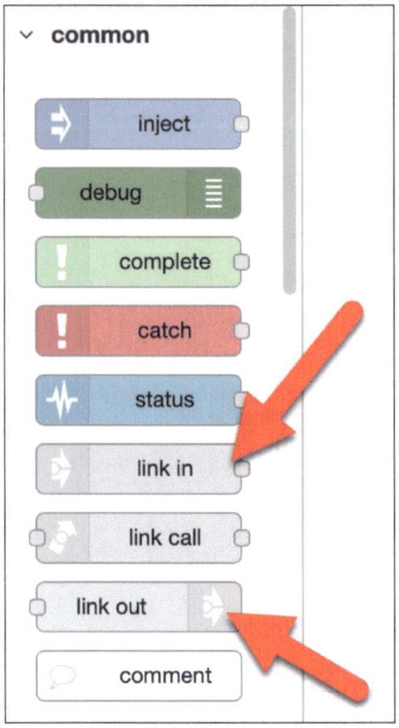

Figure 4.1.2.4: You will find the "link in" and "link out" nodes in the "common" group.

These nodes help reduce clutter in large or complex flows. They can also help encapsulate functionality, making flows more modular and easier to understand and maintain.

Examples of link node usage

The "link in" and "link out" nodes in Node-RED provide a valuable tool for managing and organising your flows. By simplifying connections, enabling data sharing across tabs, and facilitating the creation of reusable components, these nodes can significantly improve the clarity and maintainability of your Node-RED projects.

What are some use cases for "link in" and "link out" nodes? Here are some of them:

Simplifying complex flows

Consider a flow that has multiple nodes with similar functionality grouped, and these groups are wired to several different parts of the flow. Having all these wires crisscrossing your flow can make it difficult to understand and manage.

In this case, you can replace all those wires with "link out" nodes at the output of each group and corresponding "link in" nodes at the parts of the flow that need to receive those messages. This can significantly reduce the clutter and make the flow easier to understand.

Sharing data across tabs

One of the key benefits of link nodes is their ability to connect nodes across different tabs. Suppose you have a flow in one tab that fetches data from a sensor and another in a separate tab that analyses this data.

You can connect these two flows using link nodes. Adding a "link out" node at the end of the first flow and a "link in" node at the start of the second flow allows the sensor data to be passed from the first flow to the second, even though they are on different tabs.

Encapsulating functionality

Link nodes can also be used to create reusable components in your flows. For example, suppose you have a sequence of nodes that implements a particular functionality. In that case, you can encapsulate this sequence with "link in" and "link out" nodes and reuse this component in different parts of your flow or other flows.

This way, if you need to update the functionality, you only need to do it in one place, and the changes will be applied everywhere the component is used.

An example flow with the Link nodes

Let's look at a simple example of a Node-RED flow that utilises "link in" and "link out" nodes. This example demonstrates how these nodes can help simplify the flow and reduce clutter. The "link in" and "link out" nodes must be linked to each other to function as intended.

Here is how to construct the example flow:

1. **Add Nodes**: Drag and drop two "inject" nodes, two "function" nodes, two "link out" nodes, two "link in" nodes, and two "debug" nodes onto the workspace.

2. **Configure Nodes**: Double-click on each node to open its configuration:
   - "inject" nodes: Set the Payload to a string type for both nodes. For the first one, enter "Message 1"; for the second one, enter "Message 2". These nodes will be used to start the flow.
   - "function" nodes: For the first node, enter the following code to process "Message 1":

```
msg.payload = msg.payload + " has been processed.";
return msg;
```

   For the second node, enter the same code to process "Message 2".

   - "link out" nodes: You must create a new link. It is possible to link a "link out" node to a "link in" node graphically or via the "link out" node properties. You can use the graphical method if both nodes are in the same tab. Click on the "link-out" node to select it. A circular connector appears on the right side of the "link-out" node. Click in the connector to start drawing a link, and finish the connection by releasing the mouse button with the line inside the circular connector on the left side of the "link in" node. You can use the properties window if your link nodes are in different tabs. Double-click on the "link out" node to get to the properties window, and find the "link in" node you want to connect to in the list. Click the "link in" node's radio button to select it and click "done".

Use your preferred method to connect "link out 1" to "link in 1" and "link out 2" to "link in 2".

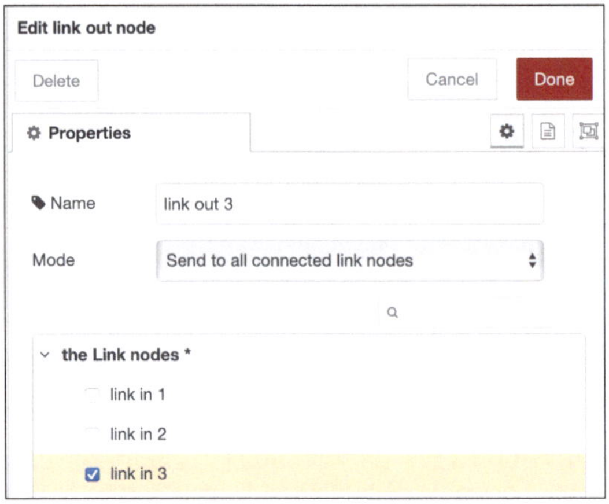

Figure 4.1.2.5: You can create a connection between links in the properties window.

- "link in" node: It is also possible to connect a "link in" node to a "link out" node. You can do this both graphically and via the properties window.
- "debug" nodes: Leave the defaults, which will output the payload of incoming messages to the debug sidebar.

3. **Wire Nodes:** Connect the nodes as follows:
   - First "inject" node → First "function" node → "link" out node
   - Second "inject" node → Second "function" node → "link" out node
   - "link in" node → First "debug" node
   - "link in" node → Second "debug" node

4. **Deploy**: Click the Deploy button to save and run the flow.

Your flow should look like this:

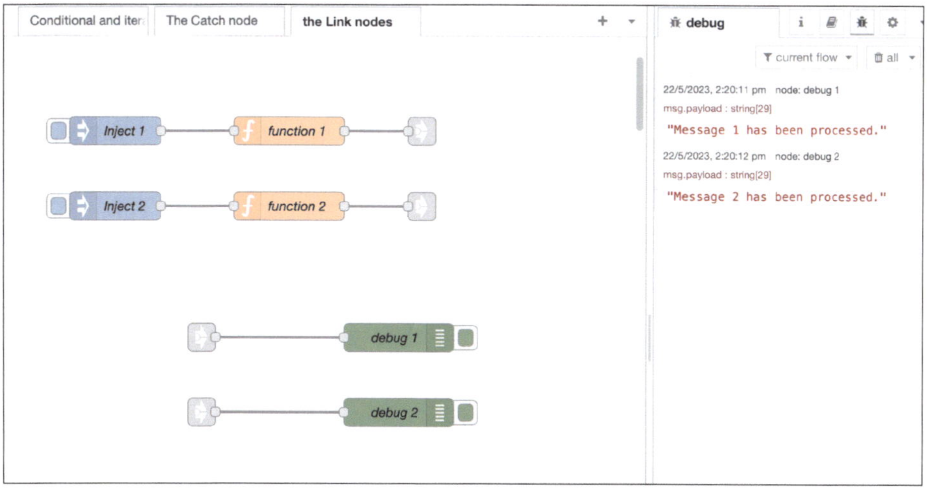

Figure 4.1.2.6: The example "link" flow, with debug output.

Now, when you inject a message using each "inject" node button, the messages will be processed by the "function" nodes, sent to the "link out" node, received by the "link in" node, and finally printed to the debug sidebar by the "debug" nodes. Even though the "link in" and "link out" are not directly connected with a wire, they are linked, and messages are successfully passed between them. This demonstrates the primary use case of the link nodes.

1.3. The "switch" Node

The "switch" node allows you to control the flow of messages based on their properties. This feature enables the creation of complex decision-making structures within a flow, enhancing the potential of your Node-RED projects. In this chapter, you will learn to use the "switch" node with the help of a practical example.

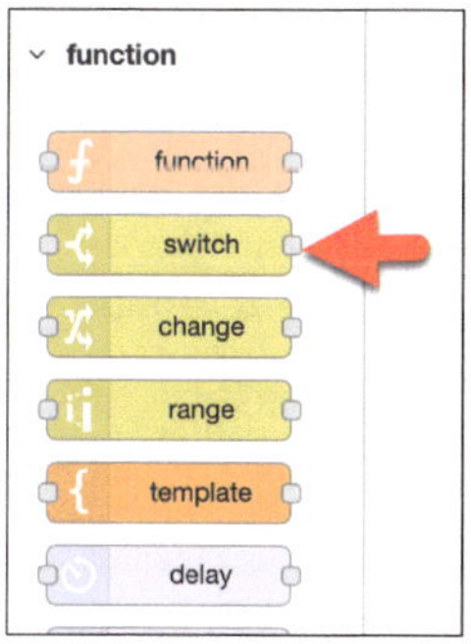

Figure 4.1.3.7: You will find the "switch" node in the "function" group.

Purpose of the "switch" node

The "switch" node in Node-RED acts as a routing mechanism, evaluating an incoming message based on its properties and routing it to an output accordingly. This allows a single input to be directed to multiple possible outputs, effectively splitting the flow based on certain conditions.

This node becomes particularly useful when performing different actions based on a message's content. Rather than using multiple "function" nodes to check other conditions, you can use a single "switch" node to evaluate multiple conditions and route messages based on the results.

You can see the properties window of the "switch" node below.

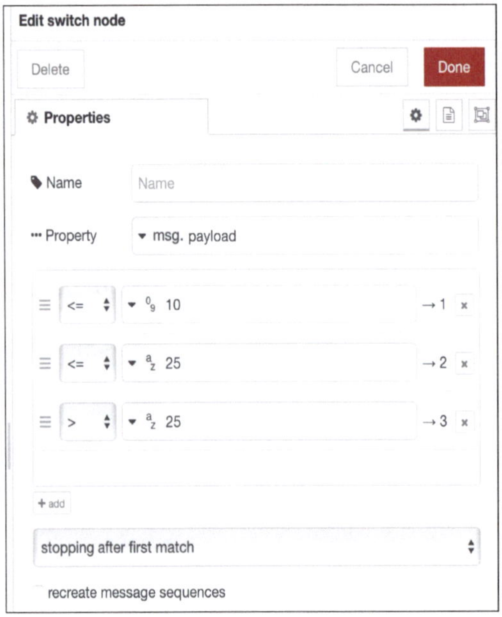

Figure 4.1.3.8: The properties of the Switch node.

As you can see from the properties window, it is possible to define multiple conditions. Each condition corresponds to a dedicated output. For example (from the screenshot above), output one will be triggered if the incoming number is less than "10".

Example uses of the Switch node

In Node-RED, the "switch" node is a powerful tool that facilitates complex decision-making structures within a flow. By allowing messages to be routed based on their properties or payload content, it provides a flexible and efficient mechanism to handle different conditions. This node is indispensable for designing intricate flows that require conditional logic.

Here are some examples of uses of the Switch node.

Routing based on payload content

Consider a flow that receives temperature data from a sensor, and you want to perform different actions depending on whether the temperature is too high, too low, or just right.

To implement this in Node-RED, you can use a "switch" node with three outputs, each corresponding to one of these conditions. The "switch" node checks the payload content of the incoming message (the temperature data) and routes it to the appropriate output based on the condition that matches.

Routing based on message properties
Another use-case of the "switch" node is routing messages based on their properties. For example, you may have a flow that handles requests from different users, and you want to route these requests based on the user role.

In this case, you can use a "switch" node to check the user role property in the message. Depending on the value of the user role, the "switch" node can route the message to an output that handles requests for that specific role.

An example flow with the switch node
Let's assume you have a weather sensor that sends a numerical value for temperature to your Node-RED flow. You want to evaluate this value and output a status message: "Cold", "Moderate", or "Hot". In this example, you will create a Node-RED flow that contains the Switch node to implement this scenario. We'll simulate the weather sensor with an inject node to keep the experiment simple. When you click the inject node button, the node will send out a number that simulates a temperature.

Here's how to create this flow:

1. **Add Nodes**: Drag and drop an "inject" node, a "switch" node, three "change" nodes, and three "debug" nodes onto the workspace.

2. **Configure Nodes**: Double-click on each node to open its configuration:
   - "inject" node: Set the Payload to a number type and input a temperature value. This will simulate your weather sensor input.
   - "switch" node: Click the Add button twice to add two new rules. Configure them as follows (see the properties screenshot earlier in the section):
     - For rule 1, select "<=" as the condition and input "10" as the value. This will check if the temperature is less than or equal to 10 degrees.
     - For rule 2, select "<=" as the condition and input "25" as the value. Then select "<=" as the second condition and input 25 as the value. This will check if the temperature is between 10 and 25 degrees.
     - For rule 3, select ">" as the condition and input "25" as the value. This will check if the temperature is above 25 degrees.
     - Set the "switch" node to stop after the first match.
   - "change" nodes: For each node, set the rules to "Set msg.payload To" a string type and input the following messages:
     - "Cold" for the first "change" node.
     - "Moderate" for the second "change" node.
     - "Hot" for the third "change" node.

3. **Wire Nodes**: Connect the nodes as follows:
   - "inject" node → "switch" node
   - "switch" node output 1 → first "change" node → "debug" node
   - "switch" node output 2 → second "change" node -> "debug" node
   - "switch" node output 3 → third "change" node → "debug" node

4. **Deploy**: Click the Deploy button to save and run the flow.

Your flow should look like this:

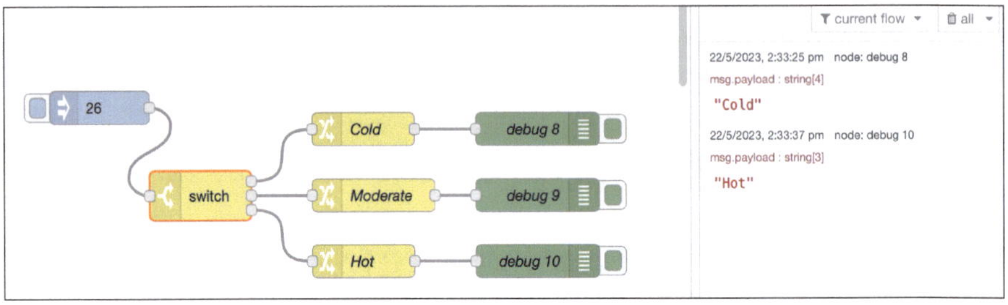

Figure 4.1.3.9: An example flow with the Switch node.

Now, when you trigger the flow using the "inject" node button, the "switch" node will check the temperature value from the "inject" node and route it to the appropriate "change" node based on the conditions set. The "change" node will then set the payload to the corresponding status message, displayed in the debug sidebar. This demonstrates how a "switch" node can route messages based on their payload content.

1.4. The "range" Node

The range node in Node-RED is a utility node that allows you to scale or map numeric input values to a new range of numbers. It performs a linear transformation of the input value, making it helpful in adapting data from different sources or preparing it for specific uses like visualisation or device control.

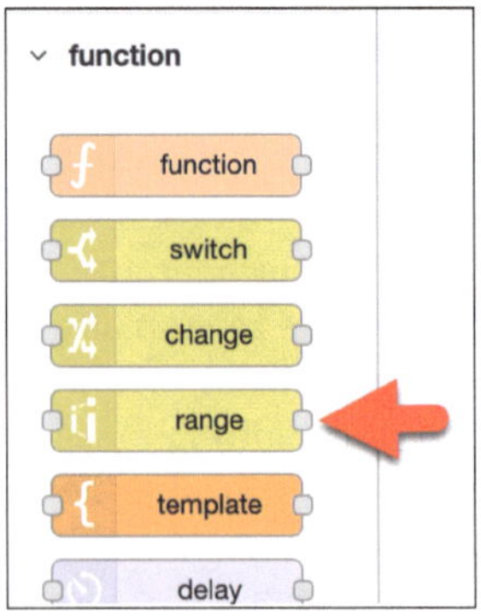

Figure 4.1.4.10: You will find the "range" node in the "function" group.

How the "range" node works

The range node takes a numeric input, typically from the msg.payload property, and scales it from a specified input range to a selected output range. You configure the input and output ranges by setting the minimum and maximum values in the node's properties panel. When a message arrives at the range node, it reads the numeric value from msg.payload, performs the scaling transformation based on the configured ranges, and then forwards the transformed value in a new message.

In the screenshot below, you can see the properties window of the "range" node. This gives you a sense of how this node works and how a number from the input range is converted to a new number in the target range.

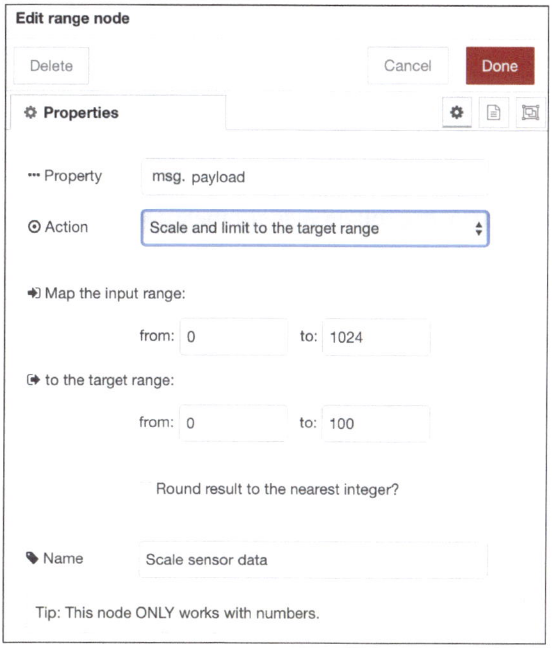

Figure 4.1.4.11: The properties of the Switch node.

Example uses of the Range node

The "range" node can be helpful in many situations and projects. Before looking at an example flow that contains a "range" node, let's look at a few typical use cases. These use case examples demonstrate the "range" node's versatility in different scenarios, from smart home applications to industrial systems. Its utility lies in its ability to adapt numerical data for specific needs easily and quickly.

Temperature scaling for HVAC control

In a smart home, temperature sensors might output values in Fahrenheit that need to be scaled to a 0-1 range to control a heating, ventilation, and air conditioning (HVAC) system. The range node could transform these temperature readings to fit the control input requirements.

Converting sensor readings for visualization
In an industrial context, pressure sensors might produce readings within a range of 0 to 150 psi. A dashboard visualisation, however, may expect these values to be normalised between 0 and 1. The range node can convert these sensor values into the appropriate scale for easy and consistent visualisation.

Adjusting light intensity
Imagine a light control system where the user inputs a light intensity value between 0 and 10 via a user interface. The actual light system, however, might accept values from 0 to 255. A range node could convert the user's desired light level to the appropriate value for the light system's API.

Battery level indicator
The voltage level for a device running on a battery could be between 3.2V to 4.2V. This range might need to be converted to a 0-100 percentage scale to give the user a clear indication of battery level. The range node can perform this conversion effectively.

Audio signal processing
In a digital audio application, the audio signal amplitude might need to be scaled to fit within a specific range to work with certain audio processing algorithms. The range node can adjust these amplitudes based on the algorithm's required input and output ranges.

An example flow with the range node
Let's create a simple flow to demonstrate the range node's functionality. The flow will consist of an inject node, a range node, and a debug node. Here's the list of nodes that this flow contains:

1. **Inject Node**: Use this node to initiate the flow. Configure it to inject a numeric value, like 500, to simulate sensor reading or user input.

2. **Range Node**: Place a range node after the inject node. In its properties, set the input range as 0 to 1024 and the output range as 0 to 100. This configuration will scale incoming values down by a factor of 100 (see the screenshot earlier in this section).

3. **Debug Node**: Attach a debug node to the range node's output. This will allow you to see the transformed value in the debug sidebar.

Link these nodes: Inject Node → Range Node → Debug Node. Your flow should look like this:

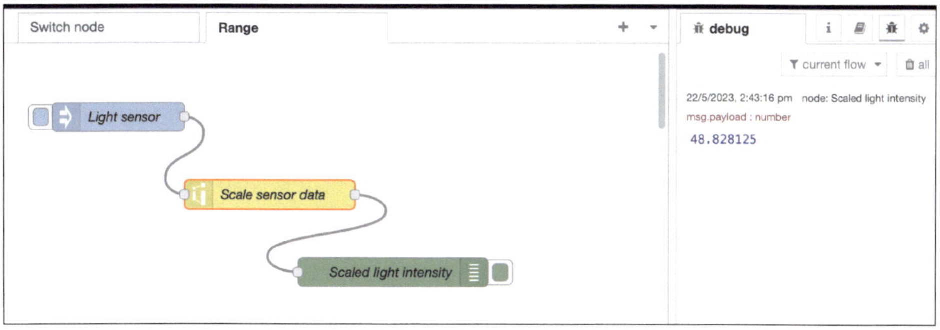

Figure 4.1.4.12: An example flow with a "range" node.

After deploying the flow, trigger the inject node by clicking its button. You'll notice that the debug node displays a value of 48.8. That's because the range node took the input value 500, scaled it down according to the set ranges, and produced an output value of 48.8.

This example illustrates the primary use of the range node, showcasing its capability to transform numerical values based on predefined ranges. Understanding this node can be particularly useful in various applications, including data normalisation and unit conversions.

1.5. The "RBE" (Report by Exception) Node

In Node-RED, the RBE (Report by Exception) node is key in managing data flow, particularly in contexts involving frequent or repetitive data. This chapter will provide an overview of the purpose and functionality of the RBE node, accompanied by practical usage examples. In the node palette of the editor, you will find the RBE node as a "filter" within the "function" group.

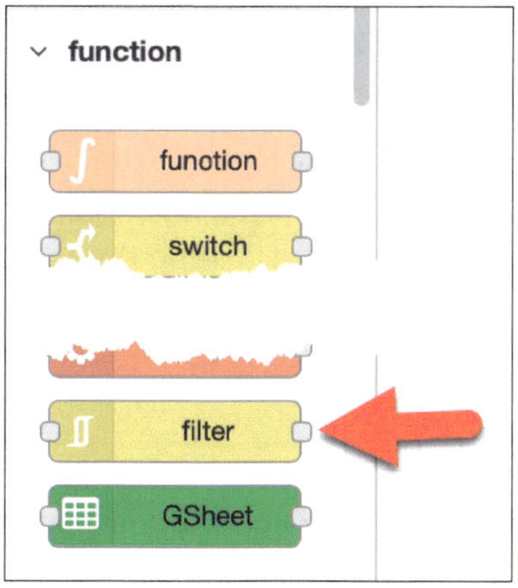

Figure 4.1.5.13: You will find the "filter" ("RBE") node in the "function" group.

How the "RBE" node works

The RBE node restricts message flow by transmitting a message only when the payload changes. When a sensor or another device sends repeated or constant values, the RBE node can help reduce network traffic and system load by only passing on new or updated data. This helps enhance the efficiency of your flows, mainly when working with devices that frequently send identical data.

In the screenshot below, you can see the properties window for the RBE node. I have also included the entire contents of the "Mode" drop-down, which reveals the available blocks.

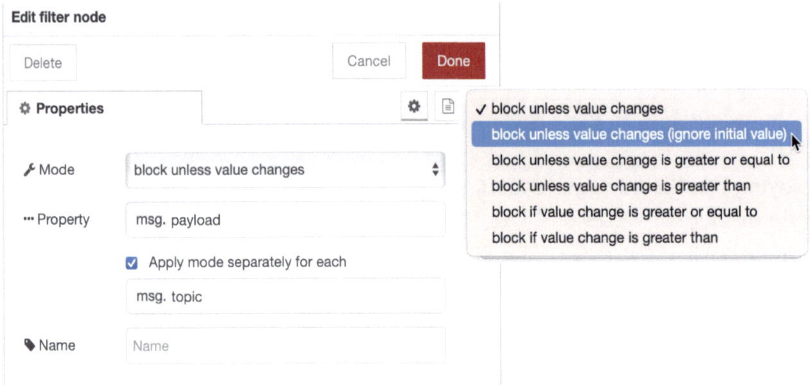

Figure 4.1.5.14: The RBE node properties and block modes.

Examples of RBE node usage

The RBE node is a fundamental tool in Node-RED for managing constant or repeated data. It helps conserve network and system resources, only transmitting messages when the payload changes. This makes the RBE node particularly useful in contexts involving frequent sensor readings or user inputs.

The examples below highlight the utility of the RBE node in various scenarios, ranging from IoT applications to industrial automation. The node is a simple yet effective way to optimise system resources by eliminating redundant operations.

Reducing network traffic in IoT applications

In an Internet of Things (IoT) environment, sending frequent sensor data updates can consume a lot of network bandwidth and energy. Using the RBE (Report by Exception) node, you can ensure that data is only sent when there's a change in the sensor reading, thereby optimising network usage.

Limiting notifications or alerts

Imagine a home security system where a motion sensor is part of the setup. If the sensor's output remains unchanged, continuously triggering notifications could be annoying. An RBE node can filter out duplicate readings, sending alerts only when a new motion is detected.

Optimising database writes

If you're logging data to a database, you don't want to write the same value repeatedly, as it would waste storage and computational resources. Using an RBE node before the database write operation will ensure that only new, unique values are written to the database.

Streamlining user interface updates
In applications where a user interface displays real-time data, avoid updating the interface unnecessarily. Using an RBE node, you can update the interface only when the underlying data changes, making the UI more responsive and efficient.

Enhancing process control
In industrial automation, you may have control loops where variable data control actuators—constantly sending control signals even when no change in value can wear out the actuators and make the system inefficient. An RBE node can be inserted to send control signals only when there's a change in the processed value.

Minimizing API calls
When integrating with third-party services through APIs with rate limits or costs associated with the number of calls, using an RBE node can help you minimise the number of API calls by only forwarding messages when the payload data changes.

Managing button presses
The RBE node can also be helpful in user input scenarios like button presses. If a button is designed to toggle a device (like a light) on and off, rapid, repeated presses could lead to unintended behaviour. An RBE node can ensure that the device only receives a command when the button state changes.

An example flow with the RBE node
Here's a simple example of how the RBE node can be used. Say you have a button that is supposed to toggle a device on and off, but the button can be pressed multiple times quickly, which might cause the device to receive repeated "on" or "off" commands. Using the RBE node, you can ensure that the device only receives a command when the button state changes.

Follow these instructions and assemble the flow:

1. **Add Nodes**: Drag and drop two "inject" nodes (to simulate button presses), an RBE node, and a "debug" node onto the workspace.

2. **Configure Nodes**: Double-click on each node to open its configuration:
   - First "inject" node: Set the Payload to a string type and input the value "on". Set the "Name" to "On Button".
   - Second "inject" node: Set the Payload to a string type and input the value "off". Set the "Name" to "Off Button".
   - RBE node: Leave the default configuration as it works out of the box. You can also see how this node is configured in the screenshot earlier in this section.
   - "debug" node: Leave the default configuration.

3. **Wire Nodes**: Connect both "inject" nodes to the RBE node and the RBE node to the "debug" node.

4. **Deploy**: Click the Deploy button to save and run the flow.

   You can see this flow in the screenshot below.

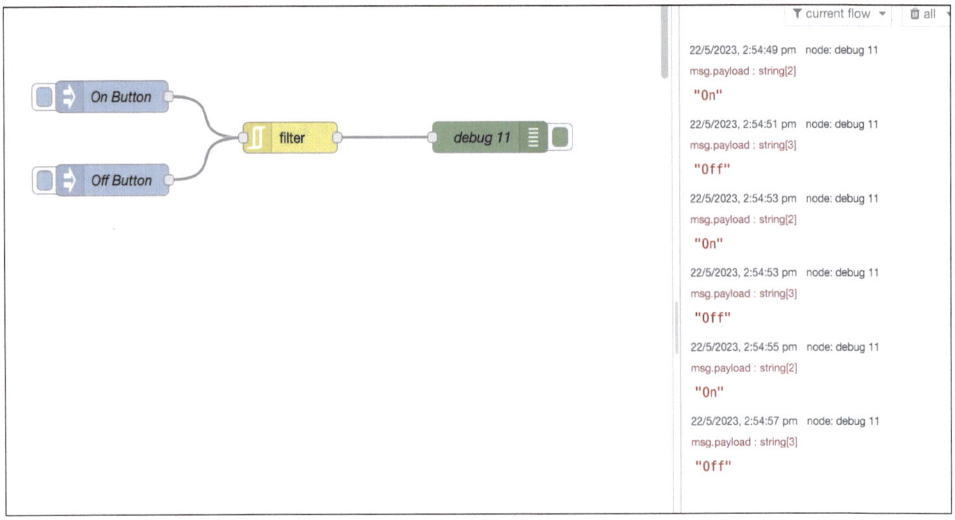

Figure 4.1.5.15: The RBE node example flow.

Now, even if you repeatedly press the same button ("inject" node), the RBE node will only pass on the message when the payload changes, i.e., when you press a different button. You can verify this by observing the messages in the debug sidebar. This simple demonstration shows how the RBE node can filter out repeated values.

1.6. The JSON Node

The JSON (JavaScript Object Notation) node is one of the parsing nodes provided by Node-RED, allowing the conversion between a JSON string and a JavaScript object. JSON has become a de facto standard for data exchange in web services due to its lightweight nature and simplicity compared to XML. In a Node-RED flow, the JSON node can be used to parse a JSON string to a JavaScript object or string a JavaScript object into a JSON string.

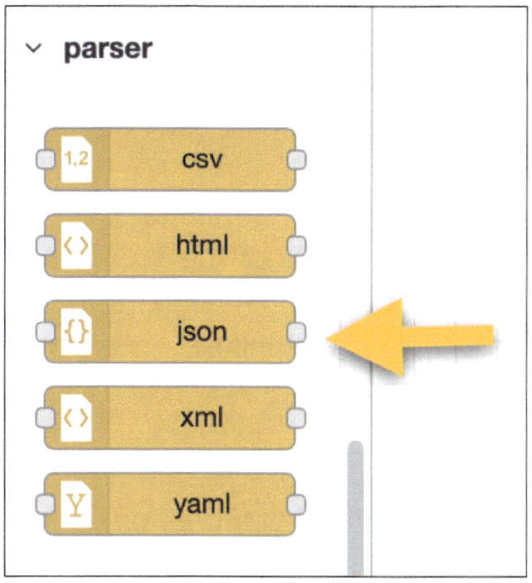

Figure 4.1.6.16: You will find the "json" node in the "parser" group.

Purpose of the JSON node

The primary purpose of the JSON node is to handle data transformation between JSON strings and JavaScript objects. This function is critical in many situations, such as when receiving data from an external API (usually in JSON format) or when you need to format data to be sent to an API.

By converting the JSON string to a JavaScript object, other nodes in the flow can easily access the data properties using JavaScript syntax. On the other hand, when you need to send data to a web service that accepts JSON, you can use the JSON node to convert a JavaScript object to a JSON string.

In the screenshot below, you can see the properties window of the JSON node as configured for the following example. I have also included an expanded "Action" drop-down menu showing the type of available conversions.

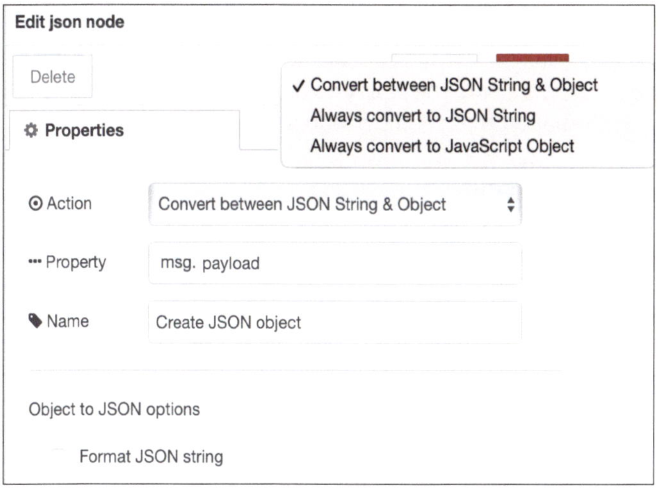

Figure 4.1.6.17: The "json" node properties and conversion actions.

Examples of JSON node usage

When working with external APIs, the data returned is often JSON. The JSON node can convert this JSON string into a JavaScript object, making it easier to manipulate and use this data within your Node-RED flow.

The examples below illustrate the versatility of the JSON node in different data manipulation scenarios. It is useful when converting between JSON strings and JavaScript objects, often required in modern data processing pipelines.

Formatting data for API requests

Conversely, you may need to send data to an external service that expects a JSON-formatted payload. The JSON node can convert a JavaScript object or array into a JSON-formatted string that can be included in an HTTP request.

Data logging and storage

You might want to store sensor data as a JSON string if you're logging sensor data or any structured data. The JSON node can convert the message payload into a JSON string suitable for storage in databases that support JSON format, like MongoDB.

Simplifying data transformation

Data coming from different sensors or inputs might be in diverse formats. Converting all this data into JSON objects using the JSON node can simplify the data transformation process. This structured format makes applying subsequent manipulations like filtering or aggregation easier.

Message routing based on payload

You may want to route messages based on the payload's content in some flows. Converting a payload to a JSON object allows you to easily inspect its properties, enabling more advanced conditional routing scenarios based on the content of the message.

Debugging and development

Viewing message payloads as easily readable JSON objects during the development and debugging phase is helpful. Using the JSON node for this purpose allows quicker identification of issues or confirmation that the flow works as intended.

Limitations of the JSON node

While the JSON node is a handy tool for handling JSON data in Node-RED, it has limitations. It works best with well-formed JSON strings. If a string is not valid JSON, the node will throw an error, which you must handle using a catch node. Also, very large JSON strings might cause performance issues and exceed the payload size limit of specific nodes or services.

Despite these limitations, the JSON node is a crucial tool in your Node-RED toolkit, especially when working with web services and APIs.

The JSON node and function node

The JSON node is often used alongside the "function" node in Node-RED. This combination allows for a flexible and powerful method of handling JSON data within a flow.

The "function" node in Node-RED allows custom JavaScript code to run within the flow. When dealing with JSON data, especially complex or nested JSON objects, using the "function" node alongside the JSON node allows for the decomposition and transformation of the JSON data into a form suitable for downstream nodes in the flow.

The JSON node is excellent for the initial conversion of a JSON string into a JavaScript object, but its capabilities stop there. If the JSON data requires further manipulation, such as extracting specific properties, handling nested objects or arrays, or transforming the data, the Function node is a powerful tool for these tasks.

An example flow with the JSON node

Let's consider an example where a JSON string represents a weather report that includes temperature, humidity, and wind speed. The JSON string is this:

```
{
  "temperature": 20,
  "humidity": 60,
  "wind": {
       "speed": 5,
       "direction": "N"
  }
}
```

The objective of the example flow is to convert this JSON string into a Javascript object so that other flow nodes can read the weather report data. To do this, the flow will implement this process:

1. Use a JSON node to convert this string into a JavaScript object.
2. Use a "function" node to extract the temperature and wind speed and pass it on to the Javascript object.

Let's set up the example flow. You can see the instructions below.

1. **Add nodes**: Drag and drop an "inject" node, a JSON node, a "function" node, and a "debug" node.
2. **Configure Nodes**: Double-click on each node to open its configuration:
   - - Debug node: Copy the JSON string from the start of this section. Remove empty spaces and make sure that the JSON string is bug-free. You can see the "inject" node properties below:

Figure 4.1.6.18: The "inject" node properties.

- JSON node: set the Action drop-down to "Convert between JSON String & Object" and the Property to "msg.payload".
- Function node: copy the Javascript code below. This code will get the temperature and wind speed values from the Javascript object that the JSON node created and create a new Javascript object with this information.

```
var temperature = msg.payload.temperature;
var windSpeed = msg.payload.wind.speed;
msg.payload = {
  temperature: temperature,
  windSpeed: windSpeed
};
return msg;
```

- Debug node: leave defaults.

3. Deploy.

This flow should look like this:

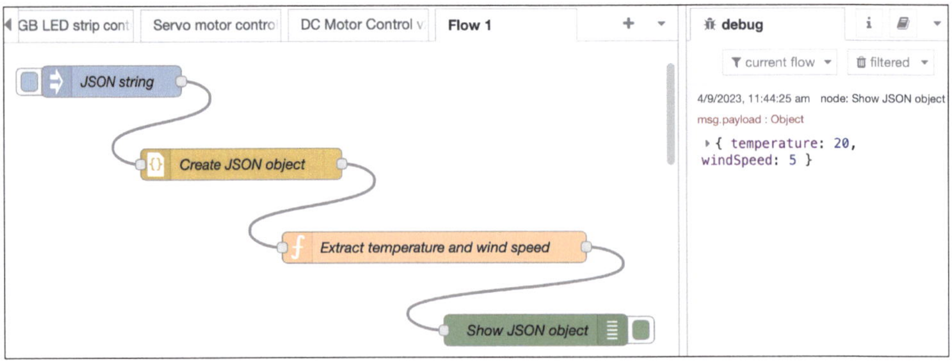

Figure 4.1.6.19: The "JSON" node example flow.

Deploy the flow and click on the Inject node button. The JSON string is injected into the JSON node and converted into a Javascript object. This object travels into the function node, where a new Javascript object is created and printed in the debug tab by the debug node.

Combining the JSON node and Function node in Node-RED offers a versatile method to handle JSON data within a flow. The JSON node is used for the initial conversion between JSON strings and JavaScript objects. In contrast, the Function node allows for customised manipulation and transformation of the data, enabling it to be used effectively in other parts of the flow.

1.7. Node Groups

A "node group" is not a node, so technically, I should not have placed this section in this chapter. But I decided to do it anyway because node groups help keep your flows tidy.

In Node-RED, a node group is a visual and functional container for organising related nodes on the workspace. Node groups offer a streamlined way to manage complexity by grouping nodes that are part of the same logical process or functionality, making flows easier to understand and manage.

Why use node groups?

Using node groups enhances the readability and maintainability of your flows. Your workspace can become cluttered and confusing when you have many interconnected nodes. Node groups allow you to section your flows into distinct parts, which makes it easier to comprehend the flow's overall logic. Additionally, node groups can simplify the export and import process, allowing you to easily reuse or share grouped parts of a flow.

How to create a node group

Creating a node group is a straightforward process. First, select the nodes you wish to group by dragging a selection box around them or clicking on them while holding down the shift key. Once the nodes are selected, right-click on one of the selected nodes and choose the "Group Selection" option from the context menu.

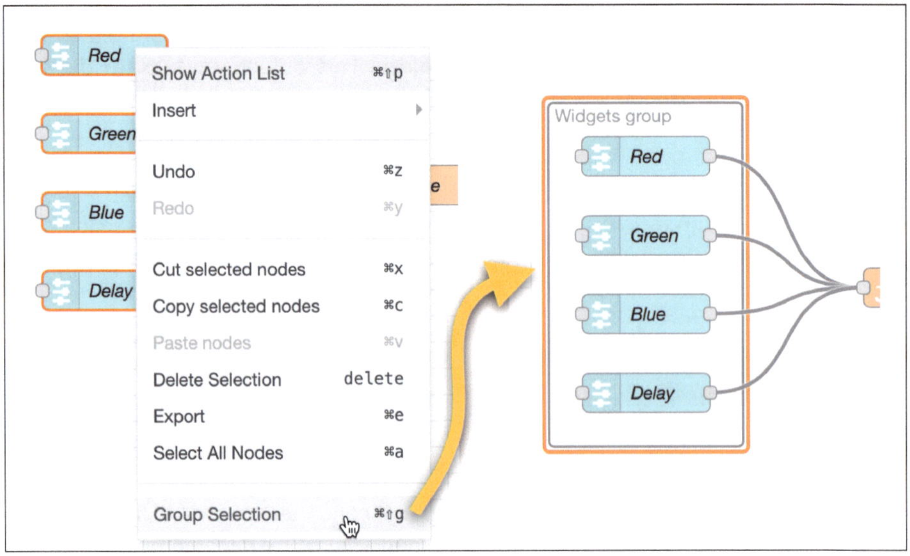

Figure 4.1.7.20: To create a node group, select the nodes you want to include and right-click on one of them, then click "Group Selection".

A new node group will be created containing the selected nodes. You can edit the group's properties by double-clicking on its title bar, including its name, label, and colour.

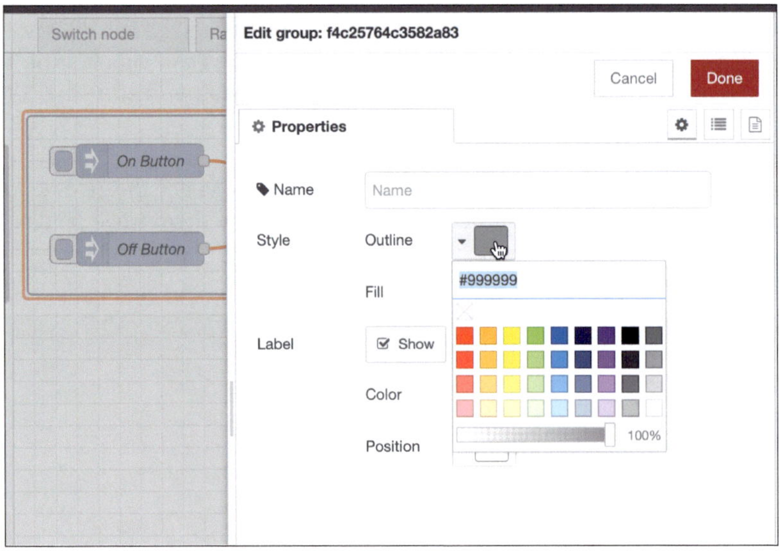

Figure 4.1.7.21: The node group properties window.

Node groups encapsulate the specific functionality related to the range node example and provide a cleaner, more organised workspace. As your flows grow more complex, the advantage of using node groups will become increasingly evident.

1.8. High-level review of other useful nodes by function

Before closing this chapter, look at a categorised list of built-in Node-RED nodes based on their primary functions. Some of those nodes you have already learned about in this book. For any nodes not covered in this book (as there are thousands of them!), you can refer to the Node-RED documentation at https://nodered.org/docs/.

Input nodes

1. **Inject**: Manually or periodically inject messages with different payloads, such as strings, numbers, or JSON objects.
2. **HTTP In**: Receive HTTP requests and generate corresponding output messages.
3. **WebSocket In**: Connect to a WebSocket server and receive messages.
4. **TCP In**: Receive messages via a TCP connection.
5. **UDP In**: Receive messages via a UDP connection.
6. **MQTT In**: Subscribe to an MQTT topic and receive messages.

Output nodes

1. **Debug**: Display messages in the debug window.
2. **HTTP Response**: Send HTTP responses based on input messages.
3. **WebSocket Out**: Connect to a WebSocket server and send messages.
4. **TCP Out**: Send messages via a TCP connection.
5. **UDP Out**: Send messages via a UDP connection.
6. **MQTT Out**: Publish messages to an MQTT topic.

Function nodes

1. **Function**: Process input messages using custom JavaScript code.
2. **Switch**: Route messages based on conditions, like comparisons or logical operations.
3. **Change**: Modify message properties by setting, deleting, or changing values.
4. **Template**: Generate message payloads using a template engine like Mustache.
5. **Delay**: Delay or rate limit messages.
6. **Trigger**: Control message flow using a trigger mechanism, such as sending an initial message and a second message after a delay.

Parsing nodes

1. **JSON**: Convert between JSON strings and JavaScript objects.
2. **XML**: Convert between XML strings and JavaScript objects.
3. **CSV**: Convert between CSV strings and JavaScript objects.
4. **HTML**: Extract content from HTML strings using CSS selectors.
5. **YAML**: Convert between YAML strings and JavaScript objects.

Social nodes

1. **Twitter In**: Receive tweets based on specific search criteria or a user's timeline. Twitter was recently rebranded to "X".

2. **Twitter Out**: Send tweets or direct messages based on input messages. Twitter was recently rebranded to "X".

Storage nodes

1. **File In**: Read data from a file and generate output messages.

2. **File Out**: Write data to a file based on input messages.

1.9. Credentials

If your flow requires credentials or other secrets, Node-RED offers several ways to manage and use such sensitive information. A good practice in such cases, whether you are working on a Node-RED flow or a program in any programming language, is to not hard-code such information in the program. If you do, then sharing the program with others, either directly or, for example, posting it on a code-sharing site like Github, would expose those secrets to the world.

In Node-RED, a simple way to protect sensitive information is to use the Credentials node (or "node-red-contrib-credentials"). This node allows you to store sensitive information at the flow level. When you export this flow to share with other people or back it up using the "json" export option, the data managed by the Credentials node are excluded. The plugin creates a separate encrypted file, "flows_cred.json" with the sensitive data.

This node adds an extra layer of security, ensuring that sensitive data isn't exposed even if someone gains access to your flow configurations. You can learn more about this node here: https://flows.nodered.org/node/node-red-contrib-credentials.

Use Cases for the Credentials node

The Credentials node in Node-RED is designed for various use cases where secure storage of credentials information is needed. I have listed several examples of such cases below to help you better understand where you might want to use this feature:

- **API Keys**: If you're using APIs that require keys for access, you can store these keys securely using the Credentials node.

- **Database Credentials**: When connecting to databases, you must often provide a username and password. Instead of hardcoding these into your flows, you can store them as credentials.

- **IoT Device Authentication**: If you're working with IoT devices that require secure tokens or passwords for communication, these can be safely stored as credentials.

- **Third-Party Services**: If your flow interacts with services like Google Sheets, AWS, or any other service requiring authentication, you can store those credentials securely.

How to use the Credentials node

Using Node-RED's Credentials feature, let's walk through securely storing sensitive information, such as API keys or passwords. Following these steps, you can keep your confidential data encrypted and separate from your flow configurations.

Step 1: Install the Credentials node

Search for "node-red-contrib-credentials" using the Palette manager and install the node. Once installed, the new node will appear in the Storage group.

Step 2: Open the node configuration

Start by double-clicking on the node where you intend to use the credentials. This action will bring up the node's configuration panel, where you can modify various settings. Drag the credentials node to the editor.

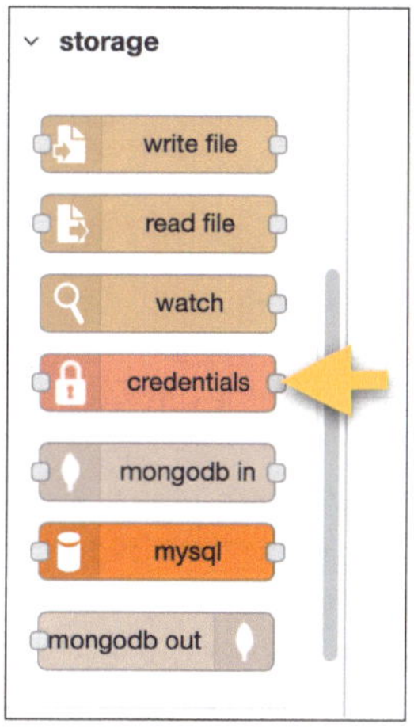

Figure 4.1.9.22: The "Credentials" node in the Storage group.

Step 3: Input your credentials

Once the configuration panel is open, look for a group labelled 'Values'. At the bottom of this group, you will see a button to add credential fields. Click on this button to add a new secret value. Add as many values as you need in your flow. In the screenshot below, I have added three secret values (the "apikey", the "username" and the "password" which is not showing in the list):

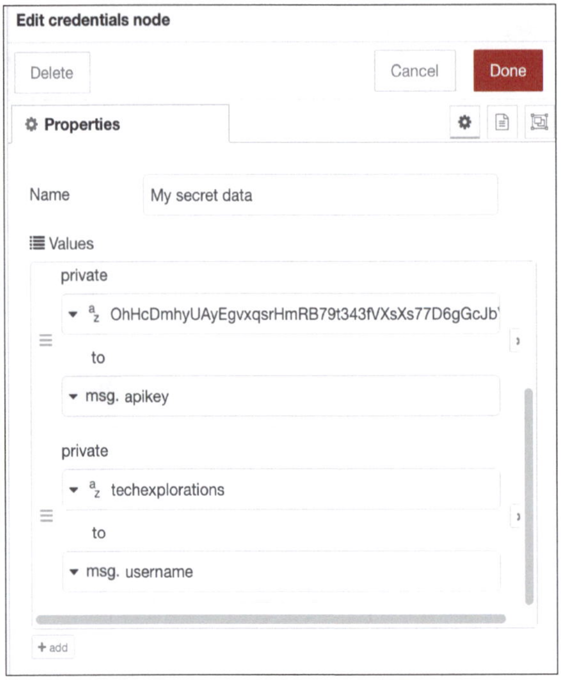

Figure 4.1.9.23: Example values in the the "Credentials" node.

As you can see, each secret value is private and set to "string" so I can see it. If you want it to be obfuscated, you can change this setting to "hidden". These values are stored encrypted outside of the flows.json file, ensuring security. Also, in the screenshot, you can see that I have made the username accessible through the "username" property of the message. Similarly, the API key is accessible through the "apikey" property of the message.

When you are finished adding secret values, click on Done.

Step 4: Utilize the stored credentials in your flow

You can now reference the credentials within your Node-RED flow with the credentials securely stored. Depending on how you've configured it, you can access these credentials using msg.credentials or flow.get("credentials"). For example, the screenshot below shows the debug node's properties connected to the Credentials node. The debug node reads the value stored in the "username" property of the message:

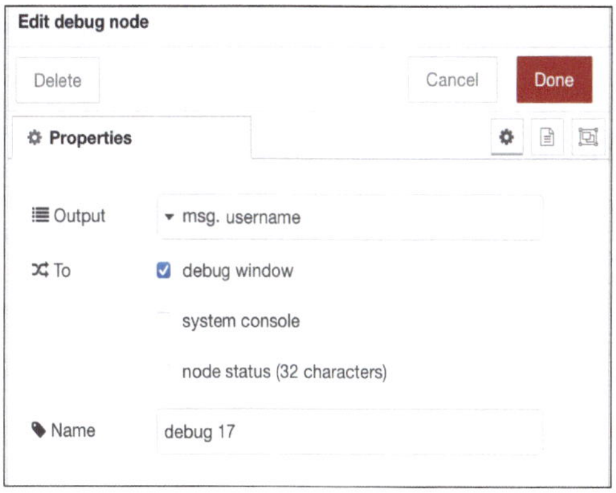

Figure 4.1.9.24: Reading an encrypted value.

Following these steps ensures that your sensitive information is securely stored and easily accessible within your Node-RED flows.

1.10. Environment Variables

In Node-RED, environment variables are a way to store data that can be accessed across different parts of your flow. They are little storage boxes that hold information you might need to use in multiple places. Unlike credentials, environment variables are not encrypted, so they're not ideal for storing sensitive information. However, they are handy for storing configuration settings or other data you want to change without modifying your flows.

Node-RED supports local (flow) and global environment variables. Local environmental variables are set and read-only within the scope of the local flow, while environment variables can be used in any flow across the Node-RED instance.

Environment variables example use cases

Here are some examples where environment variables are helpful:

1. Configuration Settings: Let's say you have a flow that interacts with a database. The database's IP address could be stored as an environment variable, making it easy to update without changing the flow.
2. Reusable Data: If a piece of data is used in multiple flows or even multiple times within a single flow, an environment variable can store that data in one place.
3. Dynamic Behavior: You can use environment variables to control the behaviour of your flows. For example, you could have a variable that switches your flow between "test" and "production" modes.

Using a variable across a flow

In Node-RED, if you want to set a variable that can be accessed by other nodes within the same flow, you can use the "flow context." The flow context allows you to store data that can be shared among all nodes in that particular flow. The setting and reading of flow environment variables are done with the help of "function" nodes.

Here are the steps required to set and read a flow environment variable:

1. **Open a Function Node**: Drag a function node into your flow and double-click it to open its configuration panel.
2. **Write the Code**: In the function field, you can set a flow context variable using the flow.set() method. For example, to set a variable named temperature, you could write:

```
flow.set("temperature", 25);
```

The "temperature" variable can now be used elsewhere in the same flow. To read its value, you will need another "function" node. Here is how to achieve this:

1. **Open Another Function Node**: Add another function node where you want to use the variable and open its configuration panel.
2. **Retrieve the Variable**: Use flow.get() to retrieve the variable. For example:

```
var temp = flow.get("temperature");
```

1. Use the Variable: Now, the variable temp contains the value from the flow context, and you can use it as needed in this function node.

```
var temp = flow.get("temperature");
if (temp > 24) {
    msg.payload = "It's warm!";
} else {
    msg.payload = "It's cool!";
}
return msg;
```

2. **Deploy**: Click "Done" and deploy the flow.

Using the flow context, you can easily share data between different nodes in the same flow, making your flows more dynamic and interactive.

Using a global environment variable

Starting with Node-RED 3.1, using global environment variables is easier than ever before. As with older versions, you can set the variable using the Node-RED user interface instead of editing the settings.js file on the command line.

To add global environment variables to Node-RED, go to the Environment tab in the User Settings window (find this under the burger icon at the top right corner of the Node-RED page). Click the "+ add" button at the bottom left corner of the Global Environment Variables box to add a variable (see the arrow in the screenshot below).

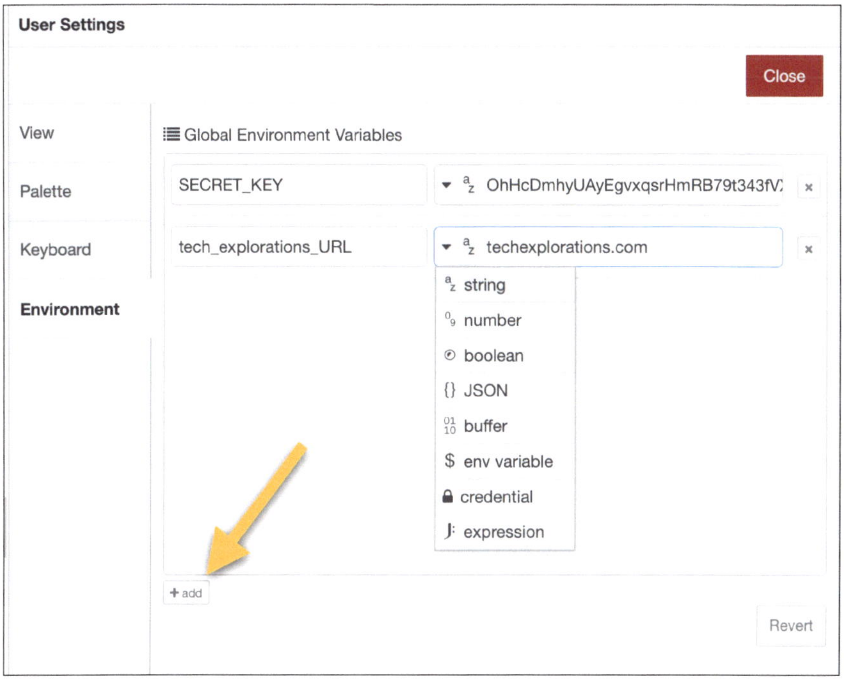

Figure 4.1.10.25: The Global Environment Variables tab in User Settings.

You can now add a key and a value for the new environment variable. The value can be one of several types: string, number, or JSON.

To use a global environment variable in any flow, you select the "env variable" data type from a node like "inject" and provide the name of the variable you want to retrieve. For example, see the screenshot below, showing the properties for an Inject node that is accessing the "tech_explorations_URL" variable:

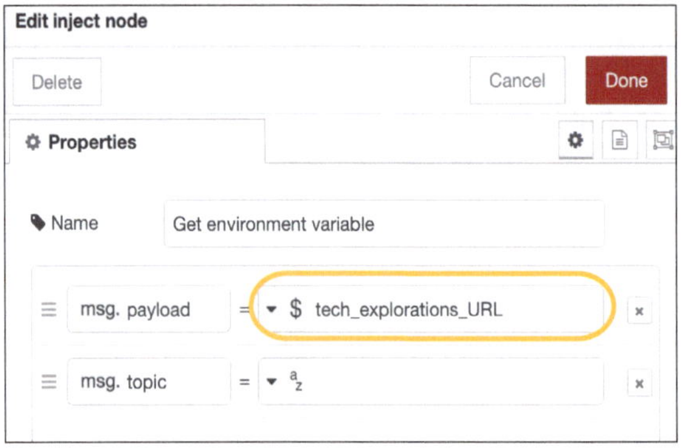

Figure 4.1.10.26: An Inject node setting the payload to the value of an environmental variable.

In the screenshot above, the environment variable type is symbolically depicted by the "$" symbol. A debug node connected to the output of the Inject node will display the value of the variable in the debug window:

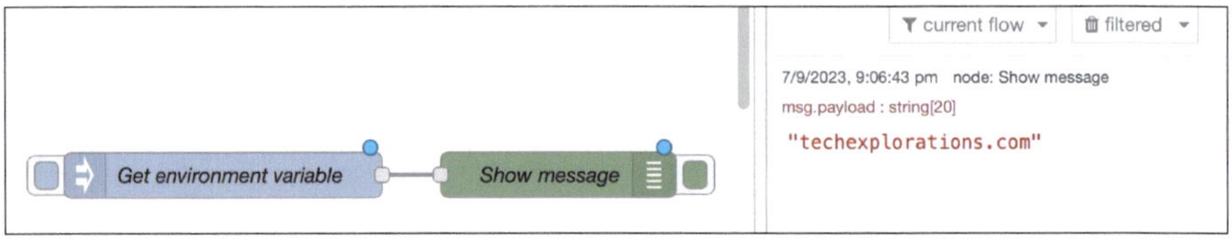

Figure 4.1.10.27: An environmental variable retrieved by an inject node, printed by Debug.

You can also retrieve global environmental variables in a "function" node. Here's an example of Javascript:

```
var message = env.get("tech_explorations_URL");
msg.payload = message;
return msg;
```

The flow below is an example of how a "function" node can retrieve an environmental variable:

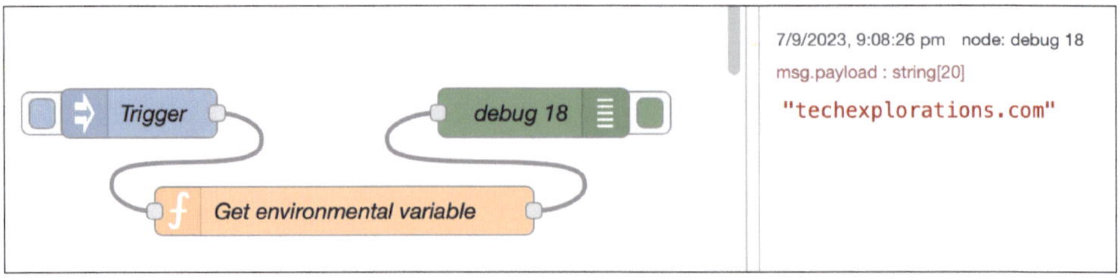

Figure 4.1.10.28: An environmental variable retrieved by a function node, printed by Debug.

By using environment variables, you're making your Node-RED flows more flexible and easier to manage. In Node-RED 3.1 and later, using environmental variables at the global scope is easier than ever before.

Chapter 4.2 • Control Structures and Loops

In the journey to master Node-RED, understanding control structures is akin to grasping the grammar of a language. It's not just about knowing the vocabulary—represented by the individual nodes—but also about understanding how to arrange them to yield robust, efficient, and manageable flows.

This chapter dives into the essence of control structures and loops in Node-RED, illuminating how they orchestrate the flow of data and events through your setups. From conditional operations like switches and gates to loops and flow controls like split and join nodes, we'll explore how these structures bring logic and versatility to your automations.

Whether you're a novice looking to build simple, conditional flows or an experienced user aiming for complex, data-driven orchestrations, the control structures discussed in this chapter will serve as indispensable tools in your Node-RED toolkit.

2.1. Conditional Nodes

In Node-RED, several nodes can be used for implementing conditional logic in your flows. Here are some of them.

Switch node

You learned about the switch node in the section "Other useful nodes" in this book at the start of Part 4, so we won't go into the details here. The switch node can route messages to different outputs based on their properties. It allows you to specify multiple conditions, each corresponding to a different output port.

Range node

You also learned about this node early in this part of the book (as you did about the RBE node next). In summary, the range node is another conditional node that routes messages based on a set range of values. This node is applicable when dealing with devices with a specific operating range, and we want to filter out either too-high or too-low values.

RBE (Report by Exception) node

The RBE node sends a message only when the payload has changed from the previous payload, reducing the number of messages in the flow. You can learn more about the RBE node in the "Other useful nodes" section at the start of Part 4 in this book.

Change node

The Change node is a powerful tool for modifying messages. It allows you to set, change, delete, or move message properties or flow variables based on rules you define.

An example application of this node: You can use a Change node to add a timestamp property to each message that passes through or to rename a message property.

Gate Node
The Gate node provides a way to control the flow of messages. It can be configured to allow or block messages based on specified conditions.

Example: Suppose you're monitoring a sensor that reports environmental data. You can use a Gate node to block all messages during night hours when data is unnecessary.

Check payload node
The Check Payload node evaluates the payload and outputs messages based on content. Example: To ensure that a payload isn't empty before sending it on, you can use the Check Payload node to divert or halt empty payloads.

Delay node (Rate Limit mode)
In "Rate Limit" mode, the Delay node can control the flow of messages based on a specified rate. It can limit the rate to a specific number of messages per time unit.

Example: If you get sensor readings every second but only want a report every minute, you can set the rate limit to one message per minute.

Traffic node
The Traffic node is similar to the Gate node but can dynamically change its state in response to Control msg inputs.

Example: If you want to stop a flow of messages based on a particular signal from another part of your flow, you can use the Traffic node to control this.

While node
The While node implements a while-loop, repeating the flow of messages while a certain condition is true.

Example: If you're receiving a series of messages and want to process them repeatedly until a condition is met (such as the payload exceeding a threshold), the While node can be used for this purpose.

Split node
The Split node divides a sequence into individual messages, like an array or string. It can be used conditionally by specifying a particular separator or matching criteria.

Example: If you have a payload containing a comma-separated string of values, you can use the Split node to create a separate message for each value.

Batch node
The Batch node is similar to the Split node but can create a batch of messages based on specified conditions.

Example: If you're receiving messages at a high rate and want to process them in groups of 10, you can use the Batch node to create these groups.

While all these nodes offer various ways to implement conditional logic, the "function" node can also be used for custom conditional logic using JavaScript. This provides the most flexibility but requires programming knowledge to use effectively.

Remember, you can also use the "function" node to implement custom conditional logic using JavaScript. It provides the most flexibility but also requires programming skills to use effectively.

2.2. Iteration Nodes

Iteration is a common requirement in data processing, and Node-RED offers a variety of nodes that can iterate over data structures or execute commands repeatedly. This chapter will discuss nodes classified as "iteration" nodes. Several nodes provide iterative functionality in Node-RED. Here, we will cover the "split", "batch", "join", "delay" (rate limit mode), and "while" nodes.

Split node
The Split node iterates over sequence data structures like arrays or strings and divides them into individual messages.

Example: Consider a situation where you have a payload containing a comma-separated string of values. You can use the Split node to create a separate message for each value. This effectively iterates the entire string, splitting it wherever it encounters a comma.

Batch node
Like the Split node, the Batch node groups incoming messages into batches based on specified conditions, effectively iterating over a stream of messages.

Example: Suppose you receive messages at a high rate and need to process them in groups of 10. In this case, you could use the Batch node to group incoming messages into batches of 10.

Join node
The Join node is the opposite of the Split node. It iterates over individual messages and joins them together into a single message with a combined payload.

Example: If you have multiple messages, each with a single word payload, you could use the Join node to create a single message with a payload of a sentence.

Delay node (Rate Limit mode)
In "Rate Limit" mode, the Delay node controls the flow of messages based on a specified rate, iterating through the incoming messages at a limited pace.

Example: If a sensor sends data every second but only wants to process one message every ten seconds, you could use the Delay node in Rate Limit mode to throttle the message flow.

While node

The While node provides a way to implement a while-loop structure in Node-RED, repeating the flow of messages while a certain condition remains true.

Example: Suppose you're receiving a series of messages and want to process them repeatedly until a certain condition is met (e.g., a payload value exceeds a threshold). The While node can be used for this purpose, effectively iterating over the messages until the condition is no longer met.

Node-RED offers a variety of nodes for implementing iterative processes. These nodes allow you to control the flow of messages, whether by dividing sequences into individual messages, grouping messages, controlling the rate of message flow, or looping over messages under certain conditions. Understanding how to use these nodes effectively is vital for building complex flows in Node-RED.

2.3. Conditional and iteration nodes example flow

Let's consider a simple flow example demonstrating iteration and conditional logic. Imagine we have a user interface that takes input of comma-separated numbers from the user through a dashboard Text Input widget. The flow will split these numbers into individual messages (iteration), check if each number is greater or smaller than 5 (conditional), and then display the results in two separate dashboard Text widgets, one for numbers that are greater or equal to 5 and one for numbers that are smaller than 5. The dashboard user interface comprises a text input node and two output nodes.

You can see the dashboard below:

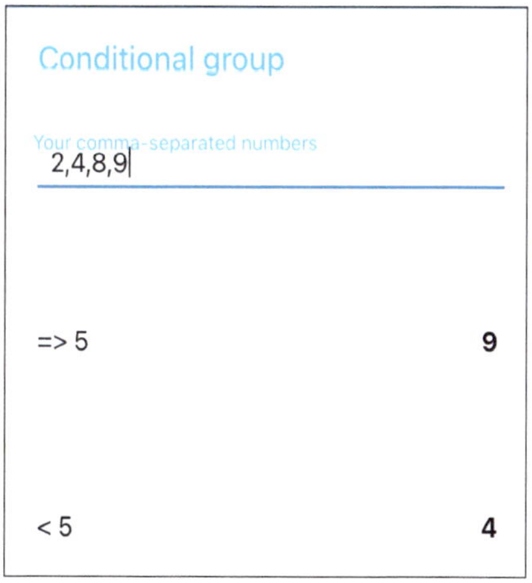

Figure 4.2.3.1: This example demonstrates iteration and conditional logic.

In the example run (see above), I have entered a few numbers in comma-delimited format in the text input widget. The flow evaluates these numbers iteratively and prints out the latest pair of smaller than five and larger or equal to five.

Here's a step-by-step guide to creating this flow:

1. **Add Input and Output Nodes**: Drag and drop a "ui_text_input" node and two "ui_text" nodes onto the workspace. The "ui_text_input" node will allow the user to input comma-separated numbers, and the "ui_text" nodes will display the results.

2. **Add Split Node**: Add a "split" node (from function nodes) to the workspace. This node will split the input string into individual numbers.

3. **Add Switch Node**: Add a "switch" node (from function nodes). This node will determine whether each number is even or odd.

4. **Configure Nodes**: Double-click on each node to open its configuration:
   - "ui_text_input": Leave the default settings. This node will create a text input field on the dashboard.
   - "split": In the properties section, set "Split using" to "Comma". This will cause the node to split the input string at each comma.
   - "switch": Click on "+ add" to create two conditions. The first should be "greater or equal to 5", and the second should be "smaller than 5".

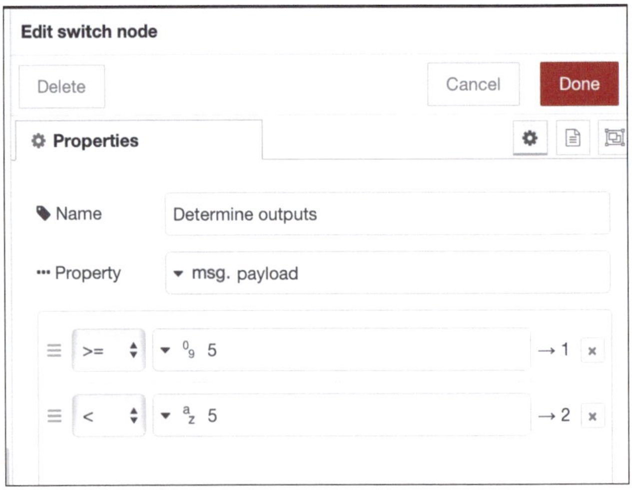

Figure 4.2.3.2: The properties of the Switch node.

   - "ui_text" (x2): Configure the first "ui_text" node for even numbers and the second for odd numbers. In the "Group" field, ensure that both nodes are in the same group so they display in the same part of the dashboard.

1. **Wire Nodes**: Connect the nodes in the following order:
   - "ui_text_input" —> "split" —> "switch" —> "ui_text". The "switch" node should connect to both "ui_text" nodes, one for each output.
2. **Deploy**: Click the Deploy button to save and run the flow.

The example flow looks like this:

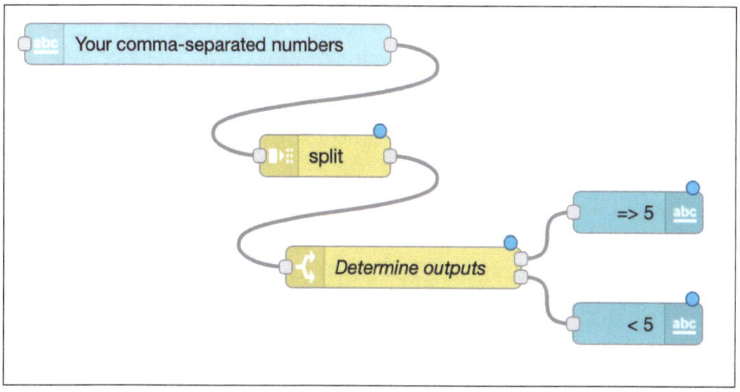

Figure 4.2.3.3: This example flow demonstrates conditional and iteration nodes.

Now, you can go to the Node-RED dashboard in your web browser. You should see a text input field where you can enter comma-separated numbers. When you enter numbers and hit enter, they will be split, classified as even or odd, and displayed in the appropriate text widget.

This simple flow demonstrates conditional logic (determining if numbers are even or odd) and iteration (processing multiple numbers from a single input string).

Chapter 4.3 • Integrating External Services and APIs

In the ever-expanding world of interconnected technologies, the true power of Node-RED lies in its ability to serve as a hub, seamlessly integrating with external services and APIs. This chapter aims to demystify extending the capabilities of your Node-RED setup beyond the confines of your local environment.

I will walk you through connecting to databases, interacting with RESTful APIs and web services, and offer guidelines on harnessing cloud platforms and IoT services for enhanced functionalities. Specific use cases, such as fetching weather information from OpenWeatherMap or utilising Google Sheets for data logging and data retrieval, will also be explored in more detail.

Whether you are building an advanced home automation system, crafting an IoT solution, or simply looking to log and manipulate data, understanding how to connect Node-RED with external entities will unlock a new level of capability for your projects.

3.1. Node-RED with MySQL

Node-red provides various nodes that can be used to interact with both SQL and NoSQL databases. This allows Node-RED to read from and write to these databases, which can be especially useful in an IoT context where data is often gathered and stored for later analysis. You can connect several SQL databases with Node-RED, such as MySQL, PostgreSQL, Microsoft SQL Server and SQLite.

This chapter teaches you to integrate your Node-RED flows with a MySQL database. The MySQL database is installed on the Ubuntu server on which the Node-RED container is also running. Of course, installing the MySQL server on a remote machine is possible if your use case requires this. For example, you can have a dedicated virtual machine or real server on the metal running a MySQL server.

One limitation of the MySQL nodes in Node-RED that I came across is that these nodes don't work with secure protocols. This means that the communication between Node-RED and the database is done unencrypted. For this reason, I decided to install the database server on the same virtual machine as the Node-RED Docker container. As I have control of the virtual machine, it is secure enough to compensate for the inability to use a secure communications protocol.

To begin with, though, I'd like to show you a simple example from the experiment you will implement later in this chapter. In this example, you can see my terminal on the top side. In this terminal session, we have issued SQL "select" instruction, which returns the available rows of data from the database table with the name "sensor data". In the bottom part of the image below, you can see a simple Node-RED flow that consists of two sub-flows.

The first subflow uses a "create" SQL instruction to record data in the database. The function node contains Javascript that will generate random numbers (to simulate values acquired from sensors) and an SQL query for storing these numbers in a new database row.

The second subflow uses a "select" SQL instruction to retrieve data from the database. The function node contains Javascript that will receive the rows from the database, ignore all except for the last row, split the three values, store them into three local variables, and export them to be printed on the debug tab.

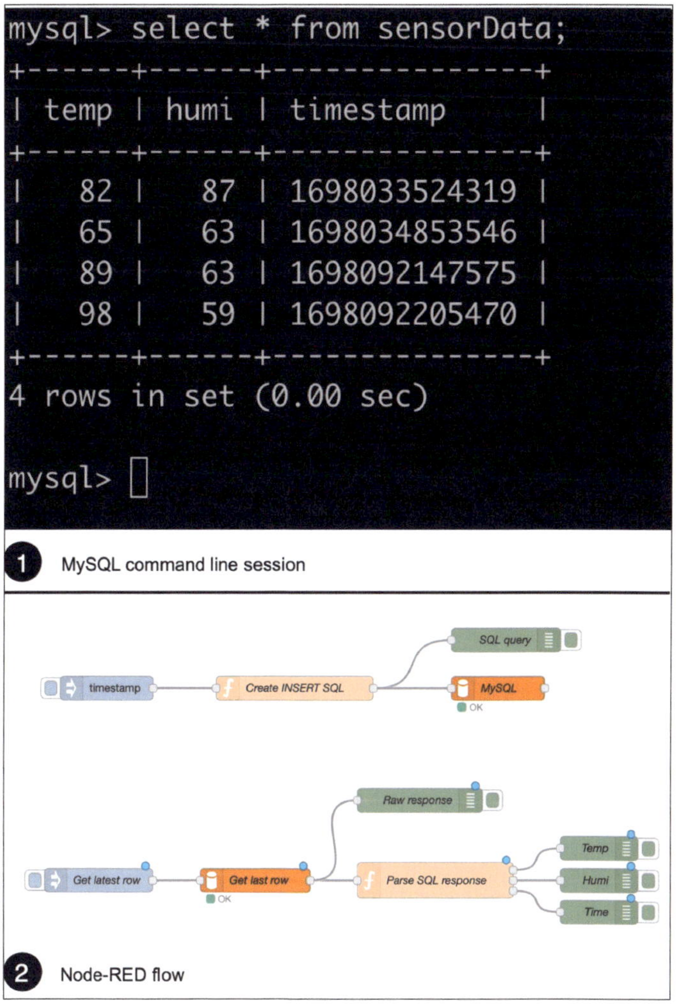

Figure 4.3.2.6: This example contains a flow (bottom) that writes and reads data to and from a MySQL database. The image on the top shows four rows of data from the sensorData table of the database.

At this point, don't be concerned about the details. Later in this chapter, you will work on the contents of the function nodes and the specialist MySQL nodes in these example sub-flows.

In this example, I have installed the MySQL server on the same virtual machine that hosts the Node-RED container.

The MySQL database uses SQL (Structured Query Language, see https://txplo.re/sql). You can use SQL on the command line to interact with the database. External systems like Node-RED also use SQL to interact with MySQL. If you are not familiar with SQL, don't worry. Even though this course is not about databases, SQL is simple enough to use immediately after seeing a few examples. Later in this chapter, I will show you everything you need to know about SQL to use MySQL with your Node-RED flows immediately. You can look for appropriate SQL resources if your project requires more complex queries. You can start here: https://txplo.re/sql_tutorial.

A database, unlike a Google sheet (about which you will learn later in this book), allows you to record a massive amount of records. This means that you can store as much data as you want, limited only by the underlying storage medium. Also, you can use the compelling SQL query language specifically designed for database interactions.

In the next section, you will learn how to install MySQL on your Ubuntu virtual machine and configure it to work well with your Node-RED instance.

MySQL for Node-RED on Ubuntu

In this section, you will install the MySQL server on the Ubuntu virtual machine—the instructions I have compiled below come from a guide on the Digital Ocean Community site. You can find the original at https://txplo.re/do_mysql.

Because of the specifics of our project in this chapter (especially about the networking and the user setup), I have made a few changes to these instructions. Let's begin.

Step 1. Install MySQL

Login to your Ubuntu server and copy the following instructions. These instructions will update the operating system and then will install MySQL.

```
$ sudo apt update
$ sudo apt upgrade
$ sudo apt install mysql-server
$ sudo systemctl start mysql.service
```

At the end of this process, you will see this text among the output of the "install mysql-server" command:

```
mysqld is running as pid 96571
```

The "pid" number ("process id") will most likely be different. However, the important thing is that your database server is running. The last command, "...start mysql.service", ensures the server is running.

Although the server is now installed and running, it isn't configured. Let's do that next.

Step 2: Set up MySQL
The MySQL installation includes a script named "mysql_secure_installation" that automates the configuration process. The script asks various questions and configures the database instance per your answers. However, as of July 2022, an error will occur when you run the mysql_secure_installation script without some further configuration. The reason is that this script will attempt to set a password for the installation's root MySQL account, but, by default, on Ubuntu installations, this account is not configured to connect using a password. Therefore, before running the script, we'll make some changes to the default root user for the database.

Let's begin by starting the MySQL command prompt:

```
$ sudo mysql
```

Change the "root" user's authentication method so it uses a password and then exit the command prompt:

```
mysql> ALTER USER 'root'@'localhost' IDENTIFIED WITH mysql_native_password BY
'password';
mysql> exit
```

With this command, the "root" user is now set to use "password" as the password. I know this is not secure, but don't worry, we'll change it soon.

Test authentication with a password by trying to log in:

```
mysql -u root -p
```

The "-p" parameter instructs the database to request a password in this command. Type "password" as the password, and confirm that you have entered the MySQL command prompt like this:

```
mysql>
```

Now, restore the default authentication method that does not require a password. This will allow us to log in to the database command line using "sudo mysql" only without providing a password. It will also restrict root login to the local host only, rejecting remote hosts' attempts.

```
mysql> ALTER USER 'root'@'localhost' IDENTIFIED WITH auth_socket;
mysql> exit
```

With this, the root user can only be used when working on the local server. You can now run the security script with sudo to continue the setup:

```
$ sudo mysql_secure_installation
```

The script will ask you various questions. Below, I have recorded my specific answers to help you complete this step of the process.

- Would you like to set up VALIDATE PASSWORD component? NO
- Remove anonymous users? Yes
- Disallow root login remotely? Yes
- Remove test database and access to it? Y
- Reload privilege tables now? Y

Let's continue with setting up a dedicated user for Node-RED.

Step 3: Create a dedicated user and grant privileges:
In this step, you will create a new database user used exclusively by your Node-RED flows. After creating the user, you will configure it to allow remote login and grant it sufficient privileges for creating tables, reading and writing.

Start by accessing the MySQL shell as root:

```
$ sudo mysql
```

Then, create a new user and grant access privileges for all databases and tables:

```
mysql> CREATE USER 'node-red'@'%' IDENTIFIED BY 'node-red-pwd';
mysql> GRANT ALL PRIVILEGES ON *.* TO 'node-red'@'%';
mysql> FLUSH PRIVILEGES;
```

Let's decipher what you just did:

- The name of this user is "node-red".
- The new user's password is "node-red-pwd".
- This user has all privlidges assigned to it.
- This user can access the database from remote hosts (this is the meaning of the "%" symbol after the "@".
- Because of the changes we made in the database's "grant" tables, we use the "FLUSH PRIVILEGES" to reload the grant tables and make these changes effective.

We now have a database user dedicated to use by Node-RED and have sufficient privileges to access the database from remote hosts. However, the database is not yet set up to accept remote connections. Let's fix this issue next.

Step 4: Allow connections from external hosts
MySQL has a configuration file that allows the fine-tuning of your MySQL instance. This configuration file is named "mysqld.cnf", and, among other things, allows tuning of parameters such as the port and binding IP address, cache, logs, maximum connections, and much more.

By default, only connections from the local host are allowed. As our Node-RED instance is handled by Docker (and the Docker container is separate and isolated from the virtual machine that hosts the MySQL instance), we have to allow connection from external hosts. This can be achieved by changing one configuration in "mysqld.cnf".

Let's go ahead and do this. Use the "nano" text editor to open this file:

```
$ sudo nano /etc/mysql/mysql.conf.d/mysqld.cnf
```

Search for a line that begins with bind-address and set its value to the IP address on which a MySQL server should listen. By default, the value is set to 127.0.0.1 (listens only in localhost). In this example, we'll set the MySQL server to listen on all IPv4 interfaces by changing the value to "0.0.0.0".

```
bind-address = 0.0.0.0
```

Save the file, and exit the text editor. Restart MySQL:

```
$ sudo systemctl restart mysql
```

That's it. Your database instance is ready to work with Node-RED. However, one more thing to do is create a new database and table for our Node-RED flows.

Step 5: Create a new database and table
Our Node-RED flows can now communicate with the MySQL instance. However, the MySQL instance is blank; there is no database to work with. We must set up a new database before Node-RED can do anything useful with MySQL. Within this database, we will also create a new table to store data from our simulated sensor. Remember, MySQL can handle many databases, each with its tables and users. This means you can create databases for your various flows as needed.

Let's set up the database and table. Log in to the MySQL command line using the "node-red" user you created in step 3. You can see my session and instructions below:

```
$ mysql -u node-red -p
mysql> CREATE DATABASE nodeRED;
Query OK, 1 row affected (0.03 sec)

mysql> use nodeRED
```

```
mysql>CREATE TABLE sensorData (temp int, humi int, timestamp varchar(255));
Query OK, 0 rows affected (0.10 sec)
```

In this session, I have used these commands:

- "mysql -u node-red -p": Login to the MySQL command line with the user name "node-red" and the password provided when prompted.
- "CREATE DATABASE nodeRED;": Create a new database named "nodeRED".
- "use nodeRED": Use database "nodeRED" going forward in this session. Any command I issue from this point forward will be applied to the "nodeRED" database.
- "CREATE TABLE sensorData (temp int, humi int, timestamp varchar(255));": Create a new table with name "sensorData". This table has three columns to store data. Column 1 is named "temp" and is an integer. Column 2 is named "humi" and is also an integer. Column 3 is named "timestamp" and is a string with up to 255 characters.

The database is now ready, and you can continue with the Node-RED flow in the next section of this chapter.

A flow that uses MySQL

At this point, you have a MySQL server running on your Ubuntu server on the same server that the Node-RED Docker container is also running. At the beginning of this chapter, I gave you a preview of the Node-RED flow you would eventually implement. This flow demonstrates the use of the "mysql" node that makes it possible to relay SQL instructions to the MySQL server. The MySQL server executes these instructions and sends the results back to the Node-RED flow for further processing.

Now, it is time for you to assemble the demonstration flow.

Nodes for MySQL

The Node-RED community has created a few nodes for MySQL. In the screenshot below, you can see two of those installed in my Node-RED instance.

Figure 4.3.2.7: Two nodes for working with MySQL.

In the example flow you will implement in this chapter section, you will use the "mysql" node, which is the simpler of the two. By "simpler", I don't mean less capable but easier to set up and use. Neither of these nodes comes with a fresh instance of Node-RED, so go ahead and install the "MySQL" node. Feel free to install the "mysql - r2" node if you want to explore that option. You can install these nodes by going to the palette manager in the burger menu (click on the top-right corner icon to show the menu that contains the palette manager).

Also, remember that you can see the documentation of any node by clicking on the node first to select it, then on the help tab, as seen below.

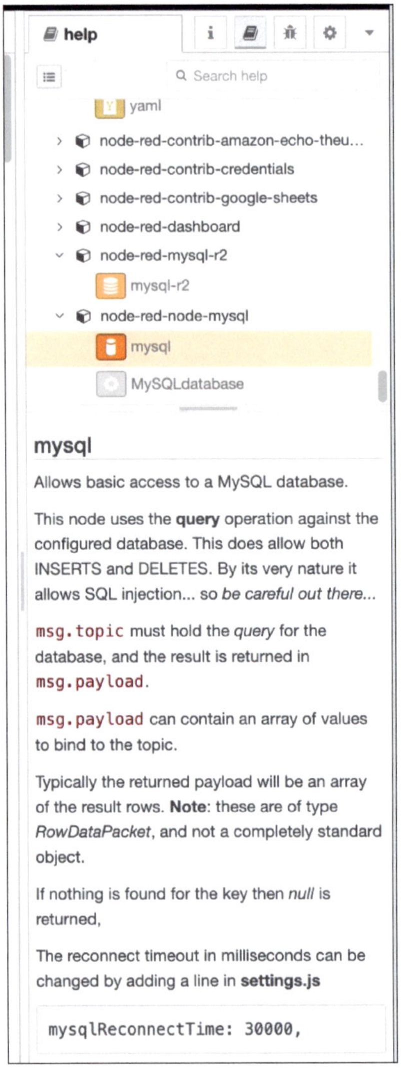

Figure 4.3.2.8: Learn more about a node by reviewing its documentation.

In the example flow below, we will explore only some of the many features of this node, so review the documentation to understand its full capabilities better.

Construct the flow

Let's focus on the example flow. This flow consists of two subflows. The first is responsible for creating new records in the database "sensor data " table, and the second is for reading the last record from the same table. You can see the flow below:

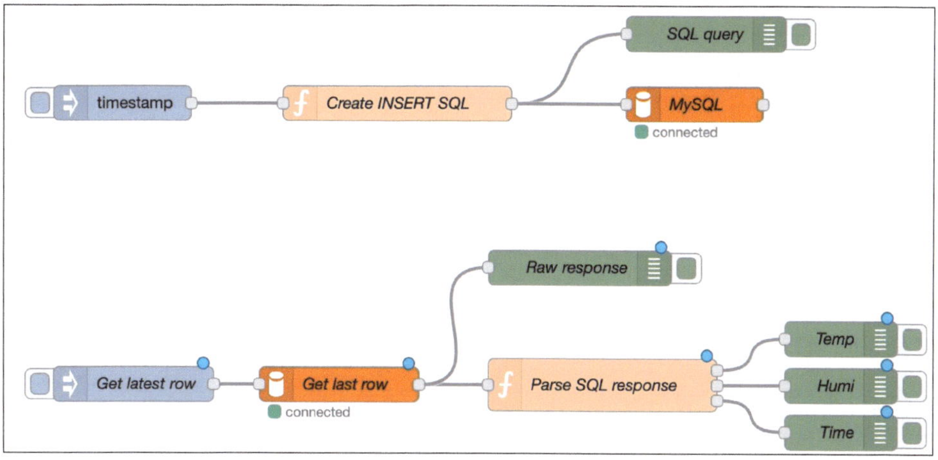

Figure 4.3.2.8: The example flow consists of two subflows: Write (top) and Read (bottom).

Let's begin with the construction of the "Write" subflow. Here is the process:

1. Drag and drop an "inject" node to the editor. This will provide a way to trigger the flow and provide a timestamp that will be one of the values recorded in the new record.

2. Add a "function" node and code it to the "inject" node. Name the function node "Create INSERT SQL", and copy this Javascript code:

```
function generateRandomNumbers() {
   let randomNumbers = [];
   for(let i = 0; i < 2; i++) {
      let randomNumber = Math.floor(Math.random() * (100 - 20 + 1) + 20);
      randomNumbers.push(randomNumber);
    }
    return randomNumbers;
}

const numbers = generateRandomNumbers();
const query   = 'INSERT INTO sensorData VALUES (${numbers[0]}, ${numbers[1]},
${msg.payload});';
msg.payload   = query;
msg.topic     = query;
return msg;
```

The code in the "Create INSERT SQL" should look like this:

Figure 4.3.2.9: The Javascript in the "Create INSERT SQL" node.

This function node is very important because it fulfils two tasks. First, it generates two random values that simulate an imaginary sensor's temperature and humidity. These values will be stored in the database in a new record along with the timestamp. Second, it generates the SQL statement that will be passed to the "mysql" node for processing by the database. You can see the SQL statement under construction in line 12 of the screenshot above. Once the SQL statement is ready and stored in a local variable, it is copied into the message object's payload. We will repeat the same process in the second subflow that reads data from the database.

Now that we have the SQL query in the message object let's continue.

3. Add a "mysql" node and connect it to the function node. This node sends the SQL query (passed along in the message object) to the database. The database will execute the SQL query and send back the results. I assume you have not yet created a database connection in Node-RED, so the first thing to do is create such a connection. Double-click on the "MySQL" node to open its properties window, give it a name (such as "MySQL), and select "Add new MySQL database...", then click on the pen button. This step looks like the screenshot below:

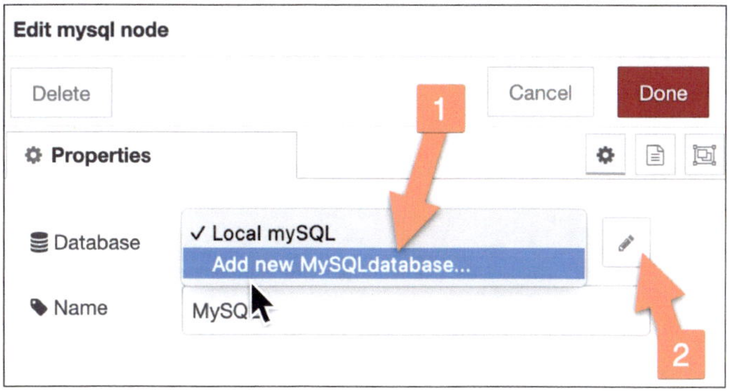

Figure 4.3.2.10: Creating a new database connection in the "mysql" node.

This will bring up the MySQLdatabase node's database editing window. In the form, complete the details for the database instance and the database you created in the previous section of this chapter. You can see my settings below. Take care to provide the correct dedicated username and password. Remember that the IP address of the database instance is assigned to your Ubuntu virtual machine. The default port of the MySQL instance is 3306.

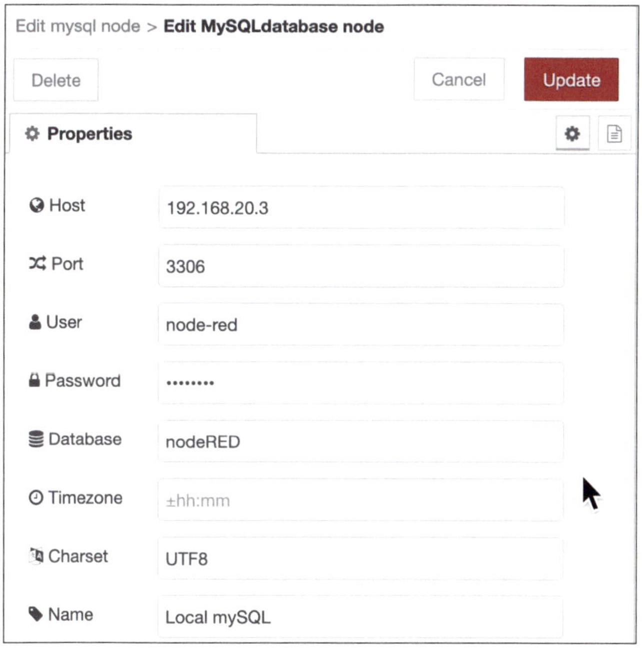

Figure 4.3.2.11: The settings for the new database connection.

When ready, click on Update, and then select the new database connection name from the "Database" dropdown menu of the "mysql" node properties window. Click "Done" to close the properties window.

4. Connect a "debug" node to the function node. This will print out the generated SQL query to see what is being passed on to the MySQL node and the database.

   This completes the setup of the "Write" subflow. Let's work on the "Read" subflow next.

5. Add a new "inject" node to the editor. We'll use this node to trigger the subflow. We will also use this node to create the SQL query passed to the "mysql" node. As this query is fixed (I.e. It does not contain any variables), we don't need a function node to generate it. Double-click the inject node to open its properties. In the "topic" parameter of the message object, copy this SQL:

```
SELECT * FROM sensorData ORDER BY timestamp DESC LIMIT 1;
```

You can see the configuration of the "inject" node below:

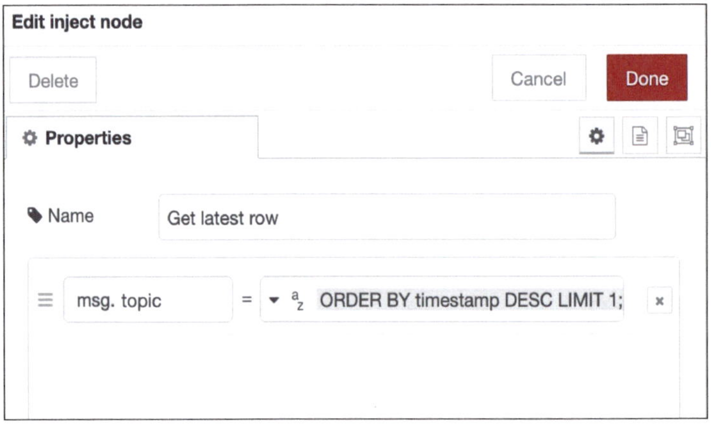

Figure 4.3.2.12: The settings for the inject node in the "Read" subflow.

Here is what this query does:

- SELECT * FROM sensorData:
  - SELECT *: This part of the query instructs the database to select all columns in the specified table.
  - FROM sensorData: This specifies the table named sensorData from which the data should be retrieved.
- ORDER BY timestamp DESC
  - ORDER BY timestamp: This clause arranges the retrieved data in an order based on the timestamp column of the sensorData table.
  - DESC: DESC stands for descending. This modifier ensures the data is sorted in descending order, meaning the newest entries (with the latest timestamps) appear first.

- LIMIT 1: This clause restricts the query output to just one row. Given the previous ordering, the query will return the row with the most recent timestamp.

This is an excellent example of how efficient SQL is. Imagine having a database table with thousands or millions of rows, and you only want to retrieve a single row. In this case, we want to retrieve the last row. The simple SQL instruction above does this. In contrast, as you will learn in the later chapter on Google Sheets, there is no efficient way to do this. In Google Sheets, you must retrieve all rows in the sheet and then use Javascript on the Node-RED side to retrieve the row you want. For small sheets, this method is tolerable; however, as the number of rows in the sheet grows to a few thousand, this inefficient method will impact the performance of your server. This is an excellent example of why and when a database is preferable to a spreadsheet.

6. Add a "mysql" node. Just like in the first sub flow, this "MySQL" node will receive an SQL query from the preceding node (in this case, the "inject" node) and send it to the connected database server for processing. In the previous sub flow, you created a database connection and can now reuse it (named "Local MySQL"). Rename the node "Get last row".

7. Add a "function" node and connect it to the "Get last row" node. This node contains Javascript that will extract the values for the three columns from the row returned from the database and output these values in three separate messages (ensure that in the Setup tab of the function node properties, you have selected three outputs). You can see the Javascript below:

```
const jsonObject              = msg.payload[0];

const tempVaIues              = {payload: jsonObject.temp};
const humValues               = {payload: jsonObject.humi};
const timestampValues         = {payload: jsonObject.timestamp};

console.log('Temperature values:', tempVaIues):
console.log('Humidity values:', humValues):
console.log('Timestamp values:', timestampValues);

return [tempValues, humValues, timestampValues];
```

Let's see what this code does.

- const jsonObject = msg.payload[0]; : The "mysql" node returns an array that contains the rows from the database. Each row contains a JSON object with the row data. As we want only one record, we use the notation "[0]" to get this record from the payload object. This line declares a constant jsonObject and initialises it with an array's first element ([0]) at msg.payload.

- The following three lines declare local variables for the temperature, humidity and timestamp. In each one, the line will create a new JSON object and store the temperature, humidity and timestamp in the payload attribute. This is important because the "function" node must output one or more JSON objects containing a payload attribute at minimum.
- The script uses the "console.log" instruction to print the database row values to the log.
- Finally, 'return [tempValues, humValues, timestampValues];' sends the three newly created JSON objects to the node output.

8. Drag four "debug" nodes. Connect one to the "Get last row" (to show the raw output of the "mysql" node) and the other three to the function node.

Deploy and test the flow

Go ahead and deploy the flow. Click on the "Write" subflow's inject node to create a new row in the database. If all goes well, you'll see the raw SQL query in the debug tab, like in the example below:

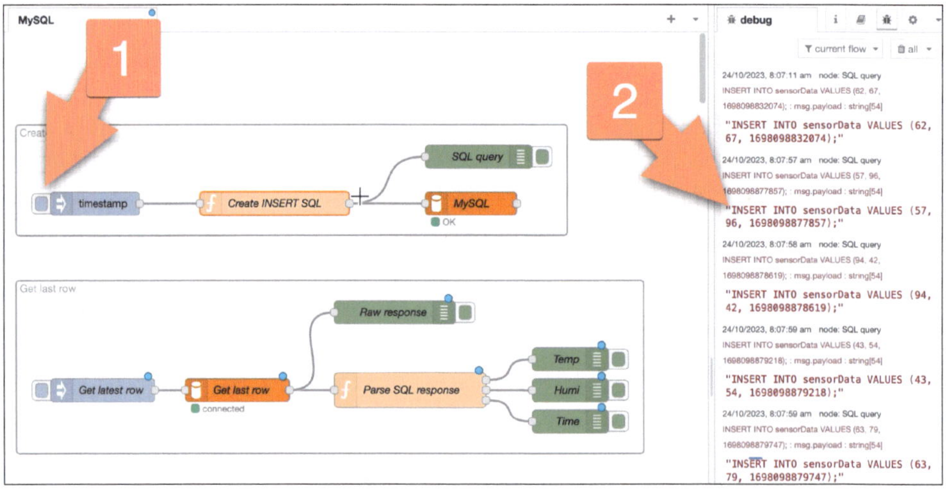

Figure 4.3.2.13: Trigger the "Write" subflow (1) to create new records in the database. The raw SQL is in the Debug tab (2).

Trigger the "Write" subflow a few more times to create more records in the database. Now, let's read the last record. Click on the "Get latest row" inject node's button to trigger the "Read" subflow.

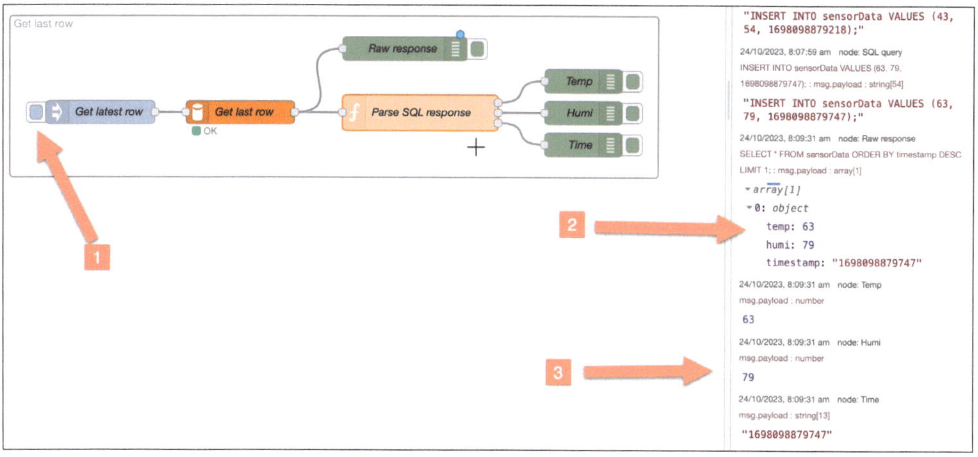

Figure 4.3.2.14: Trigger the "Read" subflow (1) and see the raw returned JSON object (2) and extracted values (3) in the debug window.

In the debug window, you can see the raw response from the "mysql" node and the individual column values extracted by the Javascript code in the "function" node.

You can also verify the new records on the command line. Connect to the MySQL command line as the "node-red" user, and use the SELECT command to list all records in the "sensorData" table. You can see the results of this command below (the command is "select * from sensorData"):

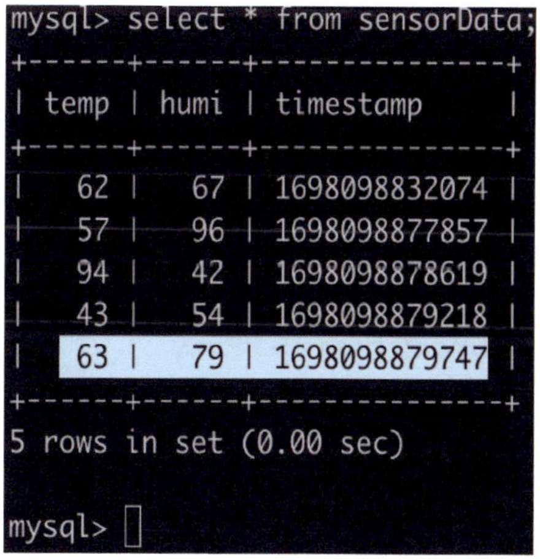

```
mysql> select * from sensorData;
+------+------+---------------+
| temp | humi | timestamp     |
+------+------+---------------+
|   62 |   67 | 1698098832074 |
|   57 |   96 | 1698098877857 |
|   94 |   42 | 1698098878619 |
|   43 |   54 | 1698098879218 |
|   63 |   79 | 1698098879747 |
+------+------+---------------+
5 rows in set (0.00 sec)

mysql>
```

Figure 4.3.2.15: A list of all current records in the database.

As you can see, working with a MySQL database as back-end storage for your Node-RED flows is simple. Because the two services are co-located on the same server, and both are fast and efficient, this setup is ideal for any application that requires the ability to handle big data sets with top-tier performance and at zero cost (at least for the software). I encourage you to do additional reading on SQL to understand the database handling options better.

3.3. Using RESTful APIs and Web Services

In the Internet of Things (IoT) and modern web applications, RESTful APIs are a powerful tool for enabling communication between devices and services. This chapter will provide an overview of how to use RESTful APIs with Node-RED from a high-level perspective and give examples of the possibilities they unlock.

RESTful APIs: an introduction

REST (Representational State Transfer) is a standard architectural style for designing networked applications. A RESTful API is a web service that uses HTTP protocols and methods (like GET, POST, PUT, DELETE) to allow communication with other software components. RESTful APIs have become increasingly popular in web development due to their simplicity, scalability, and alignment with web technologies. These APIs are particularly useful in cloud services, facilitating interaction between components and services.

When interacting with a RESTful API, you typically request specific URLs (endpoints), and the server responds with the requested data. This data is often in JSON format but can also be in XML or other formats.

The REST architecture

REST architecture is built around resources, any object, data, or service the client can access. A resource can be a single data entity (such as a number) or a collection of data entities (such as a JSON document). REST uses standard HTTP methods like GET, POST, PUT, and DELETE to perform operations on these resources. Each operation is stateless, meaning that each request from a client to a server must contain all the information needed to understand and process the request, including authentication.

Role of RESTful APIs in cloud services

In cloud computing, RESTful APIs are essential for enabling modular architecture. They allow different cloud services to interact seamlessly. For example, a cloud storage service might offer a RESTful API that allows other applications to upload or download files. Similarly, a cloud-based machine learning service could provide an API for training models, making predictions, and gathering insights.

Accessing RESTful APIs using Python and Ruby

Before we turn to examples of Node-RED flows interacting with RESTful APIs, let's look at how you can access such APIs with languages such as Python and Ruby. The objective of the following examples is to show you how easy such interactions are to do with high-level programming languages, which is a strong contributor to the popularity of RESTful APIs in cloud services.

Let's start with Python. You can use the requests library to interact with a RESTful API in Python. Here's a simple example that fetches data from an API:

```
import requests

response = requests.get('https://api.example.com/data')
if response.status_code == 200:
    data = response.json()
```

Now, let's do the same in Ruby. In Ruby, you can use the rest-client gem to make API calls. Here's how you can get data from an API:

```
require 'rest-client'
require 'json'

response = RestClient.get 'https://api.example.com/data'
if response.code == 200
    data = JSON.parse(response.body)
end
```

In both script examples, the web service will respond to the GET request from the scripts by sending back the requested data, formatted in a JSON envelope, in the response message's body. In the examples, no authentication is required, but even if authentication were needed, the additional code would only be a couple of extra lines.

Python and Ruby offer straightforward ways to interact with RESTful APIs, making it easier to integrate various services, whether they are cloud-based or not. Now, let's look at Node-RED and RESTful API, first from a high level and then with a complete example in the next section of this chapter.

Node-RED and RESTful APIs

As you might have expected, Node-RED provides built-in nodes (like the "HTTP request" node) that make it easy to interact with RESTful APIs. These nodes allow you to make HTTP requests to APIs directly from your Node-RED flows and then use the response data in your application.

In the palette, you will find the "http response" and "http request" nodes in the network group.

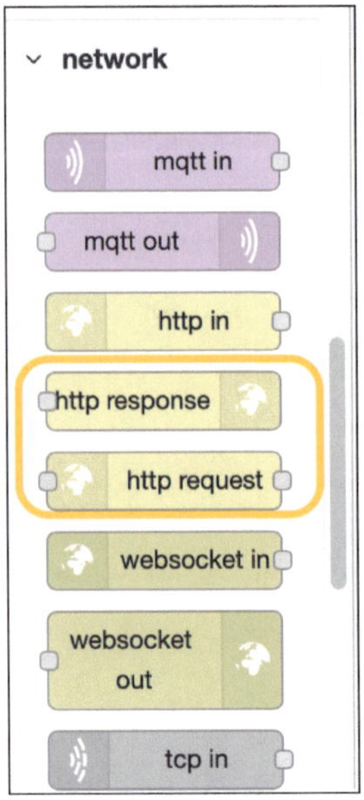

Figure 4.3.3.16: The "http response" and "http request" nodes in the Network group.

HTTP Request node

The HTTP Request node sends an HTTP request to a web server. When you place this node in your flow and configure it, you specify the URL you want to interact with and the HTTP method you want to use, such as GET, POST, PUT, or DELETE. Once the node receives an input message, it sends the HTTP request to the specified URL. The server's response is then stored in the output message payload, which can be passed on to other nodes for further processing.

You can see the properties window of the "http request" node below. This example is configured to make a GET request on the openweathermap.org API endpoint that returns the current weather information for a specific locality (you can read the details of this example in the next section). In this properties window, you can see options for authentication and security through encryption, proxies, etc.

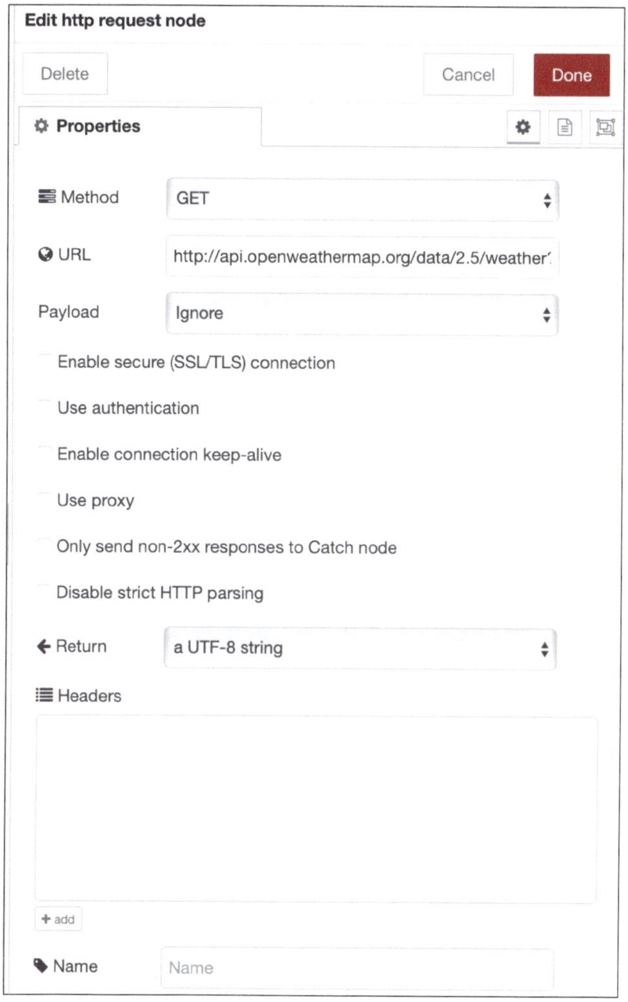

Figure 4.3.3.17: The "http request" properties.

HTTP Response node

On the other hand, the HTTP Response node is used to send back an HTTP response when Node-RED is set up as an HTTP server. This node is typically paired with an HTTP In node, which listens for incoming HTTP requests. When an HTTP request is received by the HTTP In node, it passes along the message through the flow for processing. Eventually, the message reaches the HTTP Response node, which sends back the final response to the client who made the initial request.

You can see the properties window of the "http response" node below. As you can see, there is not much to configure. Other nodes have done all the work for creating the response payload before reaching the "http response" node. The "http response" node will simply take the payload from the previous node and send it to the client that originated the requests. You can set a status code (below, I have set it to "200") and any additional headers.

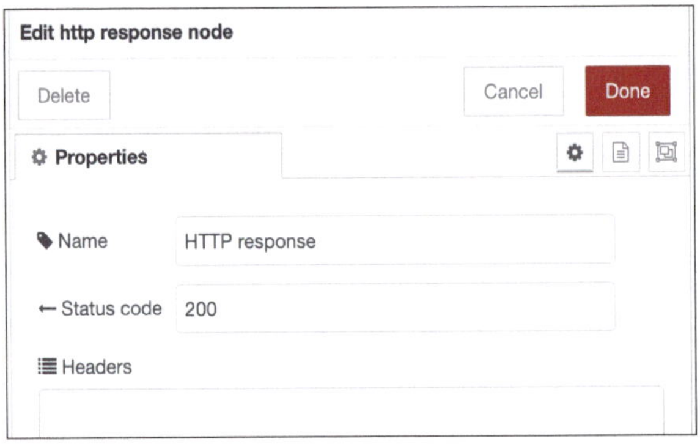

Figure 4.3.3.18: The "http response" properties.

Examples of using RESTful APIs in Node-RED

Before looking at a complete example of a Node-RED flow that interacts with a RESTful API, let's look at some of the things you can do with this knowledge.

- Fetching Data from an API: This is the classic use case. You can use the HTTP request node in Node-RED to fetch data from an API. For example, you could fetch weather data from the OpenWeatherMap API and display it on your Node-RED dashboard.
- Controlling IoT Devices: If IoT devices expose a RESTful API, you can control these devices using Node-RED. For example, you could create a flow that turns on your smart lights when a specific condition is met.
- Posting Data to an API: You can also use Node-RED to send data to a RESTful API. For instance, you could create a flow that sends sensor data from an IoT device to a cloud-based data storage API.
- Integrating with other services: Many online services offer RESTful APIs that allow you to integrate. For example, you could use the Twitter API (now "X") to post a tweet or the Google Calendar API to create a calendar event.

In these examples, you would use the HTTP request node to make the API request and other nodes (like the function node or JSON node) to process the response data. The specific details of setting up these nodes will depend on the API you're working with.

Node-RED and RESTful API best practices

When working with RESTful APIs in Node-RED, there are a few best practices to remember. Let's look at the most important of those practices now.

Keep your API keys secure

Almost all Cloud services require the use of API keys to provide authentication. Never hard-code your API keys directly in your Node-RED flows. It's a security risk. Instead, use environment variables or the built-in "Credentials" feature in Node-RED to store sensitive information like API keys. This way, even if someone gets access to your flow, they won't have your keys. To learn how to use the Credentials feature and environment variables, look in the sections at the start of Part 4 of this book under "Other useful nodes and features".

Use the right HTTP methods
When setting up an HTTP Request node, choose the appropriate HTTP method for what you're trying to do. For example, if you're fetching data, use GET. If you're sending data to create a new record, use POST. This helps maintain the semantics of the API and makes your flow easier to understand.

You can learn more about HTTP methods at this location: https://txplo.re/http_methods.

Rate limiting and error handling
APIs often have rate limits, meaning you can only make a certain number of requests in a given period. Be mindful of this when setting up your flows. Use delay nodes or rate-limiting logic to avoid hitting these limits. Also, always include error-handling nodes like "catch" and "status" to handle any issues like network errors or API downtimes gracefully.

Use headers wisely
Some APIs require specific headers, like content type or authorisation tokens. Make sure you include these in the HTTP Request node settings where needed. For example, if you're sending JSON data, you'll often need to set the "Content-Type" header to "application/json".

Keep an eye on the payload
Node-RED often works with the msg.payload object for passing data between nodes. When working with APIs, ensure you know what the API response will look like and how you'll extract the data you need. Use debug nodes liberally during development to inspect the payload and ensure it contains what you expect.

These best practices make your Node-RED flows more robust, secure, and maintainable while interacting with RESTful APIs.

Examples of RESTfull web services
What services are available on the Web that offer REST APIs? Below, I have compiled a list of some of the most widely used services. You can use most of them free of charge, from Node-RED or any programming language.

Data storage and databases

- MongoDB Atlas: MongoDB Atlas provides a hosted MongoDB service that you can easily interact with using its RESTful API. See https://www.mongodb.com/atlas.
- Firebase: Google's Firebase is a NoSQL cloud database that can sync data across clients in real time. See https://firebase.google.com/.
- AWS S3: Amazon's Simple Storage Service (S3) is an object storage service that offers industry-leading scalability, data availability, security, and performance. See https://aws.amazon.com/s3/.

Weather data

- OpenWeatherMap: Provides weather data that you can access via their API, which is helpful for various IoT applications. See https://openweathermap.org/.

IoT platforms

- IBM Watson IoT Platform: This platform provides powerful device management and analytics capabilities for IoT applications. It provides a REST API for interacting with the platform. See https://www.ibm.com/docs/en/watson-iot-platform.
- Microsoft Azure IoT Hub: Azure IoT Hub is a managed service hosted in the cloud. It acts as a central message hub for bi-directional communication between your IoT application and the devices it manages. See https://azure.microsoft.com/en-us/products/iot-hub/.

Machine Learning and AI

- Google Cloud Machine Learning Engine: This service allows you to create machine learning models for your applications. The models can be accessed and used via a RESTful API. See https://cloud.google.com/products/ai/.
- OpenAI API: The OpenAI API provides access to powerful language models like GPT-3 and GPT-4 for various natural language processing tasks. It enables developers to integrate machine learning capabilities into their applications, services, or products. The API can perform text generation, summarisation, translation, and question-answering tasks. It's commonly used in chatbots, content creation tools, and data analysis. Access to the API is generally through a paid subscription, and usage is subject to rate and data limits. See https://openai.com/product.
- IBM Watson: IBM Watson offers a variety of cognitive computing features like language recognition, image analysis, and machine learning. See https://www.ibm.com/watson.

Social media

- Twitter API: Allows you to programmatically access Twitter's vast global social media data stream. See https://developer.twitter.com/en/docs/twitter-api.
- Facebook Graph API: Allows you to read and write data to Facebook's platform. See https://developers.facebook.com/tools/explorer/.

Geolocation

- Google Maps API: Provides access to various geolocation and mapping services. See https://developers.google.com/maps/apis-by-platform.

Email services

- SendGrid: A cloud-based email service that provides reliable transactional email delivery, scalability, and real-time analytics, along with flexible APIs that make custom integration easy. See https://docs.sendgrid.com/for-developers/sending-email/api-getting-started.

Other cloud services

- Twilio: Twilio's API allows developers to programmatically make and receive phone calls, send and receive text messages, and perform other communication functions using its web service APIs. See https://www.twilio.com/docs/usage/api.

These are just a few examples. Countless other cloud services with RESTful APIs can be integrated into Node-RED workflows. It's recommended to always refer to the specific API documentation of the service for the most accurate and updated information.

3.4. Get Weather Information from OpenWeatherMap.org

In this section, you will learn how to use the OpenWeatherMap API with Node-RED. You will create a Node-RED flow that gets the current weather information for your location and displays it in the dashboard.

OpenWeather.org is a digital platform offering various weather data services, including real-time weather, forecasts and historical data. It provides accurate and reliable weather information that can be accessed via its website or APIs. The platform is popular among developers, businesses, and researchers for its ease of use and extensive data coverage.

If you're looking to integrate weather data into your Node-RED flows, the OpenWeatherMap API is a great resource. This API allows you to access various types of weather information, such as current conditions, forecasts, and even severe weather alerts. The API is well-documented, making it easier to understand how to make requests and handle responses.

It is free for up to 1000 API calls daily, which is plenty of an allowance for educational and hobbyist purposes. You can use Node-RED's HTTP Request node to fetch data from the OpenWeatherMap API and then use function nodes to process and utilise this data in your flows. Whether building a weather dashboard or automating your home based on weather conditions, the OpenWeatherMap API offers a robust solution.

The OpenWeatherMap API offers a variety of data sets, including the current weather data (which we'll use in this example), four and 16-day forecasting, 30-day climatic forecasts, global weather alerts, and much more.

You can find information about this API at https://openweathermap.org/api.

Using OpenWeatherMap API with Node-RED

In this example, we'll create a Node-RED flow that gets current weather information and shows it in the debug tab and the dashboard. When the flow is deployed and working, the dashboard will look like this:

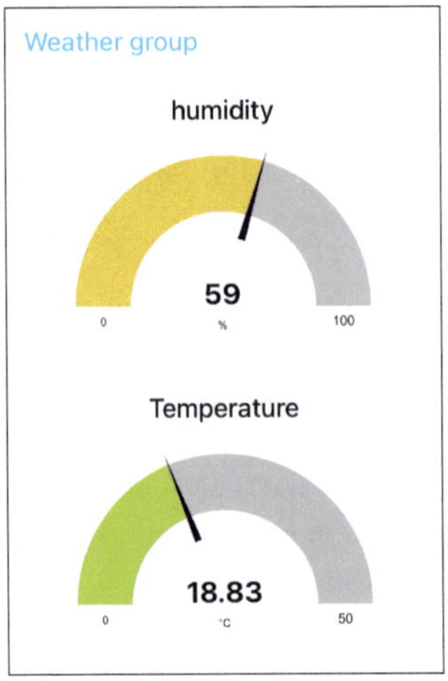

Figure 4.3.4.19: The dashboard shows current weather data from Open Weather Map.

To get the current weather information for Sydney from the OpenWeatherMap API, you'll need an API key, which you can obtain for free by creating an account on the OpenWeatherMap website. Go to https://openweathermap.org/, create a free account, log in, and go to https://home.openweathermap.org/api_keys to generate a new key. You can see my OpenWeather account API keys page below.

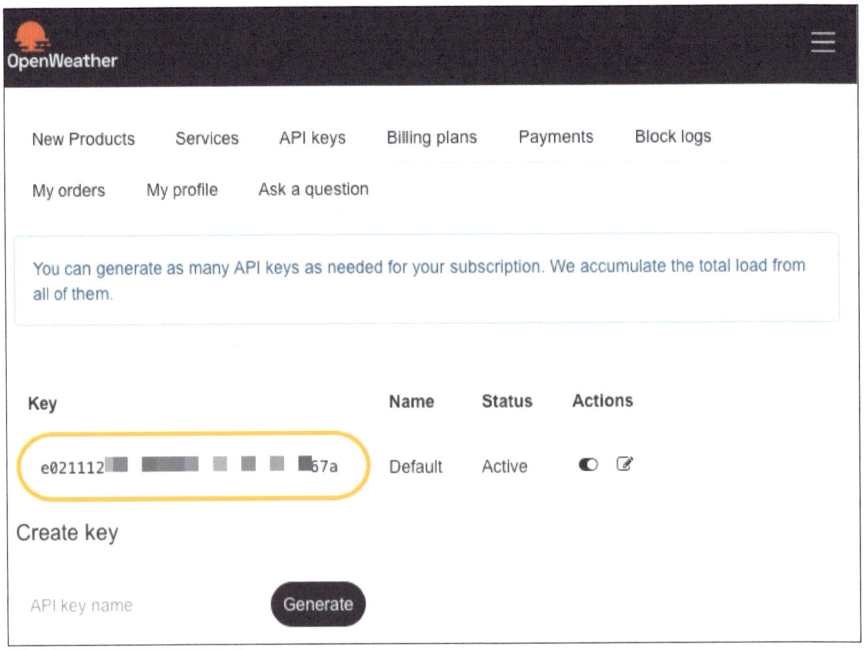

Figure 4.3.4.20: My Open Weather Map API key
(I have blurred it as it is my primary active key).

Once you have obtained your key, continue in the Node-RED flow editor. Follow these instructions to create your flow:

1. **HTTP Request Node**: Drag and drop an 'http request' node onto your workspace. Double-click the node to open its configuration settings. Set the 'Method' field to 'GET'. In the 'URL' field, enter the URL of the OpenWeatherMap API endpoint for current weather data. Replace "<API_KEY>" with your actual API key. You can also change "Sydney" to your city or locality. The URL should look like this:

```
http://api.openweathermap.org/data/2.5/
weather?q=Sydney,au&appid=<API_KEY>&units=metric
```

The "units=metric" parameter will ensure the temperature is returned in degrees Celsius. Click 'Done'.

2. **Inject Node:** Drag and drop an 'inject' node onto your workspace. This node will be used to trigger the HTTP request. Connect the output of the 'inject' node to the input of the 'http request' node. If you want to get new data in a regular interval, say, every 30 seconds, you can set up the inject node with a repeat trigger. You can see this in the properties screenshot below:

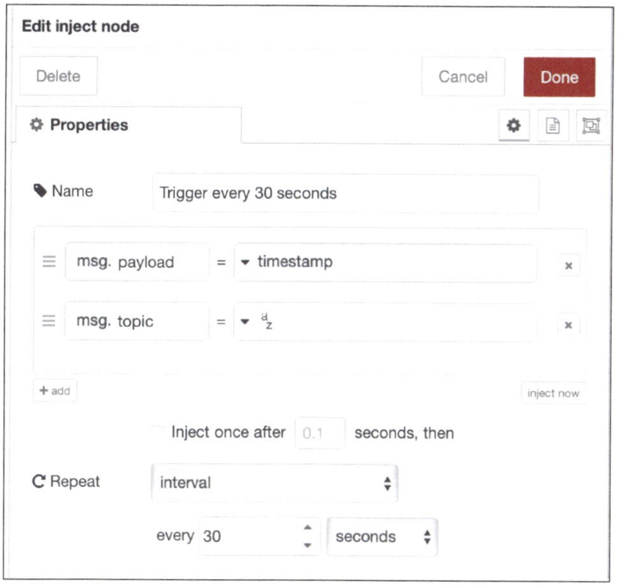

Figure 4.3.4.21: This Inject node will trigger the flow every 30 seconds.

3. **Function Node:** Drag and drop a 'function' node onto your workspace. This node will extract the temperature and humidity from the response. Double-click the node to open its configuration settings. Enter the following JavaScript code:

```
var data = JSON.parse(msg.payload);
msg.payload = {
    temperature: data.main.temp,
    humidity: data.main.humidity
};
return msg;
```

Click 'Done'. Connect the output of the 'http request' node to the input of the 'function' node.

4. **Debug Node**: Drag and drop a 'debug' node onto your workspace. This node will be used to display the temperature and humidity. Connect the output of the 'function' node to the input of the 'debug' node.

5. **Gauge nodes:** Drag and drop two "gauge" nodes onto the workspace. Assign these nodes to a new Dashboard tab and group. Set one of them to display the temperature (set its name and the range to be from 0 to 50 °C, or whatever makes sense for your location) and the other to display humidity (set its name and the range from 0 to 100 %).

6. **Change nodes**: Drag and drop two "change" nodes onto the workspace. These nodes will take the payload from the function node and extract the temperature and humidity values. You can see the properties of the temperature "change" node below:

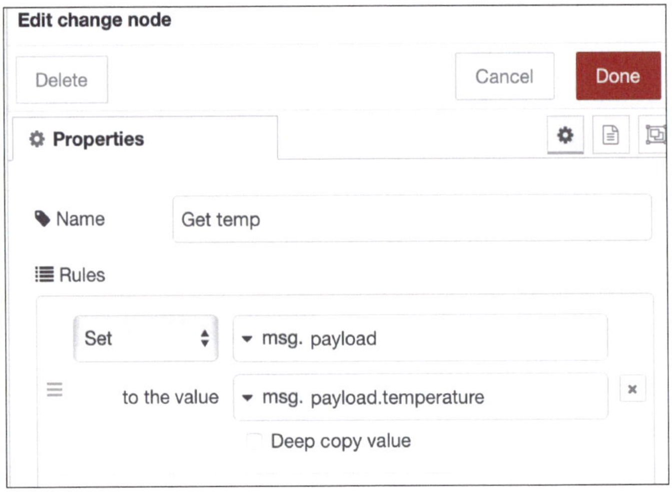

Figure 4.3.4.22: This change node extracts the temperature value from the Javascript object and sends it to the gauge node in the message payload.

7. Click 'Deploy' to deploy the flow.

Your new flow should look like this:

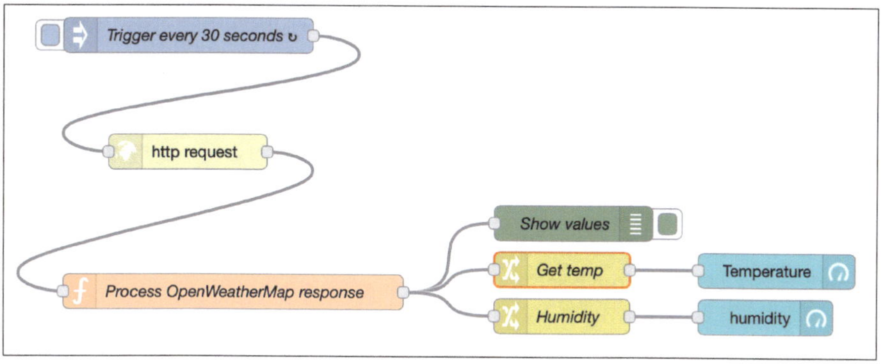

Figure 4.3.4.23: The OpenWeatherMap flow.

Now, when you click the button on the 'inject' node (and every 30 seconds after that), the following will happen:

- The 'http request' node will make a GET request to the OpenWeatherMap API,
- The 'function' node will extract the temperature and humidity from the response,
- The 'debug' node will display the temperature and humidity in the debug area.
- The Change nodes will extract temperature and humidity values from the Javascript object and pass them on to the Gauge nodes.
- The Gauge nodes will display the temperature and humidity values in the Dashboard.

Once you get this flow working, consider other data objects from the OpenWeatherMap API. For example, you can look at the 5-day, 3-hour forecast or the air pollution API. Both are free to explore and use.

3.5. Datalogging to a Google Sheet

Let's look at another example of using a web API to connect and use a cloud service. Data logging is one of the most common requirements in automation and IoT projects. Data logging allows us to store data over time, monitor system behaviour, and analyse trends.

One of the platforms that is widely used for this purpose in conjunction with Node-RED is Google Sheets. In Node-RED, there is a node specifically designed to work with Google Sheets, which makes data logging in Google Sheets a straightforward and reliable process. In this chapter, you will learn some of the benefits and limitations of using Google Sheets for data logging in a Node-RED flow, and, of course, you will create an example data logging flow.

Benefits of Google Sheets and its API

One of the best aspects of using the Google Sheets API in Node-RED is the seamless integration between your flows and a widely used spreadsheet platform. You can read from and write to Google Sheets like a database. This is handy for logging data, like sensor readings from a smart home setup. Imagine having your temperature and humidity sensors automatically log data to a Google Sheet. You can then easily share that sheet with others or use Google's built-in analytics tools to make sense of it.

Another benefit is real-time data manipulation. Let's say you're running a small e-commerce site. Set up a flow in Node-RED that updates inventory levels in a Google Sheet whenever a sale is made. This way, you can track stock levels without manually updating a spreadsheet.

Here is a short list of additional benefits:

Easy to set up: Google Sheets API is relatively straightforward. Creating a service account allows you to connect Node-RED to Google Sheets using nodes like "node-red-contrib-google-sheets".

Scalability and Accessibility: Google Sheets can handle a significant amount of data (up to 5 million cells per spreadsheet), which makes it suitable for many small to medium-sized projects. Plus, the data logged into Google Sheets is accessible from anywhere with an internet connection.

Real-Time Data Monitoring: Changes to Google Sheets are real-time and can be shared with multiple users, allowing real-time data monitoring.

Data Analysis and Visualization: Google Sheets provides built-in functions for data analysis and visualisation, reducing the need for additional tools to generate insights from logged data.

Cost: Google Sheets is a part of Google's free suite of tools, making it a cost-effective solution for data logging.

And now, let's look at some limitations to remember when creating flows.

Limitations of Google Sheets and its API

While it's a powerful tool, the Google Sheets API has limitations. For starters, you have to deal with API quotas, which means you can only make a certain number of daily requests. If you're running a large-scale operation, you might hit those limits quickly.

Another issue is the complexity of setting up authentication. Google's security protocols are robust, which is good, but you'll need to spend some time setting up API keys or OAuth credentials. This can be daunting if you're new to working with APIs. The example below shows you how to set up your Google authentication.

Google Sheets isn't a full-fledged database. While it's excellent for more straightforward tasks, it lacks a dedicated database solution's advanced querying and indexing features. So, if your project requires complex data operations, you might find Google Sheets a bit limiting.

Let's look at some more limitations to keep in mind:

Limited storage: While Google Sheets can handle millions of cells, it is unsuitable for projects that store large amounts of data or data at high frequency, as it can quickly fill up.

Rate limiting: Google Sheets API has a rate limit. You may encounter rate limit errors if you're writing data too frequently.

Security: As your data is stored in Google Sheets, it's exposed to the internet. Even though only users with the proper credentials can access the data, the risk of data breaches still exists.

Reliance on Google: The data logging solution relies heavily on Google's infrastructure. Any disruption or changes in their service or API can directly impact your project.

Limited customisation: While Google Sheets provides excellent functionality, it doesn't offer as much customisation or complexity as specialised databases, which may be required for advanced analytics or large datasets.

Google Sheets is a convenient tool for data logging in Node-RED flows, offering many advantages such as ease of setup, accessibility, real-time monitoring, and built-in data analysis tools. However, limitations exist in terms of storage capacity, rate limiting, security, reliance on Google, and customisation options. The choice to use Google Sheets should be based on project requirements, data size, and the desired level of control over the data logging infrastructure.

How to transfer data to Google Sheet

To use Google Sheets as a data logging service for a Node-RED flow, you need to set up Google Sheets API, create a service account, and use the respective Node-RED nodes for Google Sheets. Let's look at each step in detail.

Create a Google Sheets API Service Account

This is the most tedious step of the process. A Google Sheets API service account is a special type of Google account that represents your application instead of an individual user. It allows your Node-RED flows or other applications to interact with Google Sheets without needing your login details. You create this service account in the Google Cloud Console and download a JSON file containing the credentials. Once set up, you can share specific Google Sheets with the service account's email address, allowing it to read or modify those sheets. This makes it a secure and convenient way to automate tasks involving Google Sheets.

Of course, to create a Google Sheets API service account, you must have a personal Google account and be logged in. In the following steps, I assume you are already logged in to your Google account. From there, here is what to do to create a service account (these instructions are valid at the time I am writing this; Google may introduce changes as they do from time to time):

1. Go to the Google Cloud dashboard at https://console.cloud.google.com/home/dashboard.
2. Create a new project or select an existing project.
3. In the sidebar on the left, click on APIs & Services > Credentials.
4. Click on the "Create credentials" button at the top of the page and select "Service account."
5. Name the service account and grant it a "Project > Editor" role.
6. Generate a key for the new service account, which will be downloaded as a JSON file.

You can see an example of the Service account page with my new key below. Please remember that by the time you read this, this page may look different as Google updates them frequently.

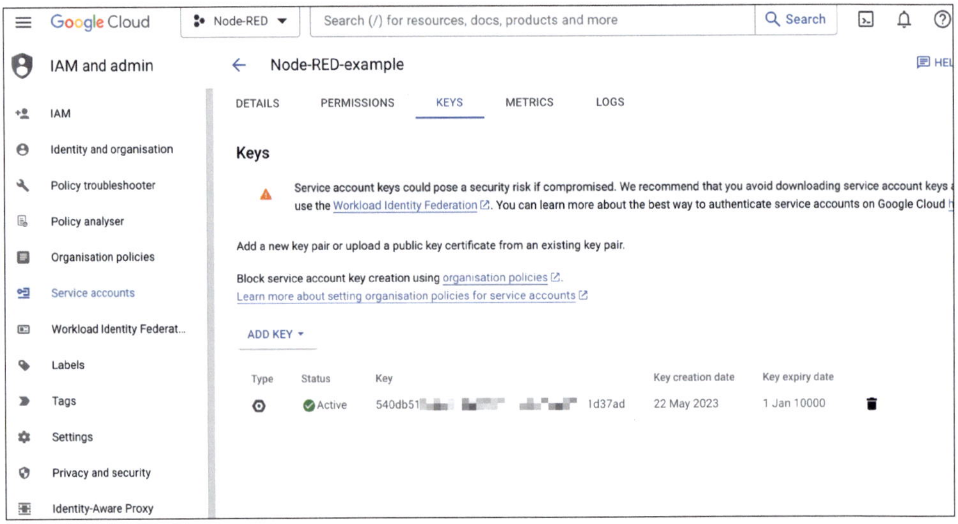

Figure 4.3.5.24: My Google Sheet service account API key.

The API key is a text JSON file; you can open it in a text editor. It will look like this:

```
{
  "type": "service_account",
  "project_id": "node-red-586301",
  "private_key_id": "540db518a407bfe3e9fbd1d37ad",
  "private_key": "-----BEGIN PRIVATE KEY-----\
nMIIEvQIBADANBgkqhkiG9w0BAQEFAASCBKcwggSjAgEAAoIBAQC+75SG453Bmbyh\
nG6osjoYhfUSxmP2bghU4WhbP8V91Iwe2aC/iIeG2oNnBCB55nrHizv2vwA+1TpLg\
niAVHernMX2bvkZC5Mj1H9KV1oh7UOUhez3N4FyltYekLaUOKVLvMtVXA887wzxvP\
nl3P7VyC0YIpL2wcxHVAFcprKB9/+9iw9rdECNfsf7nxAChdgn36CgSB+E+QJGpYi\
nWmcWWtji574rqk1CC8asseB4Xz5b1LeqFmDCdd9aWcrbEjBGhJMBNqexKCGWK9IX\
nSLBLT4ebe9gfych+Q+EnmJeCMXhspn6h22yYqeb9BtBKk5Y7csVZnNFPWAVeIfKU\
njHjlVgUjAgMBAAECggEASsmAvc0OD7TuA5TYdRdmRlvQuleYjn7HXQk8gjqjDzin\
```

```
nXrU0fDiVNnmOkWNivHSCbLpy95IdFQEv2q6skrDpcbm0kgXMUB3zjumunYLg3HOL\
ne0d6jQFVgrilFYknoVlLLgWhDzDwT7ZvyEIkdcBM2RovUUL7cCtxGpEnPBcVj8Gh\
n8ZGowUKW0tAXrqXZ0eykwEKSqdi7GoshOVe2eb1cQIPAqBJLCnRE2SNr8EXlLHiX\nzu54BDV4/
osMqeZkBpWDKt2y8mxDOrktt/sVTFUm9/1KaLDDwKBgQDAOPW2ZnyxXmOLCPzje09EdqOKbHfwb0KL\
ntUExmJ9KLHj0WIdl8eCnGD70bMIltbu54FB4yRwsZR04dijQWuJX24t4aEZR8+4F\
newg64tujefWLNvNBuLocNJlyn0Q0iqyEvrfcjeG2NT04PY1dKEYdIBxcN3U30qda\
njME2VZSMrQKBgHS2NDXIl5osZQINIKZX3A3jDqV08uARGER9ypbd44YcTyomxaIO\
nH9NtqmcmwNIaZ7LrYfYLbBJKhInWWWO8bQEE1TNvj8Tnn2IGaF3p29neyNt8yo6E\
nFPE4xIHn3t2Ga7BuccNhxQxJrxxFN0YZz/HRcq+LaKMwvT8svxZHhe0BAoGBAIdo\
n2LboVZjBDFCKQVOirwFkreo6ZPdr+S3fvpBImpWu5HoEz0TickZjjx2ya8YS0q6r\
nNSURTxZ7GG2I11GHZIFNuF9W3Yqn0pnrxK1u2Bi+NloemJhIoFp0N1CGn45/ze4p\
nR+5HyKQJtxmiSq5AS7wCvde8fYeYjqn8E7PJ7751AoGACsGgeIqxgClJkxpXgvns\
nVK2ceArvf7oP0qjNog5KT33KcqcYkNZiKst2s10Pk5q7yjqid+s/NKrodZbC1Nbc\n/
X72CsREfWgyUlCphoE0hF9oPrOUASOaRJSOW6q6ArF6MxgSte/CjrvuvBc17SOa\
nF2ZRT8T7Q5KnHkxr5XW/exY=\n-----END PRIVATE KEY-----\n",
  «client_email»: «node-red-example@node-red-387502.iam.gserviceaccount.com»,
  "client_id": "109046034126972474848",
  "auth_uri": "https://accounts.google.com/o/oauth2/auth",
  "token_uri": "https://oauth2.googleapis.com/token",
  "auth_provider_x509_cert_url": "https://www.googleapis.com/oauth2/v1/certs",
  «client_x509_cert_url»: «https://www.googleapis.com/robot/v1/metadata/x509/
node-red-example%40node-red-586301.iam.gserviceaccount.com»,
  "universe_domain": "googleapis.com"
}
```


Let's continue with step two.

Share the Google Sheet with the service account

In this step, you will share the Google sheet you intend to use for cataloguing the Node-RED flow with the new Service Account. Here's how to do this:

1. Open the JSON file you downloaded. There will be a "client_email" field. Copy the email address from there.
2. Create a new Google Sheet that you will use to log data from the Node-RED flow.
3. Open the Google Sheets document you created, click the "Share" button, and paste the service account email.

This is similar to sharing a Google Sheet (or other Google document) with a friend or colleague. This concludes the steps related to setting up the Google side of the application. Let's continue with Node-RED.

Set up Node-RED

Here, we will create the most straightforward possible flow that will allow us to test storing data to the Google Sheet. To do this, you will use the "node-red-contrib-google-sheets". Follow the instructions below:

1. Use the Node-RED Palette Manager to install the "node-red-contrib-google-sheets" node in Node-RED.
2. Once installed, you'll have a new node in the Function group of the palette titled "GSheet".
3. Drag one GSheet node to the Node-RED editor.
4. Drag an Inject node to the editor. Configure this node to send the timestamp to the message payload.
5. Connect the Inject node's output to the GSheet node's input.
6. Double-click on the Sheet node to open its configuration. Paste the entire content of your service account JSON file into the "Account" field. Then, you can select the operation you want (like "append" to add rows) and specify the document and the sheet you want to use. Here is the configuration of the GSheet node in my flow:

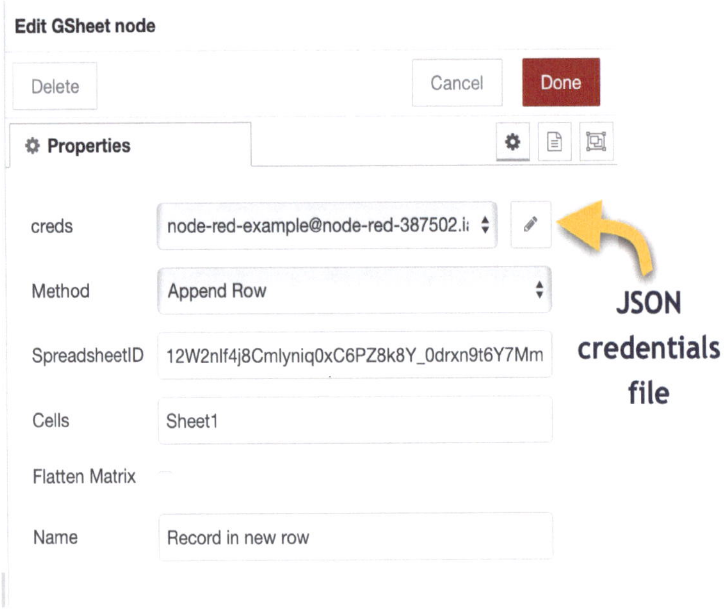

Figure 4.3.5.25: The properties window of the GSheet node.

The arrow points to the "pen" button in the screenshot above. Click this button to open a window with the text field where you can copy the credentials JSON file. In the Cells text field, ensure you have entered the name of the tab (of the spreadsheet) where you want to write the log data. In the Method field, confirm you have selected "Append Row" from the drop-down. You can get the Spreadsheet ID from the GoogleSheet URL.

The flow contains only two nodes and looks like this:

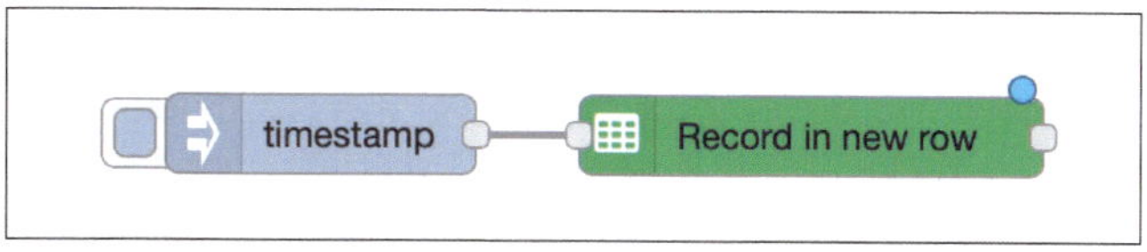

Figure 4.3.5.26: The GSheet cataloger example flow for writing data.

Deploy and use

The flow is ready now, so go ahead and deploy it. Now, every time your flow runs (i.e. when you click on the Inject node button), it will log data to your Google Sheets document.

Go ahead and click the Inject button, wait for a few seconds, and notice the timestamp appear in a new row in the Google Sheet. You can see my first such record in the screenshot below:

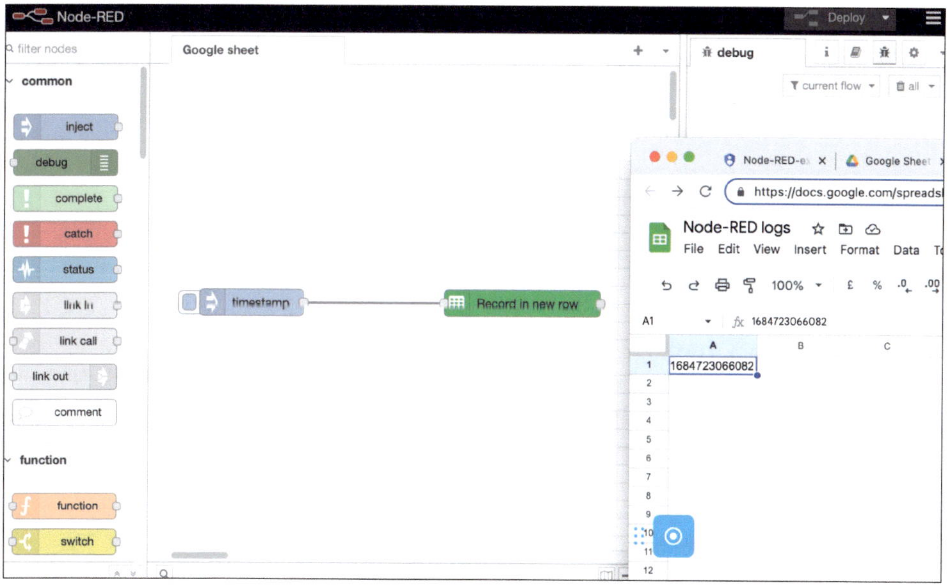

Figure 4.3.5.27: A new record in the Google Sheet.

As you can see, once you have created the Service Account for your Node-RED application, creating a flow that can record data in the Google Sheet is a matter of setting up a single node. Let's extend this example by learning how to read data from a Google Sheet in Node-RED.

3.6. Reading Data from a Google Sheet

In the previous section, you learned how to use Google Sheets as a cataloguing storage service with Node-RED. What if you want to read data from a Google Sheet? To provide a practical example, we will illustrate how to create a Node-RED flow that reads values from Google Sheets and displays the output in the debug window. Why would you want to read data from Google Sheets in a Node-RED flow? Here are some example use cases:

Event scheduling

Imagine you have a Google Sheet that contains a schedule of events for a conference or a weekly team meeting. A Node-RED flow could read this data and automatically send reminders or updates to participants via email or SMS as the event approaches.

Temperature monitoring

A Node-RED flow could read this data regularly if you log temperature data from various sensors in a Google Sheet. Based on the readings, it could trigger other actions like adjusting a smart thermostat or sending an alert if temperatures go out of a specified range.

Financial tracking

You could have a Google Sheet that tracks your monthly expenses or investment portfolio. A Node-RED flow could read this data to generate a monthly report or integrate it with other services to give you real-time financial insights.

Content management

If you're running a blog or a website, you might have a Google Sheet that lists upcoming topics or articles to be published. Node-RED could read this sheet and automatically update your content management system when a new article is due to be published.

Inventory management

A Google Sheet could hold information about product stock levels in a retail scenario. A Node-RED flow could read this data and automatically place orders for items running low or send alerts to the management team.

Customer feedback analysis

If you collect customer feedback through a form that populates a Google Sheet, Node-RED could read this data to perform sentiment analysis. Based on the results, it could trigger specific actions like sending thank-you emails or alerting customer service to negative reviews.

Reading GSheet data example

We'll continue with the "write" experiment from the previous section to learn how to read data from a Google Sheet. I assume you already have a Service Account for your Node-RED application and a Google Sheet to read from. You have also installed the "node-red-contrib-google-sheets" to your Node-RED instance. If you don't know how to do any of these tasks, go to the previous section to learn how and return here when ready.

Let's create a Node-RED flow that can read data from a Google Sheet.

1. **Google Sheets Node**: Drag a "GSheets" node to the Node-RED editor.

2. **Inject node**: Drag an "Inject" node to the Node-RED editor. We'll use this node to trigger the flow.

3. **Function node**: Drag a "Function" node to the Node-RED editor. We'll use this node to extract the last row of records returned from the GSheet node.

4. **Debug node**: Drag a "Debug" node to the Node-RED editor. We'll use this node to print the value of the last row of records returned from the GSheet node.

5. **Configure the Google Sheets Node**:
   a. Double-click on the "google sheets in" node to open its configuration.
   b. Paste the entire content of your service account JSON file into the "Account" field as you did in the "write" flow example.
   c. Specify the ID of the document and the name of the sheet you want to use in the respective fields as you did in the "write" flow example.
   d. Select the operation as "Get Cells".
   e. You can see the properties of the GSheet node below:

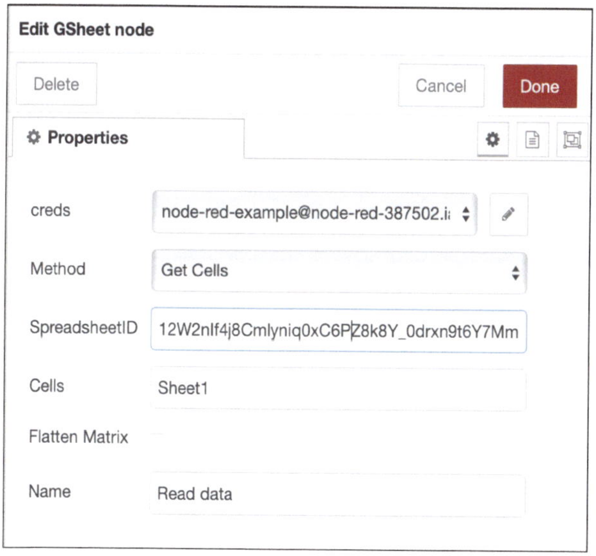

Figure 4.3.6.28: The properties window of the GSheet node.

6. **Configure the Function node**. The Gsheet node will return all Google sheet rows when set to "Get Cells". In this flow, I want to only show the last row in the debug window. To do this, we can use a simple Javascript instruction in the Function node:

a. msg.payload = msg.payload[msg.payload.length - 1];
b. return msg;

The JavaScript code modifies the msg.payload object in a Node-RED flow. Specifically, it updates msg.payload to contain only the last element of its current value, assuming that the current value is an array. After updating msg.payload, the code returns the modified msg object, making it available for the next node in the flow to use.

For example, suppose msg.payload initially contains the array [1, 2, 3, 4, 5] after this code runs. In that case, msg.payload will be updated to contain just the number 5, which is the last element of the original array.

7. **Set up a Debug Node**: Now drag and drop a "debug" node from the Node-RED's node palette onto the workspace and wire it to the "function" node. This node will display the output from the Google Sheets node in the debug window.

8. **Deploy the Flow**: Click the Deploy button to save and run the flow.

Now, if you inject a timestamp to trigger the flow (you can use an "inject" node for this purpose), you'll see the contents of the last row of your Google Sheet displayed in the debug window. You can see an example of the output in the screenshot below:

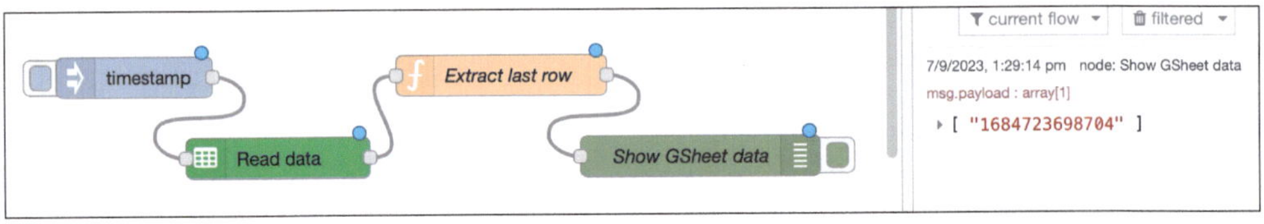

Figure 4.3.6.29: The "read" GSheet flow, with a last row in the Debug tab.

This is an example of a simple flow. The complexity of flows could be increased based on project requirements to include functions such as transforming data, filtering rows, or triggering specific actions based on the data from Google Sheets.

Index